7/8/2003

BY

COURTESY OF

KANAKARAJ

BOSTON.

Synchronization of
Digital Telecommunications Networks

Synchronization of Digital Telecommunications Networks

Stefano Bregni
Politecnico di Milano, Italy

JOHN WILEY & SONS, LTD

Other Wiley Editorial Offices

John Wiley & Sons, Inc., 605 Third Avenue,
New York, NY 10158-0012, USA

WILEY-VCH Verlag GmbH
Pappelallee 3, D-69469 Weinheim, Germany

John Wiley & Sons Australia Ltd, 33 Park Road, Milton,
Queensland 4064, Australia

John Wiley & Sons (Canada) Ltd, 22 Worcester Road
Rexdale, Ontario, M9W 1L1, Canada

John Wiley & Sons (Asia) Pte Ltd, 2 Clementi Loop #02-01,
Jin Xing Distripark, Singapore 129 809

British Library Cataloguing in Publication Data

A catalogue record for this book is available from the British Library

ISBN 0 471 61550 1

Typeset in 10/12pt Times Roman by Laserwords Private Limited, Chennai, India.
Printed and bound in Great Britain by T. J. International Ltd., Padstow, Cornwall.
This book is printed on acid-free paper responsibly manufactured from sustainable forestry, in which at least
two trees are planted for each one used for paper production.

CONTENTS

PREFACE

Network synchronization has gained increasing importance in telecommunications throughout the last 30 years, especially since transmission and switching turned digital. Infact, the quality of most services offered by network operators to their customers is affected by network synchronization performance.

Digital switching equipment requires synchronization in order to avoid slips at input elastic stores. Plain telephone conversations are not affected very much by synchronization slips, but circuit switched data services are affected significantly. Therefore, the deployment of circuit-switched data networks and of ISDN brought about the need for more stringent synchronization requirements.

Network synchronization became a thorny matter for telecommunications operators with the deployment of SDH (Synchronous Digital Hierarchy)/SONET networks, which pose new and more complex requirements on the stability of synchronization systems.

More recently, it has been also recognized that the importance of network synchronization goes way farther than SDH/SONET needs. ATM (Asynchronous Transfer Mode) and cellular mobile telephone networks such as GSM (Global System for Mobility), GPRS (Global Packet Radio Services), UMTS (Universal Mobile Telecommunications Services) are striking examples where the availability of network synchronization references has been proven to affect the quality of service.

Network synchronization addresses any distribution of time and frequency over a network of clocks, even spread over a wide area. The goal is to align the time and frequency scales of all clocks, by using the communications capacity of the links amongst them. A synchronization network is the facility implementing network synchronization. Most modern telecommunications operators have set up synchronization networks to synchronize their switching and transmission equipment.

The basic elements of a synchronization network are the nodes (autonomous and slave clocks) and the communication links interconnecting them. Two different issues must be faced: how to transfer timing from one node to the other and how to organize timing distribution among all nodes of the network (synchronization strategy).

Therefore, the good synchronization engineer should master several subjects and in particular:

- Know the subtleties of Plesiochronous Digital Hierarchy (PDH) and SDH/SONET justification processes in order to be able to understand what is really happening when synchronization problems begin to appear in transmission networks.

- Understand which system architectures are suitable and which are not for distributing synchronization, especially in the presence of various kinds of transmission systems such as PDH, SDH/SONET, radio links, satellite links, leased lines, etc.

- Be acquainted with sophisticated techniques for protection, management and performance monitoring of synchronization networks.

- Know the basics of clock modelling and characterization, in order to effectively design synchronization systems and to understand standard synchronization requirements.

- Understand the realization principles of most common clocks and synchronization sources.

- Be familiar with advanced techniques for time and frequency measurement on synchronization systems, in order to assess the performance of clocks and synchronization links and to verify their compliance with the new international standards.

It is perhaps needless to say that quality of service degradations due to a particular synchronization problem always look sudden, unexpected and of mysterious origin for almost everybody but the (good) synchronization engineer. Rather surprisingly, engineers with a solid expertise on the above-mentioned topics are not common. The results are quite evident: gross mistakes in system design and management produce quality of service degradations that unfortunately, due to ignorance, are often deemed unavoidable. On the contrary, some synchronization culture in the technical personnel can yield substantial improvements in the whole network operation.

This book provides a wealth of information needed by engineers and researchers interested in network synchronization. All the topics listed above as foundations of synchronization culture are addressed thoroughly, clearly and with detailed examples. Both theoretical and practical information is provided.

To my knowledge, this is the only book on the market to systematically deal with digital network synchronization. It covers a wide spectrum of subjects from the basics of digital multiplexing and of bit and pointer justification to advanced and specialized topics. It is well suited to a wide audience, being capable of introducing the inexperienced reader to the subject, but also providing the specialist with a deeper insight into several related advanced topics. Moreover, it features an outstanding bibliography enabling the reader to probe further.

Maurizio Dècina
Past President of the IEEE Communications Society

ACKNOWLEDGEMENTS

I wish to thank all the people who have helped me throughout the long years that I needed to complete the manuscript.

First, all members of the Italian Study Group on Synchronization, established by Telecom Italia and joined by Sirti, Fondazione Ugo Bordoni and CSELT (now Telecom Italia Lab), with whom I began to work with in the early 1990s. In particular, my former boss Luca Valtriani in Sirti; Marco Carbonelli, Domenico De Seta and Daniele Perucchini (Fondazione Ugo Bordoni); Alfonso Mariconda (Telecom Italia); Antonio Manzalin and Roberto Bonello (CSELT). Most of the experimental work presented in this book was carried out while I was with the R&D Division of Sirti, within the Italian Study Group on Synchronization.

Among them, a special mention is due to Carbonelli, De Seta and Perucchini, in particular for having derived or gathered from literature many results presented in Sections 5.10, 5.11 and 5.12, and to Luca Valtriani, who first guided me on the way of synchronization. Moreover, I want to mention Masami Kihara (NTT Network Innovation Laboratories) for his help with Section 4.8.3, Patrizia Tavella (Instituto Elettrotecnico Nazionale Galileo Ferraris) and Dominik Schneuwly (Oscilloquartz SA) for comments on clock performance data and fruitful technical discussions.

I feel also indebted to Mark Hammond, Senior Publishing Editor and Sarah Hinton, Assistant Editor of John Wiley & Sons, Ltd at Chichester, U.K. They have supported my work for years, remained friendly, and waited patiently for so long while I completed my manuscript, well beyond the deadline agreed at the beginning.

Finally, beyond this book, I wish to express my deep gratefulness to Maurizio Decina. I still remember some of his lessons when I was a student at the Politecnico di Milano and he taught me my first rudiments of telephone networks, 15 years ago. Since I have become Assistant Professor at the same university, he has been teaching me something even more difficult: the craft of the Professor.

Skill

Stefano Bregni

ABBREVIATIONS

A

AAL	ATM Adaptation Layer
AC	Alternated Current
ADEV	Allan Deviation
ADM-N	Add-Drop Multiplexer of level N
AIS	Alarm Indication Signal
AM	Amplitude Modulation
AMI	Alternate Mark Inversion
ANSI	American National Standard Institute
APLL	Analog Phase-Locked Loop
APS	Automatic Protection Switching
AS	Anti-Spoofing
AT&T	American Telephone and Telegraph Co.
ATM	Asynchronous Transfer Mode
AUG	Administrative Unit Group
AU-n	Administrative Unit of level n
AVAR	Allan Variance

B

B3ZS	Bipolar with 3 Zeroes Substitution
B6ZS	Bipolar with 6 Zeroes Substitution
B8ZS	Bipolar with 8 Zeroes Substitution
BC	Before Christ
BELLCORE	BELL COmmunications REsearch
BER	Bit Error Rate
BIP	Bit Interleaved Parity
B-ISDN	Broadband Integrated Services Digital Network
BITS	Building Integrated Timing Supply
BOC	Bell Operating Company
BPSK	Binary Phase Shift Keying
BSC	Base Station Controller
BSFR	Basic Synchronization Reference Frequency
BTS	Base Transceiver Station

BVA   *Boîtier à Vieillisement Amélioré* (packaging for improved ageing
                    performance)

C

C/A   Coarse/Acquisition (code)
CBR   Constant Bit Rate
CC   Composite Clock
CCITT   *Comité Consultatif International de Téléphonie et Télégraphie*
CD   Compact Disc
CDM   Code Division Multiplexing
CDMA   Code Division Multiple Access
CDN   *Circuiti Diretti Numerici* (Direct Digital Circuits)
CDU   Clock Distribution Unit
CEP   Circular Error Probable
CKi Clock number i out of a chain
CMI   Code Mark Inversion
C-n Container of level n
CPU   Central Processing Unit
CRC   Cyclic Redundancy Code
Cs   Caesium
CS   Convergence Sublayer
CSM   Clock Supply Module
CUT   Clock Under Test

D

DB   Data Base
DC   Direct Current
DCC   Data Communication Channel
DCF   Data Communication Function
DCN   Data Communications Network
DCO   Digital Controlled Oscillator
DCS   Digital Cross-connect System (North American abbreviation)
DGPS   Differential GPS
DMTD   Dual-Mixer Time-Difference
DoD   Department of Defense
DPLL   Digital Phase-Locked Loop
DQDB   Distributed Queue Dual Bus
DS-i Digital Signal of level i in the North American PDH
DXC   Digital Cross-Connect system (European abbreviation)

E

E-BER   Excessive Bit Error Rate
EDFA   Erbium-Doped Fibre Amplifier
E-i digital signal of level i in the European PDH
EM   Element Manager
EN   European Norm

ESF   Extended Super Frame
ETSI   European Telecommunications Standard Institute

F

FCAPS   Fault Configuration Accounting Performance Security
FDDI   Fibre Distributed Data Interface
FDM   Frequency Division Multiplexing
FEBE   Far End Block Error
FET   Field Effect Transistor
FF   Flip Flop
FFM   Flicker Frequency Modulation
FFT   Fast Fourier Transform
FLL   Frequency-Locked Loop
FPM   Flicker Phase Modulation

G

GA   Ground Antenna
GLONASS   GLObal NAvigation Satellite System
GPIB   General Purpose Interface Bus
GPS   Global Positioning System
GSM   Global System for Mobile communications

H

H   Hydrogen
HDB3   High Density Bipolar with maximum of 3 zeroes
hi-fi   High-fidelity
HMS   Hierarchical Master Slave

I

IEEE   Institute of Electrical and Electronics Engineers
IENGF   *Istituto Elettrotecnico Nazionale Galileo Ferraris*
IF   In Frame
ISDN   Integrated Services Digital Network
ISI   Inter-Symbol Interference
ISO   International Standard Organization
ISPT   *Istituto Superiore delle Poste e Telecomunicazioni* (Institute of the
                 Italian Post and Telecommunications Office)
ITU-T   International Telecommunications Union — Telecommunications
                 standardization sector

L

LAN   Local Area Network
LASER   Light Amplification by Stimulated Emission of Radiation
LOF   Loss Of Frame
LOP   Loss Of Pointer
LORAN   LOng-RAnge Navigation

LOS   Loss Of Signal
LTM-N Line Terminal Multiplexer of level N

M

M   Master
MADEV   Modified Allan Deviation
MASER   Microwave Amplification by Stimulated Emission of Radiation
MAVAR   Modified Allan Variance
MC   Master Clock
MCF   Message Communication Function
MCS   Master Control Station
MF   Mediation Function
MS   (1) Master Slave
MS   (2) Monitor Station
MS-AIS   Multiplexer Section Alarm Indication Signal
MSC   Mobile Switching Centre
MS-DCC   Multiplexer Section Data Communication Channel
MSOH   Multiplexer Section OverHead
MTIE   Maximum Time Interval Error

N

NDF   New Data Flag
NE   Network Element
NEF   Network Element Function
N-ISDN   Narrow-band Integrated Services Digital Network
NM   Network Manager
NTP   Network Time Protocol
NTT   Nippon Telephone and Telegraph Co.

O

OCXO   Oven-Controlled Crystal Oscillator
OH   OverHead
ONR   *Oscillatore Nazionale di Riferimento* (National Reference Oscillator)
OOF   Out Of Frame
OS   Operation System
OSF   Operation System Function
OSI   Open System Interconnection

P

P   Precise (code)
PABX   Private Automatic Branch eXchange
PAM   Pulse Amplitude Modulation
PCM   Pulse Code Modulation
PDH   Plesiochronous Digital Hierarchy
PJE   Pointer Justification Event
PLL   Phase-Locked Loop

POH	Path OverHead
POTS	Plain Old Telephone Service
PPS	Precise Positioning Service
PRC	Primary Reference Clock
PRS	Primary Reference Source
PSD	Power Spectral Density
PSTN	Public Switched Telephone Network

Q

QAF	Q-Adapter Function
QoS	Quality of Service
QPSK	Quaternary Phase Shift Keying
Qz	Quartz

R

Rb	Rubidium
RC	(1) Reference Clock
RC	(2) Resistor-Capacitor
RDI	Remote Defect Indication
REI	Remote Error Indication
RF	Radio Frequency
rms	root mean square
R-N	Regenerator of level N
RS-DCC	Regenerator Section Data Communication Channel
RSOH	Regenerator Section OverHead
RTG	Regenerator Timing Generator
RTIErms	root mean square of the Relative Time Interval Error (old term)
RWFM	Random Walk Frequency Modulation
RZ	Returning to Zero

S

S	Slave
SA	Selective Availability
SASE	Stand-Alone Synchronization Equipment
SC	Slave Clock
SCIU	Synchronous Clock Insertion Unit
SDH	Synchronous Digital Hierarchy
SEC	SDH Equipment Clock
SEMF	Synchronous Equipment Management Function
SEP	Spherical Error Probable
SETG	Synchronous Equipment Timing Generator
SETPI	Synchronous Equipment Timing Physical Interface
SETS	Synchronous Equipment Timing Source
SETS_MP	SETS Management Point
SF	Super Frame
SFET	Synchronous Frequency Encoding Technique

SGT	*Stadi di Gruppo di Transito* (transit office)
SGU	*Stadi di Gruppo Urbano* (local office)
SI	International System
SIS	Signal In Space
SM	Sub-Master
SNM	Synchronization Network Manager
SNMP	Simple Network Management Protocol
SOH	Section OverHead
SONET	Synchronous Optical NETwork
SPS	Standard Positioning Service
SRTS	Synchronous Residual Time Stamp
SSB	Single Side Band
SSM	Synchronization Status Message
SSU	Synchronization Supply Unit
STM	(1) Synchronous Transport Module
STM	(2) Synchronous Transfer Mode
STM-N	Synchronous Transport Module of level N
STS-N	Synchronous Transport Signal of level N
SV	Space Vehicle
SYNTRAN	SYNchronous TRANsmission

T

T	Translator
TCXO	Temperature-Compensated Crystal Oscillator
TDEV	Time Deviation
TDM	Time Division Multiplexing
TDMA	Time Division Multiple Access
TE	Time Error
TEX	Timing EXtraction unit
T-i	digital transmission system of level i carrying a DS-i
TI	Time Interval
TIE	Time Interval Error
TIErms	root mean square of the Time Interval Error
TMN	Telecommunications Management Network
TMS	Timing Monitoring System
TS	(1) Time Stamp
TS	(2) Time Slot
TSG	Timing Signal Generator
TUG	Tributary Unit Group
TU-m	Tributary Unit of level m
TVAR	Time Variance

U

UI	Unit Interval
US	United States
USA	United States of America
UTC	Universal Time Coordinated

V

VC-*n*	Virtual Container of level *n*
VCO	Voltage-Controlled Oscillator
VCXO	Voltage-Controlled Crystal Oscillator

W

WFM	White Frequency Modulation
WPM	White Phase Modulation
WS	Work Station
WSF	Work Station Function

X

XO	Crystal Oscillator

SYMBOLS

$\overline{(\cdots)}$, $\langle\cdots\rangle$	infinite time averaging: $\langle s(t) \rangle = \lim\limits_{T \to \infty} \dfrac{1}{2T} \displaystyle\int_{-T}^{T} s(t)dt$
$x(t) * y(t)$	convolution integral of signals $x(t)$ and $y(t)$:
	$$x(t) * y(t) = \int_{-\infty}^{\infty} x(\tau)y(t-\tau)\,d\tau$$
$\lfloor x \rfloor$	the largest integer less than or equal to the real number x
$\log()$	natural logarithm
$E\{\cdot\}$	expected value (ensemble average)
$\mathrm{sgn}(x)$	signum function
a	number of bits of the alignment word (frame synchronization)
$A(t)$	instantaneous amplitude of a signal
$\mathrm{ADEV}(\tau)$	Allan deviation
$\mathrm{AVAR}(\tau)$	Allan variance
b_a	number of additional bits in one frame of a digital multiplex signal
b_d	number of bits available to tributaries in one frame of a digital multiplex signal
b_m	number of bits in one frame of a digital multiplex signal (frame length)
b_s	number of bits in one sector of a frame of a digital multiplex signal
B	bandwidth
B_{IF}	equivalent noise bandwidth
B_L	bandwidth of a loop filter
B_n	occurrence time of the nth clock pulse (significant instant)
B_{PLL}	bandwidth of a PLL
$\tilde{B}_n$	occurrence time of the nth pulse of the smoothed clock $\tilde{\alpha}(n)$
$B_0^H(\rho)$	total jitter power, low-pass filtered by a desynchronizer APLL $H(f)$, resulting from a synchronizer with justification ratio ρ
$B_\varphi^H(\rho)$	total jitter power, low-pass filtered by a desynchronizer APLL $H(f)$, resulting from a synchronizer with justification ratio ρ and stuff thresholds modulated by a dither sequence $\varphi[n]$
$B_{\mathrm{RD}}^H(\rho)$	total jitter power, low-pass filtered by a desynchronizer APLL $H(f)$, resulting from a synchronizer with justification ratio ρ and stuff thresholds modulated by a random dither sequence

$B_{\mathrm{ST}}^{H}(\rho)$	total jitter power, low-pass filtered by a desynchronizer APLL $H(f)$, resulting from a synchronizer with justification ratio ρ and stuff thresholds modulated by a saw-tooth dither sequence
c	light speed in the vacuum
$c_i(t)$	code signal out of a series of several code signals
$C_x(\tau)$	autocovariance of signal $x(t)$
D	linear fractional frequency drift rate
$e(t)$	low-pass filtered error voltage in PLL theory
$e[kT]$	discrete-time, continuous-amplitude sequence of jitter values
E_i	energy of quantum level i
f	frequency (mostly denoting specifically the Fourier frequency)
$f(t)$	instantaneous frequency (mostly denoting specifically the instantaneous Fourier frequency)
f_0	nominal frequency or some centre frequency
f_{c}	cut-off frequency
f_{h}	upper cut-off frequency of power-law noise (clock hardware bandwidth)
$\hat{f}_{\mathrm{h}}$	noise upper cut-off frequency introduced by the measurement set-up (measurement hardware bandwidth)
f_{j}	jitter frequency
f_k	impulse response samples of the digital loop filter in a desynchronizer bit-leaking DPLL
f_{m}	(1) bit rate of a digital multiplex signal
f_{m}	(2) frequency of a frequency-modulating sinusoidal signal
f_{m0}	nominal frequency (bit rate) of a digital multiplex signal
f_{r}	read frequency of bits (bit transmission rate) from an elastic store
f_{r0}	nominal read frequency of bits from an elastic store
$f_{\mathrm{r}i}$	instantaneous read frequency of bits of tributary i from the elastic store of a bit synchronizer
f_{s}	sampling frequency
f_{t}	instantaneous bit rate of tributaries (same value for all) in a digital multiplexer
f_{t0}	nominal bit rate of tributaries in a digital multiplexer
$f_{\mathrm{t}i}$	instantaneous bit rate of the tributary i in a digital multiplexer
$f_{\mathrm{t}i0}$	nominal bit rate of the tributary i in a digital multiplexer
f_{VC}	average bit rate of a Virtual Container
f_{w}	write frequency (arrival rate) of incoming bits into an elastic store
f_{wi}	instantaneous write frequency of the bits of tributary i into the elastic store of a bit desynchronizer
F	magnetic moment
$F(s)$	filter transfer function (loop low-pass filter in PLL theory)
F_{slip}	slip rate in a bit synchronizer
$g(t)$	(1) symbol basic signal in base-band digital transmission
$g(t)$	(2) continuous-time signal output by the flip-flop in a desynchronizer APLL
$g_{\mathrm{F}}(t)$	low-pass filtered $g(t)$ (2) in a desynchronizer APLL
$G_0(s)$	open-loop transfer function in PLL theory

h	Planck's constant
$h(t)$	impulse response, inverse transform of the transfer function $H(s)$
$h[n]$	true jitter sequence
$\tilde{h}[n]$	true jitter sequence associated to $\tilde{B}_n$
h_{+2}	coefficient of WPM noise in the power-law model
h_{+1}	coefficient of FPM noise in the power-law model
h_0	coefficient of WFM noise in the power-law model
h_{-1}	coefficient of FFM noise in the power-law model
h_{-2}	coefficient of RWFM noise in the power-law model
$h_A(t)$	impulse response of the filter associated to the Allan variance
$h_H(t)$	impulse response of the filter associated to the Hadamard variance
$h_I(t)$	impulse response of the filter associated to the true variance
$h_{MA}(t)$	impulse response of the filter associated to the modified Allan variance
$H(f)$	(1) generic transfer function in the Fourier domain
$H(s)$	(2) closed-loop transfer function in PLL theory
$H_A(f)$	(1) transfer function associated to the Allan variance
$H_A(s)$	(2) transfer function ϕ_{out}/V_{DF} in PLL theory
$H_B(s)$	transfer function ϕ_{out}/V_{VCO} in PLL theory
$H_C(s)$	transfer function $(\phi_{out} - \phi_{in})/\phi_{in}$ in PLL theory
$H_H(f)$	transfer function associated to the Hadamard variance
$H_I(f)$	transfer function associated to the true variance
$H_{MA}(f)$	transfer function associated to the modified Allan variance
i, j, k, l, m, n	integers, mainly used as summation indices
$I^2(\tau)$	classical variance of the fractional frequency $y(t)$
k	transmission errors that can be corrected in a word of $2k + 1$ justification control bits
K	number of SSUs in the ITU-T and ETSI synchronization network reference chain
K_0	VCO gain (in PLL theory)
KA	loop gain (in PLL theory)
K_m	multiplier gain (in PLL theory)
l	(1) length (e.g., of an optical fibre)
l	(2) number of degrees of freedom of a χ^2 distribution
L	frame length [bit]
$\mathscr{L}(f)$	single-sideband measure of phase noise 'script ell'
M	(1) parameter of the M-sample variance $\sigma_y^2(M, \tau, \tau)$
M	(2) number of SECs in a subchain between two SASE clocks
M	(3) number of consecutive independent measurements of MTIE(τ) on disjointed intervals
M_p	peak overshoot
MADEV(τ)	modified Allan deviation
MAVAR(τ)	modified Allan variance
Mod σ_y^2	modified Allan variance
MTIE (τ, T)	Maximum Time Interval Error
n	(1) parameter of the modified Allan variance ($n = \tau/\tau_0$)
n	(2) refining factor in Vernier interpolation in digital time counters

$n(t)$	noise signal
n_c	group index of refraction of the fibre core
$\vec{n}_L(t)$	complex envelope of the noise process $n(t)$
N	(1) buffer size [bit] of the bit synchronizer at the input of a digital switching exchange
N	(2) hierarchical level of the SDH and of the SONET
N	(3) number of SECs in the ITU-T and ETSI synchronization network reference chain
N	(4) number of samples
N	(5) number of SASE clocks in a chain
N_0	one-sided spectral density of white noise
N_b	number of bits coding a sample in a digital signal
N_d	frequency division factor in TE direct digital measurement
N_t	number of tributaries in a digital multiplex
N_T	number of available TE samples in the measurement interval T
N_τ	number of available TE samples in the observation interval τ
$\mathcal{P}$	the probability of correct recognition of the alignment word in frame synchronization
p_{fa}	probability of fake alignment in frame synchronization
$q(t)$	quantization function
q_k	power-series coefficients of the frequency drift
Q	quality factor
r	digital multiplex signal redundancy
$r(t)$	(1) generic received signal
$r(t)$	(2) signal output by the VCO in PLL theory
$R(n)$	ratio of the modified Allan variance over the plain Allan variance
s	number of justification control bits in one frame of a digital multiplex signal
$s(t)$	generic signal
$s_i(t)$	signal out of a series of several signals
s_y^2	Allan variance estimate
$S_x(f)$	one-sided power spectral density of $x(t)$
$S_y(f)$	one-sided power spectral density of $y(t)$
$S_{\Delta v}(f)$	one-sided power spectral density of non-normalized frequency fluctuations
$S_\varphi(f)$	(1) one-sided power spectral density of $\varphi(t)$
$S_\varphi(f)$	(2) power spectral density of the dither sequence $\varphi[n]$
$S_{\varphi_{in}}(f)$	power spectral density of the phase noise at the input of a slave clock chain
$S_{\varphi_{out},j}(f)$	power spectral density of the phase noise at the output of the jth clock of a slave clock chain
$S_s^{RF}(f)$	power spectral density in radio frequency
t	time
T	(1) generic time interval (e.g., pulse repetition period)
T	(2) measurement time interval
T^*	overall measurement interval in the MTIE measurement on disjointed intervals

$T(t)$	Time function generated by a clock
t_d	delay added by a delay line
t_k	discrete-time instants
t_r	alignment recovery time (reframing time) in frame synchronization
t_{tr}	time of a trigger event
T_r	rise time
T_s	(1) sampling period
T_s	(2) settling time
T_{m0}	frame nominal duration of a digital multiplex signal
T_w	average interval between the significant instants of a clock
$\text{TDEV}(\tau)$	Time deviation
$\text{TE}(t)$	Time Error function
TE_k	Time Error samples
$\text{TIE}_{rms}(\tau)$	root mean square of the Time Interval Error
$\text{TIE}_t(\tau)$	Time Interval Error function
$\text{TI}_t(\tau)$	Time Interval function
$\text{TVAR}(\tau)$	Time variance
u_k	output sequence of the digital loop filter in a desynchronizer bit-leaking DPLL
$U(t)$	Heaviside step function
v	number of frame alignment bits in one frame of a digital multiplex signal
$V(t)$	tension signal
V_{DF}	tension noise produced cumulatively by the phase detector and the loop filter (in PLL theory)
w	number of miscellaneous overhead bits in one frame of a digital multiplex signal
w_i	relative weight in a weighted average
$x(t)$	random time deviation
$x_H(t)$	random time deviation measured on a heterodyne signal
x_k	sequence of samples of $x(t)$
$X_h(e^{j\omega T_w})$	discrete Fourier transform of the sequence $h[n]$
$X_{\tilde{h}}(e^{j\omega T_w})$	discrete Fourier transform of the sequence $\tilde{h}[n]$
$y(t)$	random fractional frequency deviation
$\{y_k\}$	first difference of the sequence $\{x_k\}$
$\overline{y}_k$	averaged sample of the normalized, fractional frequency $y(t)$
$y[m]$	sequence of stuff decisions at instant m (0: no justification; 1: positive justification; -1 negative justification)
$z_k[n]$	kth order autocorrelation sequence of $e^{-j2\pi\varphi[n]}$ (cf. Equation (3.21))
$\{z_k\}$	second difference of the sequence $\{y_k\}$
$Z_k(f)$	discrete-time Fourier transform of the kth order autocorrelation sequence $z_k[n]$

Greek letters

α	(1) number of pre-alarm states in a frame alignment strategy between the *correct alignment state* A_0 and the *out of alignment state* B

α	(2) power index in the power-law noise model
$\alpha(\Gamma)$	'phase' function (cf. Equation (3.5))
$\tilde{\alpha}(\Gamma)$	'phase function' of a smoothed clock, obtained by low-pass filtering $\alpha(\Gamma)$
β	percentile
γ	Euler's constant
Γ	buffer address variable (argument of $\alpha(\Gamma)$)
$\Gamma(t)$	gamma function
δ	number of pre-alignment states in a frame alignment strategy between the *provisional alignment state* C_0 and the *correct alignment state* A_0
$\delta(t)$	ideal pulse (Dirac's delta)
Δf_{m0}	tolerance of the multiplex signal frequency in a digital multiplexer
Δf_{r0}	tolerance of the tributary read frequency from the elastic store of the bit synchronizer in a digital multiplexer
Δf_{t0}	tolerance of the tributary frequency in a digital multiplexer
$\Delta\Phi$	total phase offset
$\Delta\varphi(t)$	instantaneous phase error
$\Delta\nu$	(1) frequency offset
$\Delta\nu$	(2) frequency uncertainty (by Heisenberg's principle)
$\Delta\nu_{\max}$	maximum frequency offset
$\Delta\omega_{HI}$	hold-in range (in PLL theory)
$\Delta\omega_{LI}$	lock-in range (in PLL theory)
$\Delta\omega_{PI}$	pull-in range (in PLL theory)
$\Delta\omega_{PO}$	pull-out range (in PLL theory)
ε	stationary rate of bit line errors (BER)
$\varepsilon(t)$	generic error function
ε_n	phase detector output sequence (in a desynchronizer bit-leaking DPLL)
ε_{tr}	error in detecting the time of a trigger event
η_H	heterodyne improving factor
ζ	damping ratio
θ	(1) temperature
$\theta(t)$	(2) phase slowly varying with respect to the initial frequency ω_0 (in PLL theory)
λ	wavelength
μ	power index in the time-domain power law of Allan variances
$\mu(x)$	input–output characteristic of a thresholder
$\nu(t)$	instantaneous frequency
ν_0	(1) starting frequency
ν_0	(2) resonance frequency in atomic clocks
ν_p	probing frequency of the resonator in atomic clocks
$\nu_a(t)$	random time-dependent component of the total instantaneous frequency
$\nu_d(t)$	deterministic (for a given oscillator) time-dependent component of the clock total frequency (modelling the *frequency drift*)
$\nu_H(t)$	beat frequency of a heterodyne signal

ν_n	nominal frequency
$\xi[n]$	jitter sequence with dithering noise added (in a synchronizer)
ρ	(1) justification ratio
ρ	(2) geometric progression ratio of the observation interval τ in the MTIE measurement on disjointed intervals
σ	standard deviation
σ^2	variance
$\sigma^2_{s(t)}$	variance of the signal $s(t)$
σ^2_x	Time variance
$\sigma^2_y(M, \tau, \tau)$	M-sample variance of $y(t)$
σ^2_y	Allan variance
τ	(1) observation interval
τ	(2) propagation delay
τ_0	sampling period
τ_b	beat period of a heterodyne signal
τ_p	clock period
$\phi(t)$	phase error between input and output signals (in PLL theory)
$\phi_{in}(t)$	phase noise on the input signal
$\phi_{out}(t)$	phase noise on the output signal
ϕ_R	phase margin
ϕ_{VCO}	phase noise generated by the VCO (in PLL theory)
$\varphi(t)$	random phase deviation in the timing signal model
$\varphi[n]$	dithering noise sequence (in a synchronizer)
$\varphi_{in}(t)$	phase noise on an input timing signal
φ_n	output phase sequence (in a desynchronizer bit-leaking DPLL)
$\varphi_{out}(t)$	phase noise on an output timing signal
φ_{VCO}	same as ϕ_{VCO}
$\Phi(t)$	total phase of a timing signal
$\Phi_{in}(f)$	Fourier transform of $\varphi_{in}(t)$
$\Phi_{out}(f)$	Fourier transform of $\varphi_{out}(t)$
$\Phi_r(t)$	buffer total read phase (in a synchronizer)
$\Phi_s(t)$	buffer phase difference (in a synchronizer)
$\Phi_w(t)$	buffer total write phase (in a synchronizer)
χ^2	probability distribution of the sum of M square normally distributed random variables (M degrees of freedom)
ω	angular frequency
ω_0	angular nominal frequency or initial frequency
ω_c	crossover angular frequency
ω_F	free-run angular frequency
ω_n	natural angular frequency

1

INTRODUCTION

Synchronization is *the act of synchronizing* (Webster's Ninth New Collegiate Dictionary [1.1]) i.e. making synchronous (cf. the Greek etymon σύγχρονος) the operation of different devices or the evolving of different processes by aligning their time scales.

Many operations in digital systems must obey a precedence relationship. If two operations obey some precedence, then synchronization ensures that operations follow in the correct order [1.2]. At the hardware level, synchronization is accomplished by distributing a common timing signal to all the modules of the system. At a higher level of abstraction, software processes synchronize by exchanging messages.

Depending on the application field, different systems of *abstractions* are adopted usefully, structured in a hierarchical fashion, where each level of abstraction relies on the features of the abstraction level below and hides unnecessary details to the higher level. Abstractions enable designers to ignore such unnecessary details and focus on the essential features, thus making easier achieving a greater complexity of the system designed.

In digital hardware systems, a common approach is to structure the system representation in abstraction levels such as the *physical level*, in which the designer is concerned about the physical laws governing semiconductor properties; the *circuit level*, where he deals with transistors, resistors, etc.; the *element level*, focused on gates, logical ports, etc.; the *module level*, where elements are grouped to form more complex entities, such as memories, logic units, CPUs, etc.

Communications protocols [1.3], on the other hand, are mostly implemented as software modules, structured according to a layer model as well. Protocol stacks are built where protocols at a given level provide services to the upper level protocols and use services of the lower level ones. In the Open System Interconnection (OSI) reference protocol model, defined by ISO, seven abstraction levels (layers) are stacked. Standards of level 1 (*physical layer*) define physical interfaces and basic bit framing, i.e. specify how logical bits (1 and 0) are transmitted over the physical media, in order to provide a raw point-to-point digital transmission channel. Standards of level 2 (*data-link layer*) define protocols which provide for example a point-to-point *error free* digital channel, basing on a layer-1 raw digital channel, through re-transmission of errored frames or error correction techniques. Protocols of upper levels provide, at increasing abstraction levels, network routing services (*network layer*), transport services through a network (*transport layer*) and so on up to services provided by applications directly to the end user.

Whichever is the abstraction criterion adopted in describing hardware and software systems, many are the entities mutually correlated, at any level, whose correct operation relies on temporal coordination. Though, entities of different abstraction levels, both in hardware and software systems, usually require different and independent synchronization functions. For example, the synchronization of protocol processes at a given level is in principle independent from the synchronization of activities of lower-level processes. Moreover, it is clear from the examples above that synchronization issues and techniques may be completely different according to the level of abstraction and to the nature of the elements or processes to synchronize.

This consideration raises some doubts on the appropriateness of adopting the elliptical term 'synchronization' to refer to a whole set of heterogeneous issues where temporal coordination is essential. Nevertheless, a thorough investigation points out some common points among the various contexts where synchronization is addressed, thus giving a reason why historically this term has been adopted comprehensively.

1.1   SYNCHRONIZATION IN TELECOMMUNICATIONS

For many digital communications engineers, the term *synchronization* is familiar in a somewhat restricted sense, meaning only the acquisition and tracking of a clock in a receiver, with reference to the periodic timing information contained in the received signal. More properly speaking, this should be referred to as carrier or symbol synchronization. On the contrary, synchronization plays an essential role in several other areas in telecommunications, at different levels of abstraction and in different contexts too. This introductory section aims at outlining the main contexts in which the word synchronization is used in telecommunications, pointing out that a wide spectrum of different timing issues may be addressed with this word.

Coherent demodulation of an amplitude-modulated signal is based on carrier reconstruction, i.e. on the extraction of a signal coherent with the carrier in frequency and phase (*carrier synchronization*).

Digital demodulation requires in any case, aside from being demodulation coherent or not, to identify the sampling and decision times in order to extract the logical information from the received analogue signal, thus deciding, in the binary case for example, whether a '1' or a '0' has been received (*symbol synchronization*).

Once the logical information has been extracted, the next step, at a higher level of abstraction, is to delineate the frames in the raw and undifferentiated stream of received bits (*frame synchronization*). Synchronizing on the frame starts allows the receiver equipment to understand the role played by the bytes at different positions in the received frames (e.g. the 30 channels allocated to different telephone calls in a European Pulse Code Modulation or PCM primary multiplex, see Section 2.2).

When the source information is split in messages or packets [1.3] transmitted or even routed independently to their destination (packet-switched networks), circuit emulation is possible if the receiver equipment is able to equalize the different delays of the received packets (packet jitter), thus rebuilding the original bit stream as if it were transmitted (more or less) across a circuit-switched network. This delay equalization (*packet synchronization*) is achieved by recovering the original timing from the received packet sequence through adaptive techniques or by processing a source timing information explicitly written in the packet headers (time stamps).

The above concepts deal with the various levels of synchronization in point-to-point transmission. Another level of synchronization is *network synchronization*: focusing on the operation of the whole system of network nodes, it may be advantageous the distribution of a common timing to every node all over the network, in order to transmit and switch in an integrated digital format, so that every Network Element (NE) can operate synchronously with the others and with all the incoming bit streams.

At the highest abstraction level, *multimedia synchronization* deals with the orchestration of heterogeneous elements (images, text, audio, video, etc.) in a multimedia communication at different levels of integration (e.g. physical and human interface).

A different kind of network synchronization is finally the *synchronization of real-time clocks* across a telecommunications network, in which the distribution of the absolute time (e.g., the national standard time) is concerned, mainly to network management purposes.

1.1.1 Carrier Synchronization

In amplitude modulation (AM) systems, modulation is given by multiplying the modulating signal $s(t)$, usually with zero mean, by the carrier wave $\cos 2\pi f_0 t$ in the form

$$x(t) = s(t) \cdot \cos 2\pi f_0 t \tag{1.1}$$

or

$$[1 + ms(t)] \cdot \cos 2\pi f_0 t \tag{1.2}$$

In the latter case, the envelope of the modulated signal $x(t)$ is proportional to $s(t)$ if $|ms(t)| \leq 1$. This allows an easier design of demodulators (*envelope demodulation*).

In the former case, instead, demodulation is possible by multiplying the modulated signal by a sine wave with the same frequency and phase of the carrier (*coherent demodulation*) and then low-pass filtering, since

$$x(t) \cdot \cos \omega_0 t = s(t) \cdot \cos^2 \omega_0 t = \frac{s(t)}{2}(1 + \cos 2\omega_0 t) \tag{1.3}$$

This kind of demodulation requires that the multiplier signal $\cos \omega_0 t$ used in the receiver (carrier reconstructed) has *exactly* the same frequency and phase of the modulated carrier as received. Any phase shift β yields an attenuated signal $(s(t)/2)\cos \beta$ at the output of the low-pass filter (a null signal if $\beta = \pi/2$!).

As can be easily shown from the formulas above, amplitude modulation is equivalent in the frequency domain to a translation of the modulating signal spectrum to the carrier frequency f_0, in case with adding the carrier itself. Actually, the spectrum of an AM signal is redundant, as made of two mirror parts around the carrier frequency f_0. Single Side Band (SSB) modulation consists of transmitting only one of the two mirror halves, thus achieving higher bandwidth efficiency. SSB demodulation must be coherent; here, the phase alignment of the carrier reconstructed is even more critical, as any phase shift yields also a distortion of the demodulated signal.

Amplitude modulation and coherent demodulation are widely treated in literature. A detailed discussion of this topic is beyond the scope of this book. The reader is thus referred to the bibliography [1.4]–[1.7] to probe further.

As stated above, coherent demodulation is based on carrier reconstruction, i.e. on the recovery of a signal coherent with the carrier in frequency and phase. This operation is known as *carrier synchronization*.

Carrier reconstruction may be easy if in the received signal spectrum there is a line at the carrier frequency f_0, as it happens when the modulating signal is not with zero mean. In this case, carrier extraction may be accomplished by means of a suitable narrow passive band-pass filter or a Phase-Locked Loop (PLL). The loop in this case is designed with a narrow band, but in such a way to allow the Voltage Controlled Oscillator (VCO) to lock and track little frequency fluctuations around the nominal frequency f_0.

Unfortunately, in many cases the spectrum line at the carrier frequency f_0 is absent. On the one hand, this is more effective from the information-transmission point of view, because the power of the f_0 line is wasted with respect to the signal-to-noise power ratio. On the other hand, this implies the need of a more sophisticated synchronizer system able to reconstruct the carrier in phase and frequency.

As a simple example of carrier synchronization, let us consider the case of Binary Phase Shift Keying (BPSK) digital transmission, where the symbols 1 and 0 are independent, have same probability and are coded with antipodal square pulses. Then, the modulated wave is of the form $\pm \cos \omega_0 t$ and the power spectrum is continuous without discrete lines at frequency f_0. Obviously, only non-linear transformations can generate the desired f_0 line from the received signal. In this simple case, squaring and frequency-dividing solve the problem (see Figure 1.1): squaring the modulated wave deletes modulation and yields $(1 + \cos 2\omega_0 t)/2$, generating the line at frequency $2f_0$ from which the desired carrier can be extracted by frequency division.

If we consider a quaternary phase modulation (QPSK system, with transmission of 2-bits symbols), the synchronizer will be based on raising the signal to the 4th power to delete the modulation and then generate a line at frequency $4f_0$.

There is a huge amount of literature published in the area of coherent reception and PLL theory. Among the others, we recommend the books [1.7] through [1.12] listed in the References. In particular, the works of Viterbi [1.7] and Lindsey [1.8] are fundamental and can be considered milestones of this field. The book by Blanchard [1.9] focuses on the application of PLLs to coherent receiver design. The books by Gardner [1.10] and Best [1.11] are aimed at practising engineers. The book by Meyr and Ascheid [1.12], finally, deals with phase and frequency control systems widely used in digital communications and provides a deep overview on the whole subject.

1.1.2   Symbol Synchronization

Digital transmission usually deals with sequences of pulses representing the transmitted symbols and sent with rate $R = 1/T$, where T is the interval between consecutive symbols, constant in isochronous transmission.

Reception requires in any case, aside from being demodulation coherent or not, to know the sequence timing, i.e. the temporal position of symbols, in order to extract the logical information from the received analogue signal. Sequence timing information allows symbol reading at the right times.

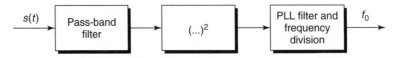

Figure 1.1.   Carrier synchronization for a BPSK system

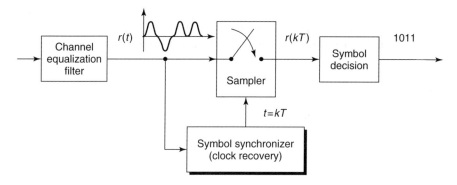

Figure 1.2. Symbol synchronization in a binary baseband receiver

The recovery of the symbol sequence timing from the received analogue signal is known as *symbol synchronization*. Sometimes it is also referred to as *clock recovery*.

Figure 1.2 outlines the scheme of principle of a binary baseband receiver. The analogue received signal $r(t)$ is sampled yielding a sequence of read values $r(kT)$ from which the sequence of bits is extracted by logical decision. The sampler is controlled by a suitable symbol synchronizer system which estimates the reading times $t = kT$ by examining $r(t)$.

Many are the techniques of symbol synchronization adopted. Sometimes the carrier synchronism is recovered first, thus allowing, in coherent band-pass systems, to transfer the signal to base band and then to recover the symbol timing. Some other times the symbol timing is recovered directly from the band-pass signal, through incoherent techniques, apart from carrier recovery.

When symbol synchronization is accomplished after signal conversion to base band, then many possible techniques adopted to recover symbol timing are not very different in principle from carrier synchronization techniques. Considering a multilevel base-band transmission, if in the spectrum of the transmitted signal

$$s(t) = \sum_k a_k g(t - kT) \tag{1.4}$$

there is a line at the symbol frequency $1/T$, with a narrow band-pass filter centred on that frequency it is then possible to recover a sine wave from which to extract the timing pulses at the symbol frequency. If such a line at frequency $1/T$ is absent, as it happens in most cases, it is still possible to generate it by means of a suitable non-linear transformation of the signal. Examples of such a non-linear transformation are a squarer $u = s^2$ or a rectifier $u = |s|$, but many other non-linear transformations generate a component with period T, making thus possible to extract the symbol frequency.

On the other hand, symbol synchronism may be recovered directly from the band-pass signal, apart from carrier recovery and conversion to base band. Considering for example a modulated signal as

$$s(t) = \sum_k a_k g(t - kT) \cos \omega_0 t \tag{1.5}$$

it is enough to take the envelope or square the signal to get a spectrum line at the symbol frequency $1/T$. This line can be used to time symbol reading.

However, there are other carrier and symbol synchronization techniques which are based on different principles from generating spectrum lines through non-linearities. To summarize, symbol-synchronizer implementations can be divided into the following three categories:

(1) based on error tracking;

(2) based on maximum seeking and post filtering;

(3) based on a non-linearity and passive filtering.

The first category is based on PLL systems. The second on comparing a training sequence signal, i.e. a sequence of know symbols sent initially by the transmitter, with a stored replica of the known symbols in order to estimate the phase shift. The third is the example already given above.

Also in this case, a detailed discussion of symbol synchronization systems is beyond the scope of this book. To probe further, the same bibliography [1.4]–[1.12] referenced in the previous section is recommended, together with [1.13] which provides a thorough overview on digital communications and clock recovery issues.

1.1.3   Frame Synchronization

Once carrier and symbol synchronizations have been accomplished and the logical information has been extracted from the incoming signal, the next step, at a higher level of abstraction, is to determine the start and the end of code words or of groups of code words, i.e. to delineate the frames in the raw undifferentiated stream of received bits (see Figure 1.3). These tasks of synchronization are called *word* and *frame synchronization* respectively.

In digital transmission, bits are usually organized in frames in order to assign different meanings to the bytes transmitted. Bytes at different positions within a certain frame may be, for example, assigned to different user channels sharing the same physical medium in Time Division Multiplexing (TDM), as in the case of the 30 channels allocated to different telephone calls in the European PCM primary multiplex (see Section 2.2), or also assigned to overhead functions (error check, transport of management and control information, etc.). Hence, frame synchronization is of primary importance in digital transmission. The *semantics* itself of the received bits, and thus e.g. proper demultiplexing of tributaries begins from the correct delineation of frames.

Any frame synchronization strategy (also referred to as *frame alignment* strategy) is made of two basic operations:

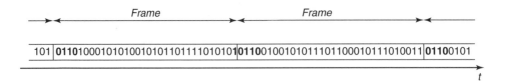

Figure 1.3.   Frame delineation in a bit stream

- *hunting*, whenever the equipment (*frame aligner*) is out of frame synchronization and the frame alignment is sought in the received bit stream;

- *maintenance*, whenever the equipment assumes to be frame-aligned and continuously checks frame boundaries where these are supposed to be.

Frame synchronization is assisted by a special *alignment word*, usually at the beginning of frames, set to special values. Hunting is performed by seeking the alignment word pattern in any position of the received bit stream, maintenance by checking that word where the frame aligner supposes that frames begin. Unfortunately, during hunting the alignment word pattern sought can be simulated in any position by the data bit stream, while during maintenance the alignment word, checked where it really is, cannot be recognized owing to line errors. Therefore, the goals in designing an effective frame alignment strategy are the following:

- under correct alignment conditions, minimize the probability of loss of frame alignment due to transmission line errors (*forced loss of alignment*);

- under out of alignment conditions, minimize the *probability of fake alignment*, owing to any alignment word pattern simulation in the received random bit stream;

- minimize the *frame alignment recovery time*.

The analysis of the stochastic process describing the alignment loss and recovery, according to a chosen frame alignment strategy, can be carried out [1.14]–[1.16] by means of a suitable Markov–Chain model [1.17] as depicted in Figure 1.4, where $\mathcal{P}$ is the probability of correct recognition of the alignment word. Obviously, $\mathcal{P}$ takes different expressions under different conditions and in different areas of the diagram.

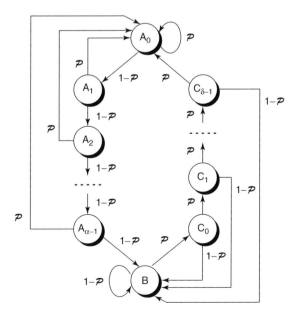

Figure 1.4. Markov–Chain model of frame alignment strategies

From the *correct-alignment state* A_0, in which the maintenance process is carried out, the aligner reaches the *out-of-alignment state* B only after α consecutive alignment words detected with errors. In the state B, the aligner carries out the hunting process and, as soon as it finds a bit pattern equal to the alignment word sought, switches to the *provisional-alignment state* C_0. From there, the aligner runs the maintenance process and reaches the normal state A_0 only after δ consecutive alignment words detected without errors, in order to cover itself against pattern simulations. Otherwise, after the first alignment word with errors detected, it gets back to the state B where the hunting process starts again.

Any transition from the state A_0 to the state B, when the aligner is in A_0 in correct-alignment state, is defined as *loss of alignment*. Such a transition may be caused by:

- line errors in the received alignment words (*forced loss of alignment*);
- loss of the bit timing or a slip of the incoming frames (*real loss of alignment*); if the bit timing is lost, in fact, the aligner operation is inhibited and, as the bit timing is restored, the aligner restarts the hunting process from the state B.

Suitable quantities characterizing the performance of an alignment strategy are therefore:

- the *mean occurrence rate* $\overline{f}_{fl}$ *of forced losses*,
- the *mean* $\overline{t}_r$ and the *variance* $\sigma^2_{t_r}$ of the *alignment recovery time* t_r (*reframing time*), defined as the interval between the start of the hunting process in the state B and the real alignment reacquisition in the state A_0;
- the *probability* p_{fa} *of fake alignment*, i.e. the probability of transition from the state B to the state A_0 owing to alignment word pattern simulations, although remaining in an out-of-alignment condition.

It is worthwhile noticing that the starting time of the hunting process coincides, in the case of forced loss of alignment, with the starting time of a frame, while, in the case of real loss, it coincides with the time of symbol timing reacquisition. Thus, the recovery time for forced losses is statistically not lower than the value for real losses.

As far as the forced losses of alignment are concerned, the probability $\mathcal{P}$ of correct recognition of the alignment word takes the following expression, assuming the system in normal alignment conditions:

$$\mathcal{P} = p_1 = (1 - \varepsilon)^a \cong 1 - a\varepsilon \qquad (1.6)$$

where ε is the stationary rate of bit line errors (assumed uncorrelated) and a is the number of bits of the alignment word (the above approximation is valid for small values of ε).

As far as the hunting process is concerned, on the other hand, the probability $\mathcal{P}$ is now the probability of simulation of the alignment word pattern by the received bit stream. Assuming for the sake of simplicity that these bits are statistically independent and having same probability[1], the probability $\mathcal{P}$ takes the following expression

$$\mathcal{P} = p_2 = \frac{1}{2^a} \qquad (1.7)$$

---

[1] This assumption is not realistic indeed when referring to the overhead bits of the transmission frames, but it works fine for the payload bits.

By analysing the model of Figure 1.4, it is thus possible to derive the following expressions for the aforementioned performance quantities:

$$\bar{f}_{fl} = \frac{f_0}{L} \frac{p_1(1-p_1)^\alpha}{1-(1-p_1)^\alpha} \simeq \frac{f_0}{L} \frac{(1-a\varepsilon)(a\varepsilon)^\alpha}{1-(a\varepsilon)^\alpha} \tag{1.8}$$

$$\bar{t}_r = \frac{L}{f_0} \left\{ 1 + \delta + (L-a)\frac{p_2}{1-p_2}\left[1 + p_2^{\delta-1}\left(\frac{1}{(1-p_2)^{\alpha-1}} - 1\right)\right] \right\} \tag{1.9}$$

$$\sigma_{t_r}^2 = \frac{L^2}{f_0^2}(L-a)\left\{ \frac{p_2^{\delta-1}}{(1-p_2)^{2\alpha}}(2 - p_2^{\delta+1}) \right.$$

$$+ \frac{p_2^{\delta-1}}{(1-p_2)^{\alpha+1}}[2(p_2^{\delta+1}-1) + p_2(1-p_2)(1+2\delta-2\alpha)]$$

$$\left. + \frac{p_2}{(1-p_2)^2}[1 - p_2^{2\delta-1} + p_2^{\delta-1}(1-p_2)(1-2\delta)] \right\} \tag{1.10}$$

$$p_{fa} = 1 - \left[(1-p_2)\sum_{i=0}^{\delta} p_2^i \right]^{L-a} = 1 - (1 - p_2^{\delta+1})^{L-a} \tag{1.11}$$

where L is the frame length (expressed in [bit]) and f_0 is the nominal bit rate of the multiplex signal under study.

As a final remark, since for $L \gg 1$ the random variable t_r tends to be normally distributed (i.e. according to a Gaussian distribution), the maximum alignment recovery time can be conventionally defined as

$$t_{r/\max} = \bar{t}_r + 3\sigma_{t_r} \tag{1.12}$$

Based on the above scheme, different frame alignment strategies have been designed for multiplex signals at the different levels of the Plesiochronous Digital Hierarchy (PDH) [1.18]. By way of example, for the PCM primary multiplex the values $\alpha = 3$ and $\delta = 1$ have been standardized [1.15][1.19].

A slightly different scheme, on the other hand, has been designed for the frame synchronizers of Synchronous Digital Hierarchy (SDH) equipment [1.20] [1.21]. Within the alignment word (up to 96 bytes for STM-16 frames), two different subsets of bytes are considered by the aligner during the hunting and the maintenance processes: seeking for a longer word on hunting reduces the probability of pattern simulation, while checking a shorter word on maintenance reduces the probability of forced loss of alignment. The state diagram is depicted in Figure 1.5. Here, the following three macro-states are considered:

- *In Frame* (IF) state, i.e. the state of normal operation under alignment conditions (corresponding to the correct alignment state A_0);

- *Loss Of Frame* (LOF) state, i.e. the alarm state of out of alignment (corresponding to the state B);

- *Out Of Frame* (OOF) state, i.e. a pre-alarm state (corresponding to the states A_i for $0 < i < \alpha$).

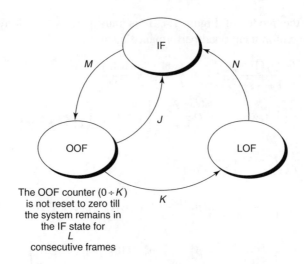

Figure 1.5.   Frame synchronization strategy in SDH systems

From the IF state, in which the maintenance process is run, the aligner reaches the OOF state after M consecutive frames with errors detected in the alignment word. After other K frames in the OOF state (i.e. other K frames with errors detected in the alignment word), where the maintenance process is still run, the frame synchronizer switches to the LOF state. In the OOF state, J consecutive frames with no errors detected in the alignment word take the aligner back to the IF state. In the LOF state, in which the hunting process is run, N consecutive frames with no errors detected in the alignment word are needed to get back to the normal alignment IF state.

Actually, the state diagram of Figure 1.5 is not a Markov model. In order to avoid intermittent OOF ↔ IF switching, the register counting the frames with errors detected in the alignment word while the system is in the OOF state (i.e. the register controlling the OOF → LOF transition and counting $0 \div K$) is not reset to zero till the system remains in the IF state for L consecutive frames.

Standard values [1.21] of the parameters of the SDH frame synchronization strategy are $M \leq 5$, $J \leq 2$, $K = 24$, $N = 24$, $L = 24$.

Finally, it is worthwhile noticing that in some transmission protocols, after frame alignment, it is then necessary to delineate further super-frames spanning several main frames already delineated. This task is called *multi-frame synchronization*. A first example is given by the bits of the PCM time slots carrying signalling, which are organized as frames spanning 16 (in the case of the European 2.048 Mbit/s PCM primary multiplex) or 24 (in the case of the North-American 1.544 Mbit/s PCM multiplex) 125-μs frames. A second example is given by the SDH Tributary Units (TU), which span four 125-μs base frames.

1.1.4   Bit Synchronization

The expression *bit synchronization* is commonly used in telecommunications with mainly two different meanings. Unfortunately, for historical reasons, it is very common in this field to face lexical ambiguities with words of the family of 'synchronous', from one context to another but especially from one author to another. Although in this book particular attention was paid to adopt consistent terms, sometimes the reader is still called to be careful with expressions used with multiple meanings, in order to avoid possible misunderstandings.

On the one hand, bit synchronization is sometimes used to refer to symbol synchronization (see Section 1.1.2) in the special case of binary symbols (bits).

More commonly, on the other hand, bit synchronization is used to denote the synchronization of an asynchronous bit stream according to the equipment local clock. This is accomplished by writing the bits of the asynchronous (i.e., with variable instantaneous frequency) bit stream into an elastic store (buffer) at their own arrival rate, and by reading them out with the frequency of the equipment local clock. Throughout this book, bit synchronization is used with this meaning.

Bit synchronization is accomplished for example to align the bits and frame starts of PCM signals at the inputs of digital exchanges, in order to allow digital switching, which is based on moving octets (speech samples) from one time slot to another (see Section 4.1.3).

Moreover, bit synchronization is accomplished in digital multiplexers in the *synchronizer* block. Here, tributaries are bit-synchronized to fit into the multiplex signal. The synchronizer may implement bit justification or slip buffering to deal with any frequency offset between the tributaries and the multiplex signal, as explained in detail in Section 2.3.

1.1.5 Packet Synchronization

Packet switching consists of splitting the source information in messages or packets [1.3] transmitted or even routed independently to their destination. Packets contain a segment of the source data (e.g., coded speech) plus some header information, and may be either fixed or varied in length. Packets of fixed length are commonly also called *cells*.

Packet switching can be an effective technology to integrate data and voice or other real-time traffic in a single network. For the implementation of the Broadband Integrated Services Digital Network (B-ISDN), a cell-switched technique was chosen by international standard bodies: the Asynchronous Transfer Mode (ATM).

A packet-switched transfer method has the following peculiar characteristics (see Figure 1.6):

- due to the stochastic nature of packet switching and in particular to the queuing inside the network, packets suffer a random delay during transport across the network and arrive at their destination with stochastic inter-arrival times;
- if the packets of a given call are routed independently (i.e., each may follow a different path through the network) packets may even arrive at their destination out of order;
- at the receiver it is impossible to recover the clock frequency of the information source (transmitter) based only on the physical incoming bit stream (OSI layer 1).

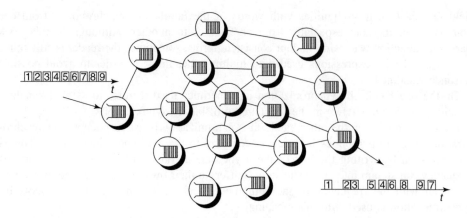

Figure 1.6. Packet jitter across a packet-switched network

The transparent transport of voice (or of any other real-time traffic inherently oriented to circuit switching) across a packet-switched network requires facing the above issues, in order to reproduce acceptable quality speech from packets arriving with varying transit delays or even out of sequence. Therefore, the characteristics above require additional functions to be implemented at the receiver interface where circuit emulation is performed. The equalization of packet random delays (*packet jitter*) is usually referred to as *packet synchronization*.

The task of packet jitter equalization, i.e. of estimating either the packet production time or equivalently its transport delay, may be achieved in several ways [1.22]–[1.27]. Such methods can be divided in two groups: those based on a non-synchronous network environment, where different nodes are timed by local independent clocks, and those based on a synchronous network environment, where a common timing signal is distributed to every node. According to this division, the main methods can be summarized as follows.

1.1.5.1   Non-Synchronous Timing Recovery

- *Blind delay estimate.* The simplest strategy for estimating the production time of an incoming packet is to make a worst case assumption: the receiver assumes that the packet on which the estimate is based arrives with minimum transit delay, and that the other packets may be delayed further not more than a given maximum amount, blindly estimated. Once the target playout time for the first packet has been estimated, the receiver uses sequence numbers in subsequent packets to determine proper playout time for each one, so that every packet results delayed of the maximum amount estimated. Packets received with higher delay are discarded.

- *Roundtrip measurement.* Although blind delay estimation is very simple, it may be not adequate in long-haul networks. A better delay estimation technique is to actually measure the roundtrip delay between the packet sender and the packet receiver on a control packet and use this value to estimate the one-way delay of other packets, assuming that delay is evenly distributed between both directions.

- *Added variable delay.* In this case, the delay experienced by a packet across the network is actually measured where it occurs. The variable delay is measured by means of a delay stamp indicating the delay accumulated by each packet. Each network element adds its delay to the delay stamp as the packet passes through, measuring it with the local clock as time difference between the entry and exit times. Knowing the packet delay allows the packet receiver to determine the playout time as the current time plus the difference between the maximum expected delay stamp value and the actual delay stamp value.

- *Adaptive strategies.* None of the above techniques produces a completely accurate delay measurement. Different algorithms have been thus conceived to adaptively change the estimated delay during packet stream reception, based for example on the receiver buffer fill level or on repeated roundtrip delay measurements.

- *PLL-based adaptive strategies.* All the above methods were conceived for voice transport on narrow-band packet-switched networks. B-ISDN ATM networks, instead, pose more stringent requirements, owing to their higher switching speed and to the wider set of services supported. In ATM networks non-synchronous techniques were proposed, based on filtering the cell jitter by means of a PLL timing recovery. This simple technique can be improved by pre-filtering, with the data buffer fill level or the cell arrival instants being used as inputs to a pre-filter. In this approach, the PLL can be viewed as a low-pass filter of the cell jitter. Although this technique is definitely better performing and more flexible than the previous ones, for circuit emulation at the output of ATM networks it is still difficult to obtain jitter performance compliant with existing standards [1.29].

1.1.5.2 *Synchronous Timing Recovery*

The fundamental concept of synchronous techniques is based on the availability of a common reference clock (synchronization network clock). This is not a problem in SDH networks, and in its North American version Synchronous Optical NETwork (SONET), which have been chosen as the underlying physical layer transport for ATM. For this reason, synchronous techniques for cell jitter reduction have been purposely designed for ATM B-ISDN networks.

- *Synchronous Frequency Encoding Technique (SFET).* The asynchronous source clock is compared to the network reference clock. The discrepancy between the two clocks is measured and encoded in the ATM Adaptation Layer (AAL) overhead [1.30] (every slip between two submultiples of the two clocks is signalled). At the receiver, the common network clock and the encoded information are used to reconstruct the source clock.

- *Time Stamp (TS).* A 16-bit counter is driven by the network reference clock. Two bytes of the Convergence Sublayer (CS) [1.30] overhead carry the current value of this counter every 16 cells. At the receiver side, the source clock is reconstructed from the received TS and the network clock.

- *Synchronous Residual Time Stamp (SRTS)* [1.26]–[1.28]. This method is essentially a modified TS and it is based on the observation that, for accurate source clocks, few least significant bits of the 16-bit TS convey all useful information. SRTS is thus made only of four bits. This allows to incorporate SRTS into the existing AAL overhead without increasing its size. SRTS has been accepted by ITU-T as the timing recovery standard technique for AAL-1 (Circuit Emulation).

A deeper discussion of the various packet synchronization methods is beyond the scope of this book. To probe further, the reader is thus referred to the relevant bibliography already referenced. In particular, we recommend as starting points of further investigation the fundamental works of Montgomery [1.22], dealing with packet voice synchronization techniques, and of Lau and Fleischer [1.26], who proposed the SRTS timing recovery standard technique for circuit emulation on ATM.

1.1.6   Network Synchronization

Network synchronization [1.31]–[1.33] deals with the distribution of time and frequency over a network of clocks, spread over an even wider geographical area. The goal is to align the time and frequency scales of all the clocks, by using the communications capacity of links interconnecting them (e.g. copper cables, fibre optics, radio links). In general, the following are only some of the better known applications:

- synchronization of clocks located at different multiplexing and switching points in a digital telecommunications network;
- synchronization of clocks in telecommunications networks which require some form of time-division multiplexing multiple access, such as satellite networks, GSM (Global System for Mobile communications) [1.34] mobile terminals, etc.;
- range measurement between two nodes in the network, and also position determination and navigation by network users;
- phased array antennas.

Many intriguing examples of synchronization of a large number of oscillators can be found in nature. Lindsey *et al.* [1.31] pointed out as one of the most spectacular ones the synchronous fireflies described by Buck and Buck in Reference [1.35]. These fireflies flash their light organs at regular but individual and independent intervals if they are not close together. Though, if many of these insects are placed in a relatively close proximity, they exhibit a synchronization of their light organs until they flash in unison. Other biological examples are the synchronization of individual fibres in heart muscles to produce the familiar heartbeat, or the resting and active periods of mammals, which exhibit rhythms.

Network synchronization plays a central role in digital telecommunications, having a determining influence on the quality of most services offered by the network operator to its customers. Digital network synchronization is the focus of this book, and it will be thus treated widely in the next chapters together with several related topics.

1.1.7   Multimedia Synchronization

Multimedia refers to the integration of heterogeneous elements such as text, images, audio and video in a variety of application environments. These data can be heavily time-dependent, such as audio and video in a movie, and can require time-ordered presentation during use. The task of coordinating such sequences is called *multimedia synchronization.*

Synchronization can be applied to the playout of concurrent or sequential streams of data, and also to the external events generated by a human user. In other words, temporal relationships between the media may be implied, as in the simultaneous playing of voice and video, or may be explicitly formulated, as in the case of a multimedia document which possesses voice-annotated text or in the case of a multimedia hypertext (*hypermedia*).

Time-dependency of data sequences may be simply linear, as in the case of an audio file played on a sequence of images (slide presentation with soundtrack), but other modes of data presentation are also viable, including reverse, fast-forward, fast-backward and random access. When non-sequential storage, data compression and random communication delays are introduced, the provision of such capabilities can be difficult.

Multimedia synchronization issues have been recently widely addressed in literature. However, a detailed discussion of this topic is beyond the scope of this book. The reader is thus referred to the bibliography [1.36]–[1.43] as a starting point to probe further.

1.1.8   Synchronization of Real-Time Clocks

A substantially different kind of network synchronization is the distribution of a reference absolute time (as the national standard time) to the equipment real-time clocks of a telecommunications network (*synchronization of real-time clocks*).

The distribution of the national standard time is to the purposes of network control and management. Any relevant event noticed by the equipment monitoring system, such as Bit Error Rate (BER) threshold overflowing, line alarms, hardware failures, etc. is recorded for future reporting. When the telecommunications network is managed by a management system (e.g., a standard Telecommunications Management Network, TMN) [1.44]–[1.46], such events are notified by the equipment itself to the Operation System (OS) by means of management messages. In both cases, the record must include, among the others, the key information *Date and Time* as read by the equipment real-time clock.

It is essential that the real-time clocks of the whole network are synchronized to the same absolute time, otherwise it would become impossible to correlate meaningfully different messages (which may be *a lot* and coming from different parts of the network) under a common label (i.e. the actual event happened, the issue that the network manager is interested to know). Only if the equipment real-time clocks have been synchronized to the standard time, it is then possible to establish temporal and logical relationships among different events, and thus to draw useful conclusions from the raw data of event records collected.

Synchronization of real-time clocks differs inherently from network synchronization as intended in Section 1.1.6 and in the main body of this book. Synchronization of real-time clocks, in particular, distributes the *absolute time* information (e.g.: '23 Dec 1998, 01.32.04 AM', or any other kind of time stamp) and poses different requirements of accuracy. For the management needs outlined above, a time accuracy of a few milliseconds is perfectly adequate, as the Date-and-Time field in the management records does not go further than specifying day, month, year and hours, minutes, seconds.

The goal of network synchronization as intended in Section 1.1.6, instead, is minimizing time error fluctuations among the clocks, regardless of the start phase offset. This implies that *synchronous physical timing signals* (e.g. sine waves) are distributed to the network clocks by suitable means. Synchronization of digital telecommunications networks mostly achieves time deviations not greater than 10 ns or 100 ns.

Synchronization of real-time clocks, on the other hand, is usually accomplished by exchanging messages that carry the time information (time stamps), according to a suitable protocol, along the communication links between the network nodes.

As an example, the Network Time Protocol (NTP) [1.47]–[1.49] is used by Internet time servers and their clients to synchronize real-time clocks, as well as to automatically organize and maintain the time synchronization subnet itself. It has evolved from previous simpler protocols, but is specifically designed for high accuracy, stability and reliability even when used over typical Internet paths involving multiple gateways and unreliable networks.

The protocol is based on messages transported over Internet Protocol (IP) and User Datagram Protocol (UDP) packets, which provide a connectionless transport service; however, it is readily adaptable to other protocol suites. Optional features include message authentication and encryption, as well as provisions for remote control and monitoring.

In NTP one or more primary servers synchronize directly to external reference sources. Secondary time servers synchronize to the primary servers according to a suitable hierarchy. Reconfiguration on alternative synchronization routes is possible to survive outages and failures. The algorithm is able to estimate and compensate with precision the random transport delay of packets across the network and thus to achieve an absolute time accuracy in the order of few milliseconds.

1.2   OUTLINE OF THE BOOK

This book surveys all the main topics concerning the synchronization of digital telecommunications networks. It focuses on technical details, when relevant, but it aims at conveying the underlying ideas first, the conceptual understanding of issues. General topics are treated in the first chapters, namely some basics of digital transmission, timing issues in SDH networks and network synchronization architectures. Then, more advanced topics are addressed, namely the characterization and modelling of clocks, their physical principles and realization technology and the time and frequency measurement techniques used in telecommunications.

More in detail, the main contents of each chapter of the book are summarized as follows.

This *Chapter 1—Introduction* presents some general information to bring the reader to the core of the subject. It is pointed out that synchronization plays a key role in several areas in telecommunications, at different levels of abstraction and in different contexts, by describing the very different processes for which the word 'synchronization' is used.

Chapter 2—Asynchronous and Synchronous Digital Multiplexing provides the basics of digital transmission systems, paying special attention to timing aspects. After having introduced some basic concepts and definitions about the timing of digital signals, three fundamental types of digital multiplexing are described: the PCM multiplexing, the asynchronous digital multiplexing with bit justification and the synchronous digital multiplexing. As practical applications of these concepts, the PCM primary multiplex, the PDH and the SDH/SONET are detailed. Throughout this chapter, it is pointed out how different techniques cope with the issue of multiplexing asynchronous tributaries: PDH (asynchronous digital multiplexing) uses bit justification, SDH/SONET (synchronous digital multiplexing) uses pointer justification. Moreover, asynchronous vs synchronous transfer modes and asynchronous vs synchronous frame structures are distinguished.

Chapter 3 — Timing Aspects in SDH Networks deals with several aspects of SDH related to synchronization[2]. In particular, this chapter addresses first the main causes of jitter and wander in an SDH transmission chain: environmental conditions, the tributary mapping structure in SDH frames, bit justification and pointer justification. Then, synchronizers and desynchronizers for the mapping and demapping of PDH tributaries into and from SDH frames are described in detail. Basic and advanced pointer processors are also treated, describing techniques to limit jitter generation due to pointer action. Finally, the functional scheme of the SDH equipment clock is outlined.

Chapter 4 — Network Synchronization Architectures addresses the core topic of this book. First, a historical perspective on network synchronization is provided, pointing out its evolution since the old analogue FDM networks up to the advent of digital switching, of digital PDH and SDH transmission and of other advanced digital techniques like ATM. Then, various network synchronization strategies are detailed, pointing out their analogy with socio-political systems. The standard architectures of synchronization networks are described. Synchronization network planning, management, performance monitoring and protection are discussed. Finally, some real examples of synchronization networks deployed and operating in several countries are provided.

Chapter 5 — Characterization and Modelling of Clocks provides the basic theory needed by engineers involved in the design and testing of network synchronization systems: clock models and the mathematical tools for time and frequency stability characterization are given. First, the timing signal model and the relevant basic quantities commonly used in telecommunications are detailed. Stability and accuracy concepts are introduced. Then, the principles of operation of autonomous and slave clocks are outlined. As far as the latter ones are concerned, Phase-Locked Loop (PLL) fundamentals are provided. Clock stability characterization is dealt with both in the frequency and time domains: all the main stability quantities are thoroughly described, including those not adopted by telecommunications standards but relevant in oscillator science. Common types of clock noise exhibited by experimental measurements are described. Finally, the behaviour of the stability quantities is studied, for both autonomous and slave clocks, depending on various types of clock noise and inaccuracy. In two appendices, fast algorithms for the computation of TVAR and MTIE standard stability quantities are provided.

Chapter 6 — Physical Principles and Technology of Clocks summarizes the principles of operation of high-precision clocks employed in telecommunications applications. All different types of quartz oscillators (namely simple crystal, voltage-controlled, temperature-compensated and oven-controlled oscillators) are described. Then, atomic frequency standards (namely caesium-beam, hydrogen-MASER and rubidium-gas-cell frequency standards) are treated. Some indicative performance data for all the clocks above are also given. Finally, the characteristics and architecture of the Global Positioning System (GPS) are outlined.

Chapter 7 — Time and Frequency Measurement Techniques in Telecommunications provides the basic knowledge needed by engineers involved in experimental measurements on network synchronization systems. Basic concepts and definitions are first introduced, pointing out especially the key aspects relevant to telecommunications applications. For example, guidelines on how to estimate the frequency offset and drift of a clock, to

---

[2] Throughout this book, most times we will simply refer to SDH, which is the ITU-T standard designed as superset of the ANSI standard SONET, understanding that most considerations apply to SONET too, *mutatis mutandis*.

assess the statistical confidence of Allan variance estimates and to separate the noise of the Clock Under Test (CUT) from that of the Reference Clock (RC) are given. Moreover, the impact of the measurement configuration and of the Time Error sampling period on the measured quantities is discussed. A brief survey on some of the most common devices and instruments used for time and frequency measurements is also provided. Then, the main techniques for measuring time and frequency stability are overviewed, from direct digital measurement to heterodyne and homodyne methods for sensitivity enhancement. Several test set-ups are described. Finally, clock stability measurement in telecommunications is examined. The standard technique, based on the acquisition of sequences of TE samples, is detailed. Moreover, MTIE measurement is addressed in particular, due to its peculiar issues. Other measurements on equipment clocks and on network interfaces, such as jitter and phase transients measurements, are described.

1.3   SUMMARY

For many digital communications engineers, the term *synchronization* is familiar in a somewhat restricted sense, meaning only the acquisition and tracking of a clock in a receiver, with reference to the periodic timing information contained in the received signal. On the contrary, synchronization plays an essential role in several areas in telecommunications, at different levels of abstraction and in different contexts too.

At different abstraction levels, the main contexts in which the word synchronization is used in telecommunications are the following:

- *carrier synchronization*, i.e. the extraction of the carrier from a modulated signal in coherent demodulation;

- *symbol synchronization*, i.e. the identification of the sampling and decision times in digital demodulation, in order to extract the logical information from the received analogue signal;

- *word and frame synchronization*, i.e. the identification of the start and end of code words or of groups of code words (frames), or also the delineation of the frames in the raw and undifferentiated stream of received bits;

- *packet synchronization*, i.e. the delay equalization of packet arrival times in order to reconstruct a user circuit with constant bit rate over a packet-switched network;

- *network synchronization*, i.e. the distribution of a common timing over a network of clocks, spread over an even wider geographical area;

- *multimedia synchronization*, i.e. the orchestration of heterogeneous elements (images, text, audio, video, etc.) in a multimedia communication at different (e.g. physical and human interface) levels of integration;

- *synchronization of real-time clocks*, i.e. a substantially different kind of network synchronization in which the distribution of the absolute time (e.g. the national standard time) across a telecommunications network is concerned, mainly to network management purposes.

1.4 REFERENCES

[1.1] *Webster's Ninth New Collegiate Dictionary*. Merriam-Webster Inc., 1986.

[1.2] D. G. Messerschmitt. Synchronization in digital system design. *IEEE Journal on Selected Areas in Communications* 1990; **SAC-8**(8).

[1.3] A. S. Tanenbaum. *Computer Networks*. Englewood Cliffs, NJ, U.S.A.: Prentice Hall Inc., 1996, 3rd edn.

[1.4] S. Benedetto, E. Biglieri, V. Castellani. *Digital Transmission Theory*. Englewood Cliffs, NJ, U.S.A.: Prentice Hall Inc., 1987.

[1.5] J. G. Proakis. *Digital Communications*. New York: McGraw-Hill, 1989.

[1.6] S. Haykin. *An Introduction to Analog and Digital Communications*. New York: John Wiley & Sons, 1989.

[1.7] A. J. Viterbi. *Principles of Coherent Communications*. New York: McGraw-Hill, 1966.

[1.8] W. C. Lindsey. *Synchronization Systems in Communications and Control*. Englewood Cliffs, NJ: Prentice Hall Inc., 1972.

[1.9] A. Blanchard. *Phase-Locked Loops*. New York: John Wiley & Sons, 1976.

[1.10] F. M. Gardner. *Phaselock Techniques*. New York: John Wiley & Sons, 1979.

[1.11] R. E. Best. *Phase Locked Loops*. New York: McGraw-Hill Book Company, 1984.

[1.12] H. Meyr, G. Ascheid. *Synchronization in Digital Communications. Vol. 1: Phase-, Frequency-Locked Loops, and Amplitude Control*. New York: John Wiley & Sons, 1990.

[1.13] Y. Takasaki. *Digital Transmission Design and Jitter Analysis*. Norwood, MA: Artech House, 1991.

[1.14] O. Brugia, M. Decina. Reframing Statistics of PCM Multiplex Transmission. *Electronic Letters* 1969; **5**(24): 625–627.

[1.15] D. T. R. Munhoz, J. R. B. De Marca, D. S. Arantes. On frame synchronization of PCM systems. *IEEE Transactions on Communications* 1980; **COM-28**(8): 1213–1218.

[1.16] P. F. Driessen. Improved frame synchronization performance for CCITT algorithms using bit erasures. *IEEE Transactions on Communications* 1995; **COM-43**(6).

[1.17] L. Kleinrock. *Queueing Systems. Vol. 1: Theory*. New York: John Wiley & Sons, 1975.

[1.18] ITU-T Rec. G.702 *Digital Hierarchy Bit Rates*. Blue Book, Geneva, 1988.

[1.19] ITU-T Rec. G.706 *Frame Alignment and Cyclic Redundancy Check (CRC) Procedures Relating to Basic Frame Structures Defined in Recommendation G.704*. Geneva, April 1991.

[1.20] ITU-T Rec. G.707 *Network Node Interface for the Synchronous Digital Hierarchy (SDH)*. Geneva, March 1996.

[1.21] ITU-T Rec. G.783 *Characteristics of Synchronous Digital Hierarchy (SDH) Equipment Functional Blocks*. Geneva, April 1997.

[1.22] W. A. Montgomery. Techniques for packet voice synchronization. *IEEE Journal on Selected Areas in Communications* 1983; **SAC-1**(6).

[1.23] J. Y. Cochennec, P. Adam, T. Houdoin. Asynchronous time division networks: terminal synchronization for video and sound signals. *Proc. of IEEE INFOCOM '85*, Nov. 1985, pp. 791–794.

[1.24] M. De Prycker, M. Ryckebusch, P. Barri. Terminal synchronization in asynchronous networks. *Proc. of IEEE ICC'87*, Seattle, WA, USA, June 1987, pp. 22.7.1–22.7.8.

[1.25] R. P. Singh, S.-H. Lee, C.-K. Kim. Jitter and clock recovery for periodic traffic in broadband packet networks. *IEEE Transactions on Communications* 1994; **COM-42**(5).

[1.26] R. C. Lau, P. E. Fleischer. Synchronous techniques for timing recovery in BISDN. *IEEE Transactions on Communications* 1995; **COM-43**(2/3/4).

[1.27] K. Murakami. Jitter in synchronous residual time stamp. *IEEE Transactions on Communications* 1996; **COM-44**(6).

[1.28] K. Murakami. Waveform analysis of jitter in SRTS using continued fraction. *IEEE Transactions on Communications* 1998; **COM-46**(6).

[1.29] ITU-T Recs. G.823 *The Control of Jitter and Wander within Digital Networks which are Based on the 2048 kbit/s Hierarchy*; G.824 *The Control of Jitter and Wander within Digital Networks which are Based on the 1544 kbit/s Hierarchy*; G.825 *The Control of Jitter and Wander within Digital Networks which are Based on the Synchronous Digital Hierarchy*, Geneva, March 1993.

[1.30] ITU-T Rec. I.363 *B-ISDN ATM Adaptation Layer (AAL) Specification*, Geneva, 1996/2000.

[1.31] W. C. Lindsey, F. Ghazvinian, W. C. Hagmann, K. Dessouky. Network synchronization. *Proceedings of the IEEE*, vol. 73, no. 10, Oct. 1985, pp. 1445–1467.

[1.32] P. Kartaschoff. Synchronization in digital communications networks. *Proceedings of the IEEE*, vol. 79, no. 7, July 1991, pp. 1019–1028.

[1.33] S. Bregni. A historical perspective on network synchronization. *IEEE Comm. Magazine* 1998; **36**(6).

[1.34] P. Mouley, M. B. Pautet. *The GSM System for Mobile Communications*. Lassay-les-Château: Europa Media Duplication S. A., 2000.

[1.35] J. Buck, E. Buck. *Synchronous Fireflies*. Scientific American, May 1976.

[1.36] T. D. C. Little, A. Ghafoor, C. Y. R. Chen, C. S. Chang, P. B. Berra. Multimedia synchronization. *IEEE Data Engineering Bulletin* 1991; **14**(3): 26–35.

[1.37] R. Steinmetz. Synchronization properties in multimedia systems. *IEEE Journal on Selected Areas in Communications* 1990; **SAC-8**(3): 401–412.

[1.38] T. D. C. Little, A. Ghafoor. Multimedia synchronization protocols for broadband integrated services. *IEEE Journal on Selected Areas in Communications* 1991; **SAC-9**(9): 1368–1382.

[1.39] C. L. Hamblin. Instants and intervals. *Proc. of the 1st Conference of the International Society for the Study of Time*, by J. T. Fraser *et al.* New York: Ed. Springer-Verlag, 1972, pp. 324–331.

[1.40] S. R. Faulk, D. L. Parnas. On synchronization in hard-real-time systems. *Comm. of the ACM* 1988; **31**(3): 274–287.

[1.41] T. D. C. Little, A. Ghafoor. Synchronization and storage models for multimedia objects. *IEEE Journal on Selected Areas in Communications* 1990; **SAC-8**(3): 413–427.

[1.42] T. D. C. Little, A. Ghafoor. Interval-based temporal models for time-dependent multimedia data. *IEEE Transactions on Data and Knowledge Engineering* 1993; **5**(4): 551–563.

[1.43] *IEEE Journal on Selected Areas in Communications*, Special Issue on 'Synchronization Issues in Multimedia Communications', 1996; **SAC-14**(1).

[1.44] U. D. Black. *Network Management Standards — The OSI, SNMP and CMOL Protocols*. New York: McGraw-Hill Series on Computer Communications, 1992.

[1.45] U. D. Black. *Network Management Standards — SNMP, CMIP, TMN, MIBs and Object Libraries*. New York: McGraw-Hill Series on Computer Communications, 1995.

[1.46] W. Stallings. *Network Management*. IEEE Computer Society Press, 1993.

[1.47] D. L. Mills. *Home page of NTP*. http://www.eecis.udel.edu/intp.

[1.48] RFC 1305, *Network Time Protocol (Version 3)-Specification, Implementation and Analysis*. By D. L. Mills, Univ. of Delaware, USA, March 1992.

[1.49] D. L. Mills. Internet Time Synchronization: the Network Time Protocol. *IEEE Trans. on Commun.* **COM-39**, Oct. 1991, pp. 1482–1493.

2

ASYNCHRONOUS AND SYNCHRONOUS DIGITAL MULTIPLEXING

Any continuous-time analogue signal can be converted into a digital signal by making discrete both the time and the amplitude axes. The process of making the time axis discrete is called *sampling* and consists of substituting the whole continuous-time analogue signal with a series of its analogue values (samples) taken at particular instants. This process is reversible only if the original signal has limited bandwidth. A well-known result, the fundamental *Sampling Theorem* [2.1]–[2.3], commonly attributed to Shannon or Nyquist[1], assures that the series of (analogue) samples taken with sampling frequency f_s is perfectly equivalent to the original signal if

$$f_s \geq 2B \tag{2.1}$$

where B (bandwidth) is the maximum Fourier frequency in the signal spectrum.

The process of making the amplitude axis discrete is called *quantization*. It consists of dividing the amplitude axis in contiguous intervals and in associating to all the amplitudes within any interval a single amplitude value chosen among them. In practical applications, the number of intervals is finite and the quantized amplitude values can be thus expressed in a numerical form, with a fixed number of digits depending on the total number of intervals chosen.

Through the joint processes of sampling and quantization, therefore, any continuous-time analogue signal is converted to a sequence of numbers or binary digits (*digital signal*). The sampling frequency f_s (or its reciprocal quantity the sampling period T_s) and the number of binary digits (bits) N_b expressing every sample (word) are the two key parameters yielding the quality of reproduction of the original signal. For example, in the case of the telephone voice channel (PCM) the parameters take the values $f_s = 8$ kHz, $N_b = 8$ bit, while for the hi-fi audio Compact Disc (CD) the values are $f_s = 44.1$ kHz, $N_b = 16$ bit (for each channel of the stereo audio signal).

Digital transmission deals with transferring digital signals from one point in a network to another far one. The fundamental issue in digital transmission is *multiplexing*, i.e.

---

[1] As reminded by S. Haykin [2.3], the Sampling Theorem was introduced to communication theory by C. Shannon in 1949. However, the interest of communication engineers in this theorem may be traced back to H. Nyquist (1928). Indeed, the Sampling Theorem was known to mathematicians much earlier. In particular, the corresponding mathematical theorem was first proved by E. T. Whittaker in 1915. For a tutorial review of the Sampling Theorem and some historical notes, see Reference [2.4].

transmitting several signals (e.g., telephone signals) on a single shared transmission channel. Digital multiplexing implies some key issues related to the timing of the digital signals involved. Indeed, the taxonomy itself of multiplexing techniques is based on timing criteria (see Section 2.1.5).

This chapter deals with the basics of digital transmission systems, providing the fundamentals of digital multiplexing by paying special attention to the timing aspects. After having introduced some basic concepts and definitions about the timing of digital signals, the three fundamental types of digital multiplexing are described: the PCM multiplexing, the synchronous digital multiplexing and the asynchronous digital multiplexing with bit justification. As practical applications of these concepts, the PCM primary multiplex, the PDH and the SDH/SONET are detailed. Throughout this chapter, it is pointed out how different techniques cope with the issue of multiplexing asynchronous tributaries: asynchronous digital multiplexing (PDH) exploits bit justification, synchronous digital multiplexing (SDH/SONET) pointer justification. Moreover, asynchronous vs synchronous transfer modes and asynchronous vs synchronous frame structures are distinguished.

2.1   BASIC CONCEPTS

This first section introduces some fundamental concepts and definitions about the timing of digital signals and their multiplexing. In particular, basic definitions about the timing relationships between digital signals are provided. Then, the phenomena of jitter and wander are addressed and the difference between asynchronous and synchronous transfer modes is pointed out. Finally, the taxonomy of multiplexing techniques is outlined paving the way to the rest of the chapter.

2.1.1   Timing Signals and Digital Signals

A *timing signal*, or *chronosignal*, or *clock signal* may be defined as a periodic or pseudo-periodic signal used to control the timing of actions (e.g., the symbol decision at reception) on digital signals. The *timing recovery* (clock recovery) is thus the process of extraction of the timing signal associated to a digital signal received (see Section 1.1.2).

Two examples of timing signals with frequency v_0 and period T are shown in Figure 2.1. Typical waveforms are sine and square waves. In the former case, for the sake of simplicity, a pseudo-periodic timing signal is usually modelled as

$$s(t) = A(t) \sin \Phi(t) \tag{2.2}$$

where $A(t)$ is the instantaneous amplitude, $\Phi(t)$ is the total phase and the instantaneous frequency $v(t)$ is given by

$$v(t) = \frac{1}{2\pi} \frac{d\Phi(t)}{dt} \tag{2.3}$$

Obviously, in the ideal case, when the pseudo-periodic timing signal is periodic indeed, the total phase is linearly increasing with time, i.e.

$$\Phi(t) = 2\pi v_0 t \tag{2.4}$$

where v_0 is the nominal frequency. In the actual case, other components affect the total phase increasing, including frequency offset, drifts and purely random fluctuations. This topic is thoroughly addressed in Chapter 5.

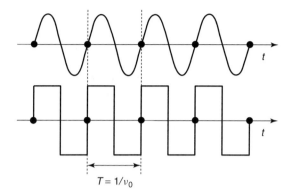

$T = 1/v_0$

Figure 2.1. Timing signals (sine and square waves) and significant instants identified at signal zero crossings

A chronosignal fulfils its duty by triggering events, i.e. timing the controlled process. From this point of view, a timing signal can be modelled also by a series of pulses spaced T, at special instants called *significant instants*. The timing signal triggers the controlled process at those instants. Suitable significant instants can be identified, for example, at the signal zero-crossing instants, for ease of practical realization of electronic circuitry.

An *isochronous*[2] digital signal is a digital signal in which the time intervals between significant instants have, at least on the average[3], the same duration or durations which are integer multiples of the shortest one. Standard digital signals are always isochronous (e.g., the HDB3-coded 2.048 Mbit/s [2.5]).

Conversely, an *anisochronous* digital signal is a non-isochronous digital signal. For example, a hypothetical digital signal, where the symbols 1 and 0 are coded with pulses of different variable length, is anisochronous.

With *regular* timing signal, it is usually denoted an isochronous timing signal with all its expected significant instants evenly spaced in time (see Figure 2.2(a)). All digital signals transferred along network links are timed by such regular clocks.

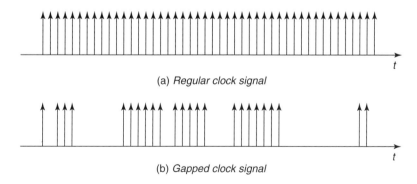

(a) *Regular clock signal*

(b) *Gapped clock signal*

Figure 2.2. Regular and gapped clock signals

---

[2] From the Greek etyma ἴσος = *equal* and χρόνος = *time*.
[3] Here and in the following definitions, the expression *at least on the average* denotes that little fluctuations around the mean value are tolerated, as unavoidable in real systems.

Conversely, a *gapped* timing signal is generally derived from a regular clock, by leaving gaps in the clock signal where the regular clock would yield a pulse (see Figure 2.2(b)). A gapped clock signal is still an isochronous timing signal. It is of lower average frequency than the regular clock signal from which it is derived, yet it has the same basic clock period (i.e. the same peak clock rate). In digital technology, it is very easy to generate a gapped clock signal from a regular clock, by gating out the unwanted pulses. Gapped clock signals are used for instance to time the insertion of tributary bits into a multiplex signal (see Section 2.3), according to certain mapping rules.

The *Unit Interval* (UI) is defined as the nominal time interval between consecutive significant instants of an isochronous signal, or rather the shortest interval if intervals are integer multiples of it. To provide some examples, Table 2.1 summarizes the UI durations (rounded to five digits of mantissa) of the most common digital signals of PDH and SDH hierarchies (see Sections 2.4 and 2.5).

2.1.2   Timing Relationships between Digital Signals

Two *synchronous*[4] digital signals are isochronous digital signals whose respective timing signals have the same frequency, at least on the average, and a phase relationship controlled precisely (i.e., with phase offset $\Delta\Phi = $ constant).

Conversely, two digital signals are *asynchronous* if they are not synchronous.

Two *mesochronous*[5] digital signals are isochronous, asynchronous digital signals, whose respective timing signals have the same frequency, at least on the average, but no control on the phase relationship. It is worthwhile noticing that, being the phase fluctuation function the integral of the frequency fluctuation function, in this case the phase error $\Delta\Phi$ is not theoretically limited over an infinite time interval even for small zero-mean frequency fluctuations (thus, $\Delta\Phi$ can take any value).

UNIT
INTERVAL

Table 2.1 UI durations for some common digital signals

Digital Signal Rate	UI Duration
64 kbit/s	15.625 μs
1.544 Mbit/s	647.67 ns
2.048 Mbit/s	488.28 ns
6.312 Mbit/s	158.43 ns
8.448 Mbit/s	118.37 ns
34.368 Mbit/s	29.097 ns
44.736 Mbit/s	22.353 ns
139.264 Mbit/s	7.1806 ns
155.520 Mbit/s	6.4300 ns
622.080 Mbit/s	1.6075 ns
2488.320 Mbit/s	0.40188 ns
9953.280 Mbit/s	0.10047 ns

---

[4] From the Greek etymon σύγχρονος, built by σύν = *with* and χρόνος = *time*.
[5] From the Greek etyma μέσος = *medium* and χρόνος = *time*.

Two *plesiochronous*[6] digital signals are isochronous, asynchronous digital signals, whose respective timing signals have the same frequency values only nominally. Actually, their frequency values are allowed to be slightly different, within a given tolerance range.

Two *heterochronous*[7] digital signals are isochronous, asynchronous digital signals, whose respective timing signals have different nominal frequencies.

To give sound examples of the above abstract concepts, a locked PLL outputs a timing signal which is synchronous with the input signal, owing to the feedback control on the phase error between them (see Section 5.5.1). A Frequency-Locked Loop (FLL), i.e. a feedback system operating like a PLL but controlling the frequency error instead between the input and the output signals, outputs a signal which is mesochronous with the input. Two oscillators, even designed and built as equal by the same supplier, output two plesiochronous timing signals, owing to unavoidable manufacturing tolerances. In the same way, for example, two 2.048 Mbit/s PCM digital signals are always plesiochronous, whenever generated by two independent pieces of equipment. Finally, two digital signals with different rates (e.g. a 2.048 Mbit/s PCM and a 8.448 Mbit/s PDH multiplexes) are heterochronous.

To summarize, the taxonomy of the different timing relationships between digital signals is outlined in Figure 2.3.

2.1.3   Jitter and Wander

Jitter and wander are fundamental impairments affecting digital signals. These two names are used to refer extensively to any kind of *phase noise* of digital signal symbols (i.e., fluctuations of transmitted symbols along the time axis). Phase and amplitude noises are the two basic orthogonal components of the noise affecting a digital signal and accumulate along transmission lines.

Jitter and wander are a transmission anomaly that must be considered in designing, developing, deploying, interconnecting and maintaining any digital system and network. Indeed, they have been extensively studied since the first digital systems were conceived, beginning in the 1950s.

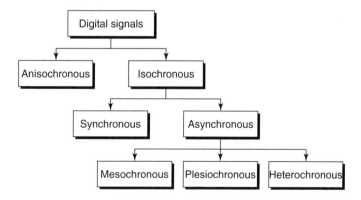

Figure 2.3.   Taxonomy of the different timing relationships between digital signals

---

[6] From the Greek etyma $\pi\lambda\acute{\eta}\sigma\iota o\varsigma = close$ and $\chi\rho\acute{o}\nu o\varsigma = time$.
[7] From the Greek etyma $\acute{\varepsilon}\tau\varepsilon\rho o\varsigma = different$ and $\chi\rho\acute{o}\nu o\varsigma = time$.

In an ideal digital transmission system, the pulses of the digital pulse stream would arrive at times which are integer multiples of the pulse repetition period T, i.e. at times $t_k = kT$ ($k = \ldots, -2, -1, 0, +1, +2, \ldots$). However, in real systems, pulses arrive at times that differ from ideal integer multiples of T. We call this (unwanted) pulse position modulation of the pulse stream *jitter*. Therefore, jitter is defined as the

phase/time deviation of the received digital signal, at every time $t_k = kT$, *from the expected pulse times* $t_k = kT$ *of the ideal signal.*

These time deviations from ideal times kT form a discrete-time, continuous-amplitude sequence $e[kT]$[8]. The sequence $e[kT]$ is the mathematical description of jitter: it has dimension of time or phase in amplitude and it takes value only at integer multiples of T. It is defined positive for pulses that arrive earlier than due time kT.

This concept is clarified by Figure 2.4, where a train of unit pulses represents a digital signal. In an ideal (expected) digital signal pulses are spaced equally in time, and they thus come at $t_k = kT$. In an actual digital signal, affected by jitter, pulses become spaced irregularly in time and thus come earlier or later than due times kT. The jitter affecting the actual digital signal is depicted by the third plot, according to the convention of defining positive the jitter of pulses that come earlier.

In digital transmission systems $e[kT]$ is a random function of time, a stochastic process. Several statistical quantities can be adopted and evaluated to characterize quantitatively $e[kT]$. Among them, for example, the power spectrum and the root mean square (rms) value of $e[kT]$ are often measured or assessed.

As made clear by Figure 2.4, the nominal frequency $1/T$ of the digital signal and the frequency f_j of the phase-modulating jitter entail two different concepts. For example, let a 2.048 MHz digital signal (pulses coming every about 488 ns) be phase-modulated by a 10 Hz sine-wave jitter; in that case, pulses are still spaced about 488 ns, on the average, but move back and forward along the time axis every 100 ms. Jitter can be

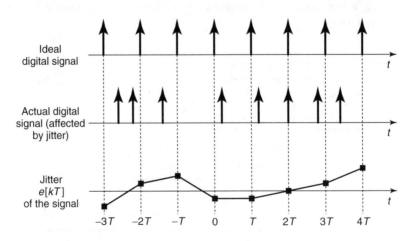

Figure 2.4.   Jitter affecting a digital signal

---

[8] In the literature about jitter, the notation of using square brackets to denote discrete-time functions and round brackets to denote continuous-time functions is somehow common.

for instance low-pass filtered. Obviously, low-pass filtering jitter has nothing to do with low-pass filtering the digital signal itself. An RC low-pass filter simply low-pass filters the digital signal, but it does not affect substantially the phase-modulating jitter. A way to low-pass filter jitter, but not affecting otherwise the digital signal, is using a PLL instead.

Low-frequency jitter is often called the specific name *wander*. The border between jitter and wander is purely conventional and is located around 10 Hz: phase fluctuations are referred to as jitter if $f_j > 10$ Hz, as wander otherwise. Jitter frequencies of interest can be as high as 100 kHz, while wander can be so slow to have f_j in the order of μHz (diurnal wander) or even nHz (annual wander).

Jitter and wander are commonly measured and expressed in terms of [UI], although from a 'physical' point of view a measurement unit with dimension [time] would be more appropriate.

Just to take a glance, international standards on jitter and wander [2.6] specify limits for equipment and network digital interfaces. As far as equipment interfaces are concerned, the following three jitter and wander limits are specified:

- *jitter tolerance*, i.e. the minimum jitter amplitude that the equipment should be able to tolerate at digital input interfaces for error-free reception (higher jitter may cause errors in symbol decision due to inaccurate clock recovery and thus to time-misplaced sampling);

- *jitter transfer function*, i.e. the ratio between the output jitter amplitude and the input jitter amplitude over a jitter frequency range (this test is accomplished by feeding the equipment input with a signal affected by a sinusoidal test jitter at given frequencies);

- *output jitter*, i.e. the maximum jitter allowed at digital output interfaces (this is thus the limit to the jitter generated by the equipment).

The jitter tolerance and the jitter transfer function are specified over a wide frequency range, i.e. from few μHz up to 100 kHz. Jitter tolerance masks are decreasing functions of the frequency f_j, as slower jitter is easier to track by clock recovery circuits. These masks go from jitter peak-to-peak amplitudes in the order of tens of UI, for $f_j < 10^{-5}$ Hz, to tenths of UI, for $f_j = 10^5$ Hz and above. Jitter transfer functions, on the other hand, are decreasing functions of the frequency f_j as well, since any PLL works as a low-pass filter of the input jitter towards the output [2.7]–[2.13].

Primary sources of jitter and wander in digital transmission systems are regenerators, multiplexers and the transmission lines themselves.

A regenerator receives the incoming pulse stream and transmits a new regenerated pulse stream that replicates the original sequence as closely as possible. To regenerate the signal, timing information must be known so that the regenerated pulse stream can be transmitted with the proper intervals between the pulses. The most commonly used regenerators in digital transmission systems are self-timed regenerators that extract timing information directly from the incoming pulse stream using a timing extraction circuit as explained in Section 1.1.2. Because the timing extraction process is imperfect, the transmitted pulse stream is not an exact replica of the original pulse stream, but contains the unwanted pulse position modulation that we call jitter.

Usually, the pulse stream arrives at the receiver corrupted by additive noise and with pulse shapes that spill over into adjacent time slots, resulting in Inter-Symbol Interference

(ISI). Thus, the timing extraction circuit must work on a noisy, dispersed pulse stream. This results in a regenerator output which is inherently jittered. Jitter accumulates along chains of cascaded line-regenerator systems, and may even cause symbol decision errors.

As it will be better shown in the rest of this chapter, digital multiplexers combine several lower rate pulse streams into a higher rate pulse stream using time division multiplexing. To accomplish this task, the lower speed pulse streams must be bit-synchronized to a common rate. There are three bit synchronization schemes in use in modern networks:

- *slip buffering*, used for example at input stages of digital switching exchanges (see Section 4.1.3);

- *bit justification*, or *bit stuffing*, or *pulse stuffing*[9], used in PDH asynchronous digital multiplexers (see Section 2.3.4) and in the adaptation of asynchronous tributaries into SDH Containers (see Section 2.5);

- *pointer processing*, used in SDH network elements where multiplexer sections are terminated (see Section 2.5.8).

When lower rate tributaries are demultiplexed, they result affected by jitter due to the above processes. Also in this case, cascaded multiplexer/demultiplexer systems make jitter accumulate in transported tributaries so that controlling jitter may be a real issue.

As far as the transmission lines are concerned, the main cause of jitter and wander is the temperature variation, as the propagation speed of signals in transmission media depends on the medium temperature (this holds especially for copper cables). Diurnal and annual temperature variations follow pseudo-periodic trends, which yield the so-called diurnal and annual wander.

The aim of this short section was definitely not to provide an exhaustive treatment of jitter and wander theory. Readers looking for further information will find all the details desired in the specific literature. Surprisingly, although many excellent technical papers have analysed and discussed jitter (interested readers may wish to read for example the fundamental papers of Duttwailer [2.15] and Chow [2.16] on the jitter due to bit justification), few books feature a unified and comprehensive treatment of jitter. Among them, we recommend for example the book of Trischitta and Varma [2.17], which provides a detailed treatise on all aspects of jitter in digital transmission systems. Besides that, we mention also the book of Takasaki [2.13], which features other topics related to jitter and clock recovery.

2.1.4   Asynchronous vs Synchronous Transfer Modes

As already noted, it is rather easy for the inexperienced reader to get confused with the words 'synchronous' and 'asynchronous', as they may address different concepts depending on the context and the abstraction level.

First, 'synchronous' may refer to the *physical level of information transfer*. In this case, synchronicity regards the timing of sine waves, bits, frames, etc. The concepts of carrier, symbol, frame synchronization, etc. dealt with in Section 1.1 fall within this case.

---

[9] According to Rec. G.701 [2.14], ITU-T recommends the use of the term *justification* and deprecates the term *stuffing* in the same sense, although in common usage since the 1960s. In this book, the term justification has been preferred in most cases.

Nevertheless, 'synchronous' may also refer to the information *transfer mode* instead, at an upper level of abstraction (*logic level of information transfer*). In this case, the concern is about how the various sources send information, if they are mutually dependent in deciding the instants of starting transmission of information units or not.

The two fundamental information transfer modes are therefore:

- the *Asynchronous Transfer Mode* (ATM), where the information sources are mutually asynchronous and the information is segmented in information units, which are sent by sources in independent instants, with interval dependent on source demand (see Figure 2.5(a)).

- the *Synchronous Transfer Mode* (STM[10]), where the information sources are mutually synchronous, i.e. they can start sending their information units only in preassigned time-slots (see Figure 2.5(b)); a special case of STM occurs when fixed-size words are sent periodically (see Figure 2.5(c)).

ATM gives its name to the cell-switched transport technique chosen for B-ISDN support. A sound example of STM with fixed-size words sent on a periodical base, on the other

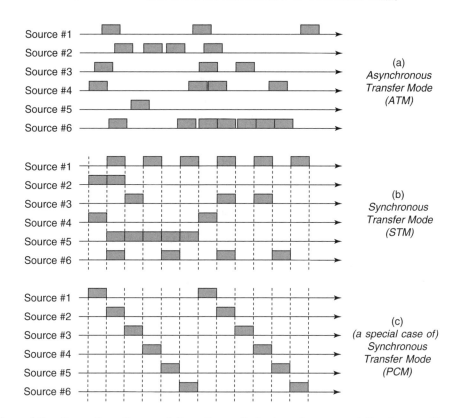

Figure 2.5. Generation of source information units in Asynchronous and Synchronous Transfer Modes

---

[10] Not to be confused with the same acronym STM meaning *Synchronous Transport Module* (SDH frame).

hand, is the PCM primary multiplex, where each frame includes a fixed number of time slots assigned to user channels (30 in the case of the European 2.048 Mbit/s PCM Primary Multiplex), as explained in Section 2.2. There, each speech source generates one sample (byte) every 125 μs during its assigned time slot (see again Figure 2.5(c)).

2.1.5 Taxonomy of Multiplexing Techniques

Transmission relies on multiplexing, i.e. transmitting several telephone[11] signals on a single shared transmission medium. Referring to Figure 2.6, the various multiplexing techniques are divided in three basic kinds, depending on the domain where different signals are multiplexed: frequency, time and code division multiplexing.

2.1.5.1 *Frequency Division Multiplexing*

Frequency Division Multiplexing (FDM) has been the foundation of analogue telephone networks for a long time. The principle of FDM consists, as the name itself says, in shifting every tributary channel in the frequency domain to different sub-bands in the spectrum of the multiplex signal. If sub-bands are well separated, there is no channel interference and it is still possible to extract the single channels from the multiplex signal by band-pass filtering.

FDM applies directly to telephone continuous-time analogue signals (gross bandwidth $B = 4$ kHz). Thus, sub-bands are 4 kHz wide, while the single-signal net bandwidth is about 3 kHz (300 Hz $\div$ 3400 Hz), in order to make easier channel separation by band-pass filtering (see Figure 2.7).

In principle, each signal can be moved to its sub-band through one SSB suppressed-carrier amplitude-modulation stage; in practice, the multiplex signal is formed through

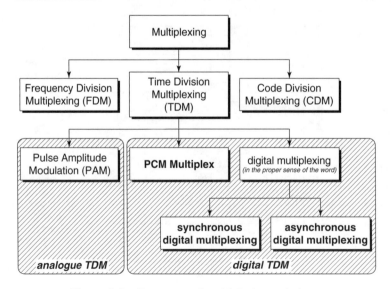

Figure 2.6. Taxonomy of multiplexing techniques

---

[11] Here and in the following, we focus in particular on telephone signals for the sake of simplicity, although FDM and TDM techniques are applied to other types of signals as well.

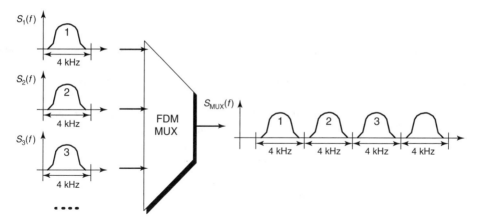

Figure 2.7.   Principle of Frequency Division Multiplexing on telephone signals

several modulation stages, according to standard multiplexing hierarchies defined by CCITT Recommendations (now ITU-T) [2.18]. The 10 800 channels of the 60 MHz FDM system, designed for transmission on the 2.6/9.5 mm coaxial cable, represent the greatest capacity achieved with FDM, before the digital technology put an end to further growth.

To summarize, as shown by the graph in the time-frequency plane of Figure 2.8(a), multiplexed channels share the same transmission time interval T, but are separated in different sub-bands within the frequency interval F.

2.1.5.2   *Time Division Multiplexing*

Time Division Multiplexing (TDM) is the dual technique of FDM in the time domain: several signals are interleaved in time for transmission over a common channel.

The principle of TDM applies to sampled telephone signals with sampling rate $f_s = 8$ kHz (samples spaced $T_s = 125$ μs) and consists in allotting the sampling period T_s into as many *time slots* as the telephone channels to multiplex are. In each time slot, one sample of the corresponding channel is transmitted, so that all the input lines are repeatedly served by the TDM multiplexer and each of them sends one sample on the output line every T_s.

To summarize, as shown by the graph in the time-frequency plane of Figure 2.8(b), multiplexed channels share the same frequency interval F, but are transmitted in different time slots within the time interval T.

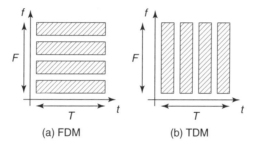

(a) FDM   (b) TDM

Figure 2.8.   Time and frequency shared in frequency- and time-division multiplexing

2.1.5.3   Code Division Multiplexing

Code Division Multiplexing (CDM), also known as spread-spectrum, is a third transmission technique that in the last years gained much importance for various applications (one above all: cellular mobile telephone networks).

It consists of transmitting simultaneously and in the same frequency band a set of N signals $s_i(t)$, for example by multiplying each of them by a particular code $c_i(t)$ which enables separation of the single signals from the multiplex signal (direct-sequence CDM). If the N codes $c_i(t)$ are orthogonal (i.e. with null mutual correlation) or approximately so, then demultiplexing achieves null mutual interference between signals $s_i(t)$, as it happens in FDM and TDM systems, which are inherently orthogonal.

Alternatively, in band-pass transmission, it is possible to change the carrier frequency during the transmission time interval of each information symbol, according to proper codes (frequency-hopping CDM).

2.1.6   Time Division Multiplexing (TDM)

TDM may be analogue, if samples take analogue values, or digital, if samples are coded in a digital form. The various TDM techniques are outlined as follows.

2.1.6.1   Analogue TDM of Pulse Amplitude Modulation (PAM) Signals

TDM may apply to *Pulse Amplitude Modulation* (PAM) signals, i.e. to analogue sampled telephone signals where samples $s(kT)$ take analogue values equal to the actual values of the continuous-time signal at sampling times. PAM–TDM signals are thus analogue signals where N PAM signals are multiplexed in TDM.

The principle of operation of a PAM–TDM multiplexer is shown in Figure 2.9. The input telephone signals $s_i(t)$ are sampled with sampling period T_s in subsequent time instants kT, indicated by the black arrows. The analogue samples $s_i(kT)$ drive a transmission filter with impulse response $h(t)$ and then form the output multiplex signal.

The PAM–TDM technique is not suited for direct transmission, due to the issue of dealing with analogue sampled signals. It was adopted, instead, as intermediate stage in PCM Primary Multiplex Equipment before analogue-to-digital coding (see Section 2.2).

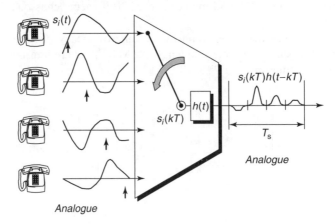

Figure 2.9.   Analogue time-division multiplexing of PAM telephone signals (PAM–TDM multiplexer)

2.1.6.2 PCM Multiplexing

PCM multiplexing consists of TDM multiplexing N analogue telephone signals, having converted them to PCM digital form. The principle of a PCM multiplexer is shown in Figure 2.10.

The main examples of PCM multiplex signals are the PCM Primary Multiplexes in the European (2.048 Mbit/s) and North American (1.544 Mbit/s) digital hierarchies. The former carries 30 telephone channels (over 32 time-slots including frame alignment and signalling information), the latter 24. The PCM Primary Multiplex is described forth in Section 2.2.

2.1.6.3 Digital Multiplexing

Digital multiplexing (in the proper sense of the word) means specifically TDM-multiplex N digital signals (tributaries), which usually have the same (but in general different) nominal bit rate. The distinctive feature of digital multiplexing, compared to PCM multiplexing, is that the tributary signals to multiplex are already in digital form. Thus, there are no intermediate PAM stage or PCM coders. The principle of a digital multiplexer is shown in Figure 2.11.

Digital multiplexing is further divided into:

- *synchronous digital multiplexing*, where the tributary signals are assumed to be synchronous (described forth in Section 2.3.1);

- *asynchronous digital multiplexing*, where the tributary signals are assumed to be plesiochronous instead and the *bit justification* technique is adopted to synchronize tributaries into the multiplex signal (described forth in Section 2.3.4).

The rest of this chapter deals with the three most important types of digital multiplexing (highlighted in boldface in Figure 2.6): the PCM multiplexing, the synchronous digital multiplexing and the asynchronous digital multiplexing with bit justification. As practical applications of these concepts, the PCM primary multiplex, the PDH and the SDH/SONET will be detailed.

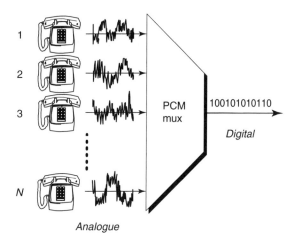

Figure 2.10.   PCM multiplexing of N analogue telephone signals (PCM multiplexer)

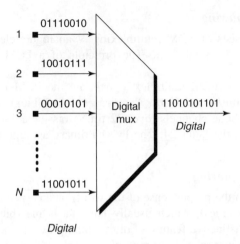

Figure 2.11. Digital multiplexing of N digital tributaries (digital multiplexer)

2.2   THE PCM PRIMARY MULTIPLEX

The principle of PCM multiplexing can be summarized in four fundamental functions: *sampling*, *quantization*, *coding*, *multiplexing*. Hence, PCM multiplex equipment basically may perform the following tasks in transmission:

- it takes as inputs N analogue telephone signals (each with gross bandwidth $B = 4$ kHz),
- it samples and TDM-multiplexes them thus producing a PAM–TDM signal,
- it quantizes and PCM-codes its amplitude at each time slot,
- it outputs the resulting digital TDM signal (PCM multiplex signal).

The equipment that performs this PCM-multiplexing task in transmission, and the inverse task in reception (demultiplexing), is called *PCM multiplexer*.

The distinctive feature of such a PCM multiplexer, compared to the other types of digital multiplexers, is thus the analogue input and the intermediate sampling and coding stage. A PCM multiplex signal has a frame length equal to the sampling period $T_s = 125$ μs. Frames are divided in time-slots carrying one octet (coding one speech sample) and allocated to different channels, so that every frame carries one sample per channel.

Two standard PCM multiplex signals [2.19]–[2.25] are the first levels of PDH digital hierarchies and the foundations of digital networks worldwide:

- the European E1 signal, with rate 2.048 Mbit/s carrying 30 telephone channels over 32 time-slots including frame alignment and signalling information (non-uniform quantization of the analogue speech signals according to the so-called *A-law* is adopted);
- the North American DS1 signal, with rate 1.544 Mbit/s carrying 24 telephone channels (non-uniform quantization of the analogue speech signals according to the so-called *μ-law* is adopted).

8 KHZ sampling Frequency

Beyond formal aspects, the above signals reflect the same concept. For this reason, and for the sake of simplicity, we will focus only on the European E1 2.048 Mbit/s signal detailing its main features [2.19].

As shown by Figure 2.12, frames have a duration equal to the sampling period $T_s =$ 125 μs and are divided in 32 Time Slots (TS), numbered 0 through 31. Among them, most commonly only 30 TS carry speech information, i.e. one speech sample (PCM-coded as one byte), while the TS0 carries the frame alignment word and miscellaneous overhead and the TS16 carries signalling information. The overall frame length is thus 256 bits.

8 bits

The time slot TS0 carries the frame alignment word and some overhead. In particular, two kinds of words alternate in the TS0 of odd and even frames:

- in the frames containing the frame alignment signal, TS0 carries the octet $S_i 0011011$, where S_i is a bit reserved for international use (mostly, it carries a Cyclic Redundancy Code, CRC4 bit for BER assessment) and 0011011 is the frame alignment word;

- in the frames not containing the frame alignment signal, TS0 carries the octet $S_i 1 A S_{a4} S_{a5} S_{a6} S_{a7} S_{a8}$, where S_i is the same as before, A is a remote alarm indication bit and $S_{a4} S_{a5} S_{a6} S_{a7} S_{a8}$ are additional spare bits reserved for other service functions or national use (one of them can be used to convey Synchronization Status Messages, cf. Section 4.7).

The TS16, on the other hand, usually carries the signalling channel (if not needed for signalling, in some cases it may be used for another 64 kbit/s channel as for TS1 through TS15 and TS17 through TS31). The use of the 64 kbit/s channel in TS16 is recommended for either *common channel* or *channel-associated* signalling as required.

In the case of channel-associated signalling, the bit stream carried by the TS16 is organized in frames: its 64 kbit/s capacity is sub-multiplexed into lower-rate signalling channels. The signalling frame is called *signalling multiframe*, as it spans 16 basic PCM frames (2 ms), and is 128 bits long: 4 signalling bits for each channel plus 8 bits of alignment and miscellaneous overhead. Therefore, a gross signalling capacity of 2 kbit/s is available to each channel.

In more detail, the signalling multiframe comprises 16 basic PCM frames, numbered 0 through 15. The multiframe alignment pattern is 0000 and occupies digits 1 to 4 of the TS16 in frame 0. Bits 5 to 8 of the same octet are reserved (miscellaneous overhead). Each telephone channel has four 500 bit/s signalling channels designated a, b, c and d. Signalling bits *abcd* associated with channel TS1 and *abcd* associated with channel TS17

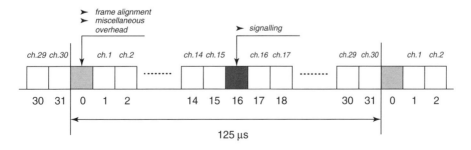

Figure 2.12.   Frame structure of the European 2.048 Mbit/s PCM Primary Multiplex

are allocated in the byte of TS16 in frame 1 as *abcdabcd*. In the same way, bits *abcd* associated with channel TS2 and *abcd* associated with channel TS18 are allocated in the byte of TS16 in frame 2, and so on.

As far as the North American DS1 signal is concerned, frames are divided in 24 time slots, each of which may carry a telephone channel, and one additional header bit (F bit). The overall frame length is thus 193 bits, transmitted within 125 μs at the bit rate 1.544 Mbit/s. The F-bit constitutes a subrate channel of 8 kbit/s, which have three basic functions: to mark frame alignment, to provide a data link for service purposes and to carry error-control bits.

Two multiframe structures have been standardized to allow interpretation of the F-bit channel: the *Super Frame* (SF) over 12 DS1 basic frames and the *Extended Super Frame* (ESF) over 24 DS1 basic frames. In the SF structure, overhead F-bits are used for frame and signalling phase alignment only. In the ESF structure, overhead F-bits are shared among an ESF alignment signal, a cyclic redundancy check for error control and a data link to service purposes. This data link may be used to convey Synchronization Status Messages (cf. Section 4.7).

Early PCM multiplexers were designed to perform exactly the tasks listed at the beginning of this section, in that sequence. The functional scheme of such PCM multiplexer is thus depicted in Figure 2.13 (European standard). The input signals (numbered #1 through #30) are analogue continuous-time telephone signals and are PAM-multiplexed. The PAM signal is then converted to digital form by the PCM coder, which operates directly on the speech samples in the PAM frames. The PCM samples are then copied onto the output PCM multiplex signal with the alignment and the signalling words. The signalling multiplex signal is built basing on the signalling information taken from each input analogue channel by devices called *translators* (T). Let us note the essential role of the timing generator and distributor, driven by the equipment clock, which synchronizes every block ensuring that words are transmitted in the proper time slot. Needless to say,

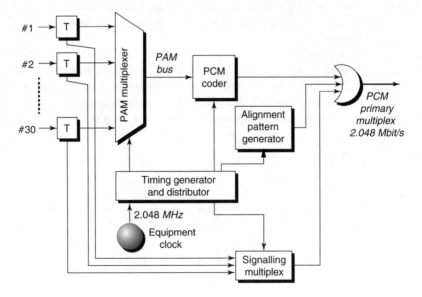

Figure 2.13.   Functional scheme of the European PCM primary multiplexer

Table 2.2 Summary of characteristics of the European 2.048 Mbits/s PCM primary multiplex signal

Nominal bit rate of the multiplex signal	2.048 Mbit/s
Frequency tolerance	± 50 ppm
Nominal bit rate of the PCM channels	64 kbit/s
Number of available PCM channels	30
Frame length	256 bit
Frame period	125 μs
Frame repetition rate	8 kHz
Frame alignment word	A (even frames): C0011011 B (odd frames): C1AXXXXX
Rate of frame alignment bits[12]	32 kbit/s
Frame alignment strategy parameters[13]	$\alpha = 3, \delta = 1$
Signalling multiframe period	2 ms
Signalling multiframe alignment word	0000XAXX
Gross signalling bandwidth per channel	2 kbit/s
Intra-office physical interface signal peak level	HDB3 2.37 V

we do not detail the functional scheme of the receiver side, as analogous tasks are carried out in the reverse direction.

In the scheme of Figure 2.13, it is worthwhile to notice that analogue input signals are first PAM–TDM multiplexed and then PCM coded. This scheme was adopted by designers of traditional PCM multiplexers (around the 1960s and 1970s), due to the high cost of early digital circuitry. Nowadays, on the contrary, digital circuits are the cheapest part of such equipment. Therefore, PCM coding moved left: in more recent PCM multiplexers, analogue input signals are first PCM coded to yield 64 kbit/s channels, which are then digitally multiplexed.

In this section we pointed out the first use of the PCM primary multiplex, i.e. putting together 30 (or 24) telephone channels in one digital multiplex signal. Nevertheless, in modern networks, the same signal is commonly used to carry data circuits as well. The PCM primary multiplex should then be seen simply as a multiplex of several 64 kbit/s circuits, which may code speech or carry data circuits of various kinds. Moreover, standard mappings have been defined which specify multiplexing of several digital subrates — ranging for example from 400 bit/s to 28.8 kbit/s, 33.6 kbit/s and above — into one basic channel 64 kbit/s. E1 or DS1 are the main standard interfaces of both telephone digital switching exchanges and Digital Cross Connect (DXC) 1/0 systems, which support digital circuit provisioning in circuit-switched data networks.

For ease of reference, finally, Table 2.2 summarizes the most relevant data of the European 2.048 Mbit/s PCM primary multiplex signal.

2.3 DIGITAL MULTIPLEXING

Digital multiplexing (in the proper sense of the word) means specifically to TDM-multiplex N_t digital signals (called *tributary signals*), which in telephone applications

---

[12] The significant alignment bits are 7 in even frames and 1 in odd frames.
[13] See Section 1.1.3.

have the same nominal bit rate f_{t0}. The resulting multiplex signal has thus a nominal bit rate $f_{m0} > N_t f_{t0}$ (some frame alignment information must be added). Generally speaking, the tributary signal may be either binary or multilevel; nevertheless, here we focus on the binary case, which is the most relevant to telephone network applications.

The piece of equipment that performs this digital-multiplexing task in transmission, and the inverse task in reception (demultiplexing), is called *digital multiplexer*.

The distinctive feature of a digital multiplexer, compared to PCM multiplexers, is that the tributary signals to multiplex are already in digital form and thus TDM multiplexing is not based on intermediate sampling and coding.

The operation of digital multiplexing can be outlined in two basic processes, functionally distinct but temporally superposed, as shown in Figure 2.14:

- in the *writing phase*, the binary digits of the ith tributary are written in its buffer (elastic store), with write frequency equal to the tributary instantaneous bit rate f_{ti} (in general, the tributary instantaneous bit rates have the same nominal value f_{t0} but actual different values f_{ti}.);

- in the *reading phase*, the elastic stores associated with the N_t tributaries are read cyclically, by extracting, from each buffer and in every cycle, one bit or a set of bits; the read frequency f_{r0} is the same for all the N_t buffers, is derived from a local equipment clock and must be chosen properly.

The bits read, with other *additional bits* carrying various information, are transmitted in sequence by structuring the resulting signal in frames made of b_m bits. The output multiplex frames are denoted *bit-interleaving* frames if reading is made bit-by-bit, or *channel time-slot interleaving* frames if reading is made word-by-word.

Letting b_a be the number of additional bits (viz. the frame alignment word, miscellaneous overhead, fixed-stuff bits, etc.) inserted in each frame and b_d the number (multiple of N_t) of bits available to tributaries[14], the multiplex signal redundancy is

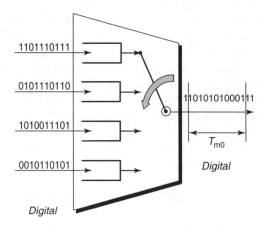

Figure 2.14. Operation of a digital multiplexer

---

[14] These b_d bits are available to tributaries, but not necessarily entirely filled by tributary bits, as they may also be justification opportunity bits in the case of asynchronous digital multiplexing (see Section 2.3.4).

defined as

$$r = \frac{b_a}{b_d} \tag{2.5}$$

Then, the nominal bit rate f_{m0} of the digital multiplex signal resulting from multiplexing N_t tributaries with nominal read frequency f_{r0} is

$$f_{m0} = f_{r0} N_t (1 + r) \tag{2.6}$$

The frame of this multiplex signal is made of $b_m = b_d + b_a = b_d(1 + r)$ bits and has nominal duration

$$T_{m0} = \frac{b_m}{f_{m0}} = \frac{b_d}{f_{r0} N_t} \tag{2.7}$$

which in general *does not coincide, unlike the PCM multiplex signals, with the sampling period T_s and is fully independent on both T_s and the code used* (e.g. the number of bits per sample).

At this point, it is worthwhile remarking that the instantaneous frequency of an input tributary signal depends on the upstream equipment clock (as well as on possible line jitter), but is obviously out of any control by the receiving equipment. This concept is well expressed by what we humorously like to call the

Fundamental Law of Input Buffers

BITS ARRIVE TO EQUIPMENT INPUT PORTS WITH THE RATE THEY CHOOSE
AND WHICH NONE CAN AFFECT

which is the basis of the design of any digital equipment block based on elastic stores. Therefore, two different cases must be considered:

- the N_t tributaries have the same instantaneous bit rate f_t, in fixed ratio with the multiplex signal bit rate $f_m = N_t f_t(1 + r)$ (*synchronous* tributaries);
- the N_t tributaries have the same nominal bit rate f_{t0}, but actually have different and independent bit rates f_{ti}, within a given tolerance range (*plesiochronous* tributaries).

The former case takes place in *synchronous digital multiplexers*, the latter in *asynchronous digital multiplexers*.

2.3.1 Synchronous Digital Multiplexing

In synchronous digital multiplexers, the timing signals of the tributary sources are synchronous, i.e. they all work at the same instantaneous frequency f_t, at least on the average, in fixed ratio with the multiplex signal frequency f_m.

Little fluctuations of the instantaneous frequency around the mean value are unavoidable, owing for example to jitter accumulated along transmission chains, and must therefore be compensated by suitable means. In synchronous digital multiplexers, relative phase fluctuations between the various tributaries and the multiplex signal are supposed null-mean and are absorbed by the buffers or elastic stores, which indeed owe such a name to this damping action.

The buffer read frequency f_{r0} is equal — on the average — to the write frequency f_t. In this particular case, therefore, Equation (2.6) becomes

$$f_m = f_t N_t (1 + r) \tag{2.8}$$

In synchronous digital multiplexes, the additional bits are of just two kinds, namely:

- the frame alignment word;
- miscellaneous overhead bits, i.e. to service purposes like the maintenance and management of the multiplex equipment and of the overall transmission system.

The fact that $f_m > N_t f_t$, owing to these additional bits, implies itself indeed the necessity of an input elastic store per each tributary, since the unevenness of the read timing signal due to the additional bits insertion (gapped clock) must be compensated. In normal operation, the phase relative error between memory write and read signals follows a saw-toothed trend, due to the gaps in the reading timing signal, left free for the additional bits. Dimensioning the buffer, if aiming at compensating such unevenness, depends on the number of tributaries, the type of interleaving and the number and position of the additional bits (the worst case is when they are concentrate, i.e. in consecutive positions).

The inadequate dimensioning of input elastic stores causes, as well as any asynchrony between tributaries and the multiplexer clock, buffer overflow or underflow, and thus the loss or repetition of some tributary bits (*slip*). Such events are particularly troublesome, as a slip on a multiplex signal may yield avalanche losses of alignment in the multiplexed tributaries.

2.3.2   Slips

The process of writing and reading tributary bits into and from an elastic store, before their retransmission in the multiplex signal, is called *bit synchronization*, because the multiplexer receives a bit stream with a given rate and retransmit it with another rate, driven by the local equipment clock.

By detailing further the scheme of Figure 2.14, the principle of operation of a bit synchronizer is shown in Figure 2.15. For each input line, the tributary bits are written into an elastic store (buffer) at *their* own arrival rate f_w, but are read with the equipment local clock frequency f_r. The elastic store absorbs any random zero-mean frequency fluctuation between the write and read clocks, within given bounds due to buffer limits. Of course, nevertheless, any frequency offset $|f_w - f_r|$ between the write and read clocks will make the buffer empty or overflow, sooner or later.

The elastic store is implemented as a circular memory with cyclic access. If the buffer empties (i.e., the write and read memory addresses coincide because the read address overtakes the write address), some bytes are repeated in transmission. If the buffer overflows (i.e., the write and read memory addresses coincide because the read address is overtaken by the write address), some bytes are deleted and lost. Such events are called *slips* (hence the name of *slip buffering* for this technique of bit synchronization). Repetitions or losses

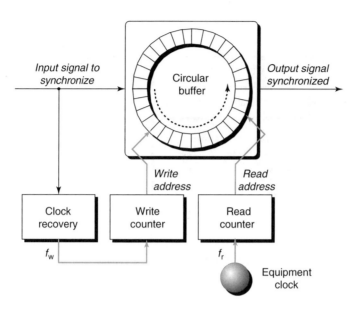

Figure 2.15.   Scheme of principle of a bit synchronizer

of an integer number of frames, thus maintaining frame alignment and limiting data loss, are called *controlled slips*.

The slip rate F_{slip} is function of the number N of bits repeated or lost in one slip (obviously, the buffer size cannot be smaller than $2N$) and of the frequency offset $|f_w - f_r|$. Expressing the frequency offset in [Hz], the following simple relationship holds

$$F_{slip} = 86\,400\frac{|f_w - f_r|}{N} \quad \text{[slips/day]} \tag{2.9}$$

where the number $86\,400$ is the number of seconds in one day. In other words, a larger buffer allows reducing the slip rate, for any given clock accuracy.

However, for example in the case of the 2.048 Mbit/s European PCM primary multiplex signal, a slip size equal to one frame (256 bits) and a frequency tolerance of 50 ppm with respect to the nominal value $f_0 = 2.048$ MHz, as specified by the ITU-T Rec. G.703 [2.5] for multiplexers, would yield an unrealistic slip rate as high as 24 slips per minute! It is clear that this mechanism for bit synchronization is not suited for operation in a plesiochronous network environment.

2.3.3   Synchronous Digital Multiplexer and Demultiplexer

The functional scheme of principle of a synchronous digital multiplexer is depicted in Figure 2.16, where only the ith tributary input port is detailed for the sake of simplicity. The bits incoming from the tributary input port are written into the buffer with their arrival rate f_t (regular clock) at the address selected by a write counter. These bits are read from the buffer at instants triggered by a timing generator and distributor, driven by the local equipment clock, which synthesizes the regular clock signal f_m (multiplex frequency) and generates the buffer read timing signal f_r by properly gapping the f_m signal, so that the buffer read frequency f_r (gapped clock) is equal — *on the average* — to

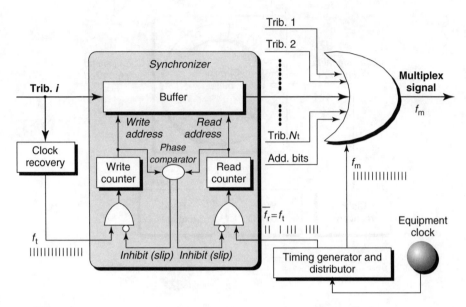

Figure 2.16.   Functional scheme of principle of a synchronous digital multiplexer

the write frequency f_t. More precisely, denoting with the upper line the time-averaging, the average of the read frequency f_r is equal to

$$\overline{f_r} = \overline{f_m/N_t(1+r)} = f_t \tag{2.10}$$

Moreover, the buffer write and read addresses are compared to detect slips, which may be due to excessive relative phase fluctuations as well as to any frequency offset between the tributary and the multiplexer clocks. When a slip is detected, reading or writing is inhibited (AND gates at counter inputs).

The block performing the mapping of the tributary into the multiplex signal, thus comprising mainly the buffer and the read and write controls, is usually called *synchronizer*, as it accomplishes the synchronization of the tributary bits into the multiplex frame available positions.

The bits read from the buffer (synchronizer output) are finally interleaved with the bits coming from other tributaries and with the additional bits, to form the output multiplex signal. The role of the timing generator and distributor is thus to generate synchronously N_t clock signals f_{ri} (one per each tributary) plus the clock signal for the additional bits, by properly gapping the common clock signal f_m so that every bit fed into the OR gate fits in its own position in the output multiplex frame.

On the reception side, demultiplexing takes place after frame alignment, as described in Section 1.1.3. Once the additional bits have been extracted, the tributaries are separated and written into their own demultiplexing buffers with average frequency

$$\overline{f_m/N_t(1+r)} = f_t \tag{2.11}$$

These buffers are then read with frequency f_t, so that to fill the gaps corresponding to the additional bits and to output digital signals with instantaneous nominal frequency f_t. Since the demultiplexer is synchronous, the frequency f_t can be synthesized directly from the multiplex frequency f_m.

The functional scheme of principle of a synchronous digital demultiplexer is depicted in Figure 2.17, where only the ith tributary output port is detailed for the sake of simplicity. The tributary bits incoming from the multiplex input port are written into the buffer at the address selected by a write counter. The write frequency f_w (gapped clock) is derived by inhibiting the multiplex clock f_m for the additional bits and the other tributaries, so that f_w is equal — *on the average* — to the tributary frequency f_t. The read frequency f_t is synthesized directly from the multiplex frequency f_m.

The block performing the demapping of the tributary from the multiplex signal, thus comprising mainly the buffer and the read and write controls, is usually called *desynchronizer* and fulfils the complementary function of the synchronizer of Figure 2.16: it smoothes the gaps in the output tributary clock due to mapping thus returning a bit stream with regular clock.

2.3.4   Asynchronous Digital Multiplexing: Bit Justification

The optimal situation of synchronism among all the tributaries assumed in the operation of synchronous digital multiplexers is not easy to achieve nor has even been practically feasible in many cases for a long time. Even though it is possible to transfer timing from one multiplexer to another one in both directions via one tributary signal, yet it results more difficult to design a comprehensive plan of synchronization of all the digital multiplexers so that *all* the tributaries in *each* multiplexer are synchronous. Moreover, the jitter accumulated along transmission chains may be enough to disturb the operation of synchronous digital multiplexers. Dimensioning elastic stores depends on the peak-to-peak amplitude of phase fluctuations to compensate and may become difficult in wide area networks.

For these reasons, historically, synchronous digital multiplexing has been adopted quite seldom in digital transmission networks, at least till the introduction of SONET/SDH which follows this principle but with some important innovations (see Section 2.5).

In asynchronous digital multiplexers, the timing signals of the tributary sources are asynchronous. Precisely speaking, they are plesiochronous, i.e. they have the same nominal

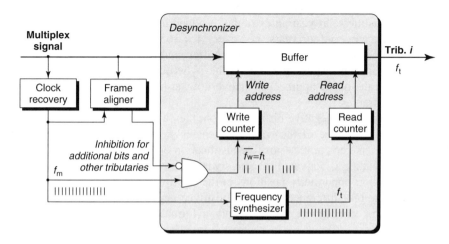

Figure 2.17.   Functional scheme of principle of a synchronous digital demultiplexer

bit rate f_{t0} but actually different and independent bit rates f_{ti}, within a given tolerance range. Moreover, the multiplex signal frequency f_m is not in fixed ratio with the tributary frequencies. This situation of plesiochrony is because each piece of multiplexing equipment works under the control of a local independent clock.

In asynchronous digital multiplexers, bit synchronization of tributary signals into the multiplex signal is performed by means of a *bit justification*[15] (historically known also as *pulse stuffing or bit stuffing*, see footnote 9 in this chapter) technique. Three fashions of bit justification have been defined: positive, negative and positive/negative bit justification. They are not conceptually different. They differ solely because of the tributary transmission rate in the multiplex signal when justification is not taking place: slightly higher than the tributary maximum rate in positive justification, lower than its minimum rate in negative justification or same as its nominal rate in positive/negative justification. In the following description, we will always refer to positive bit justification, which has been standardized in North American and European PDH (see next section), unless otherwise stated.

As it happens in synchronous digital multiplexers, the binary digits of each tributary are written in their buffer, with write frequency equal to the tributary instantaneous bit rate f_{ti}. Here, nevertheless, the buffer read frequency f_{r0}, the same for every tributary, is synthesized from the local clock and, in the case of positive bit justification, it is chosen slightly higher than the maximum value tolerated for the N_t write frequencies f_{ti}.

The tributary bits are then read and multiplexed by bit interleaving. Since the input elastic stores are read faster than written, they tend to progressively empty. To cope with this, reading is inhibited when the buffer approaches its depletion, at certain bit positions within the output multiplex signal (*justification* or *stuffing opportunity bits*), where non-meaningful, dummy bits are inserted. Such bits are called *justification* or *stuffing bits*. Usually, one justification opportunity position is reserved per tributary in each multiplex frame; such positions are filled with meaningful tributary bits in case of no stuffing.

The term 'justification' originated in the printing industry, where it describes the process of adjusting the spaces between printed words so that all the lines of print have the same length. Another practical example of justification is embodied by the concept of leap year. A nominal length calendar year is 365 days long, but to make the calendar year nearly the same as the solar year an extra day is added to the year at the end of February approximately once every four years (i.e., every 4×365 days, a day is justified).

In the case of negative bit justification, the buffer read frequency f_{r0} is chosen slightly lower than the minimum value tolerated for the N_t write frequencies f_{ti}. Excess reading is activated when the buffer approaches its overflow, at justification opportunity positions in the multiplex frames.

In the case of positive/negative bit justification, the buffer read frequency f_{r0} is chosen equal as the nominal value of the write frequencies f_{ti}. Reading is inhibited when the buffer approaches its depletion, at positive-justification opportunity positions in the multiplex frames, or reading is activated in excess when the buffer approaches its overflow, at negative-justification opportunity positions in the multiplex frames.

It is clear that the difference between positive, negative and positive/negative bit justifications is not conceptual, but it is only a formal aspect. Regardless of the type

---

[15] The specification *bit* is often omitted but helps to distinguish from the pointer justification technique, adopted in SDH and SONET, which is sometimes also referred to as *byte justification*.

of bit justification, in all cases the number of justification opportunity bits stuffed with dummy bits is adaptively regulated so to adjust on the actual tributary rate.

The presence or absence of stuffing bits in each multiplex frame must be properly signalled to the far-end demultiplexer, in order to allow it to neglect the dummy bits in the reconstruction of tributary signals. To this purpose, special additional bits are inserted into the multiplex frame, called *justification* or *stuffing control bits*.

This Boolean information of either presence or absence of stuffing in the controlled opportunity bit is coded with words of $2k + 1$ bits, with the aim at allowing the correction of up to k transmission errors in the control word. Indeed, the protection against transmission errors is critical and necessary, as any error would yield the interpretation of a dummy bit as meaningful, or vice versa, and thus the insertion of dummy bits or the deletion of meaningful bits in the demultiplexed tributaries. Such an event would yield a non-controlled slip of the tributary frame and thus its real loss of alignment. Therefore, in each multiplex frame

$$s = N_t(2k + 1) \qquad (2.12)$$

justification control bits are inserted.

To summarize, in asynchronous digital multiplexes there are $b_a = v + w + s$ additional bits. They are the following:

- v frame alignment bits (A bits);

- w miscellaneous overhead bits (O bits), i.e. to service purposes like the maintenance and management of the multiplex equipment and of the overall transmission system;

- $s = N_t(2k + 1)$ justification control bits (C bits).

Moreover, in every frame there are:

- N_t justification opportunity bits (S bits), that may be filled by tributary bits (I bits) or by non-meaningful stuffing bits (X bits).

The justification control scheme is as follows (for positive bit justification):

- all C bits equal to '1' mean that the controlled justification opportunity S bit of the corresponding tributary is non-meaningful (X bit, i.e., in that frame justification took place);

- all C bits equal to '0' mean that the controlled justification opportunity S bit of the corresponding tributary is meaningful and carries a tributary I bit (i.e., in that frame justification did not take place).

Majority voting is used to make the justification decision for protection against k bit errors in the C bits.

Focusing in particular on the case of the European PDH standard[16], for the sake of simplicity, the multiplex frame is made of b_m bits and is structured in $2(k + 1)$ sectors

---

[16] The North American and Japanese PDH formats are more complex, since frames of some signals are organized in multi-frames and the scheme of allocation of the additional bits to the sectors itself is different. Nevertheless, the principle is the same and the concepts expounded above still apply.

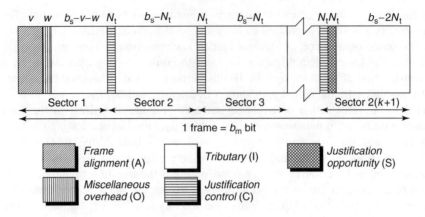

Figure 2.18. Frame structure of an asynchronous digital multiplex signal

(see Figure 2.18), each made of

$$b_s = \frac{b_m}{2(k+1)} \tag{2.13}$$

bits.

The first sector includes the alignment word, made of v consecutive A bits, followed directly by w overhead O bits and by $b_s - v - w$ tributary meaningful I bits. The last sector includes N_t justification control C bits (each of them is associated with a different tributary), followed by N_t justification opportunity S bits (each dedicated to a different tributary) available to be filled with dummy stuffing X bits or with meaningful tributary I bits, followed finally by $b_s - 2N_t$ meaningful tributary I bits. The $2k$ intermediate sectors are structured as the last one, except for the N_t justification opportunity S bits substituted by tributary I bits.

In the above scheme, it is worthwhile noticing that the $N_t(2k+1)$ justification control C bits are evenly distributed among all the sectors but the first one. This aims at making those bits less vulnerable to bursty line errors.

Example: European PDH Frame Format

Let us consider $N_t = 4$ tributaries (denoted with A, B, C e D) and suppose that we want to be able to correct up to $k = 2$ errors in the justification control word.

The total number of sectors in the asynchronous multiplex signal results $2(k+1) = 6$.

In each of the $2k + 1 = 5$ sectors following the sector 1, there are $N_t = 4$ C bits (justification control bits), forming the word $C_A C_B C_C C_D$ (each control bit is associated to one tributary).

In the last sector, there are $N_t = 4$ S bits (justification opportunity bits), forming the word $S_A S_B S_C S_D$ (each opportunity bit is associated to one tributary).

Putting together the $2k + 1 = 5$ words of $N_t = 4$ bits $C_A C_B C_C C_D$, we get the $N_t = 4$ words

$C_A C_A C_A C_A C_A$,
$C_B C_B C_B C_B C_B$,

$C_C C_C C_C C_C C_C C_C,$
$C_D C_D C_D C_D C_D,$
indicating the stuffing status of the respective bits S_A, S_B, S_C and S_D while correcting up to $k = 2$ errors.

This bit justification technique works fine, i.e. it is able to compensate the plesiochrony of tributaries, only if the justification opportunities are enough to accommodate excess bits from tributaries. In other terms, the following inequalities must hold (written for positive bit justification)

$$
\begin{cases}
(f_{r0} - \Delta f_{r0}) = \dfrac{f_{m0} - \Delta f_{m0}}{N_t(1 + r)} > (f_{ti0} + \Delta f_{t0}) & \forall i \\[4mm]
(f_{r0} + \Delta f_{r0}) - (f_{ti0} - \Delta f_{t0}) < \dfrac{f_{m0}}{b_m} & \forall i
\end{cases}
\tag{2.14}
$$

where $f_{ti0}(\forall i)$, f_{r0} and f_{m0} are the nominal values of the ith tributary write frequency, the tributary read frequency and the multiplex signal frequency, respectively, and Δf_{t0}, Δf_{r0} and Δf_{m0} are their tolerances.

The former inequality guarantees that in the output multiplex frame there is enough room to accommodate the input tributary bits, even at the extreme limits of frequency tolerance ranges. The latter states that the maximum difference between the read and write frequencies, i.e. the *maximum justification demand*, must be lower than the *maximum justification frequency* (in the frame format described above there is one justification opportunity bit per frame per each tributary).

The *nominal justification (stuffing) ratio*[17] ρ_{nom} is the ratio between the nominal justification frequency and the maximum justification frequency ($0 \leq \rho_{nom} \leq 1$). For example, in the case of positive justification, it is defined as

$$
\rho_{nom} = \frac{f_{r0} - f_{ti0}}{\dfrac{f_{m0}}{b_m}} = \frac{\dfrac{f_{m0}}{N_t(1 + r)} - f_{ti0}}{\dfrac{f_{m0}}{b_m}} \qquad \forall i
\tag{2.15}
$$

In case of positive justification, the nominal justification frequency is the frequency at which justification opportunity bits are not assigned to tributary bits, in nominal conditions. The opposite holds for negative justification, which happens when justification opportunity bits are assigned to tributary bits. The maximum justification frequency, on the other hand, is the overall frequency of justification opportunity bits.

In other words, in the case of positive justification, the nominal justification ratio ρ_{nom} is the fraction of justification opportunity bits assigned to dummy bits in nominal conditions, over the total number of available opportunities. In the case of negative justification, conversely, the ratio ρ_{nom} is the fraction of justification opportunity bits used to carry tributary bits, over the total number of available opportunities. In all cases, its reciprocal $1/\rho_{nom}$ denotes every how many frames, on the average, justification takes place.

---

[17] As noticed previously, ITU-T [2.14] recommends the term *justification* instead of *stuffing*. Nevertheless, the expression *stuffing ratio* has been very popular historically and is still in wide use.

When the multiplex and tributary frequencies are not set exactly to their nominal values, as it happens normally, the justification ratio ρ departs from ρ_{nom} in the interval $0 \le \rho_{nom} \le 1$.

2.3.5 Asynchronous Digital Multiplexer and Demultiplexer

The functional scheme of principle of an asynchronous digital multiplexer (with positive justification) is depicted in Figure 2.19 (analogous to Figure 2.16), where only the ith tributary input port is detailed for the sake of simplicity. The bits incoming from the tributary input port are written into the buffer with their arrival rate f_{ti} (regular clock) at the address selected by a write counter. These bits are read from the buffer at instants triggered by a timing generator and distributor, driven by the local equipment clock, which synthesizes the regular clock signal f_m (multiplex frequency) and generates a buffer read timing signal f_{r0} by properly gapping the f_m signal. The average frequency $\overline{f_{r0}}$ is slightly greater than the maximum value tolerated for all the write frequencies f_{ti}.

The buffer write and read addresses are compared, so that when the buffer gets empty, reading is inhibited (AND gate at the read counter input) at justification opportunity positions. Therefore, the actual buffer read frequency f_{ri} results equal — *on the average* — to the write frequency f_{ti}.

The block performing the mapping of the tributary into the multiplex signal, thus comprising mainly the buffer together with the justification control, is usually called *synchronizer*, as it accomplishes the synchronization of the tributary bits into the multiplex frame available positions (cf. the synchronizer of Figure 2.16).

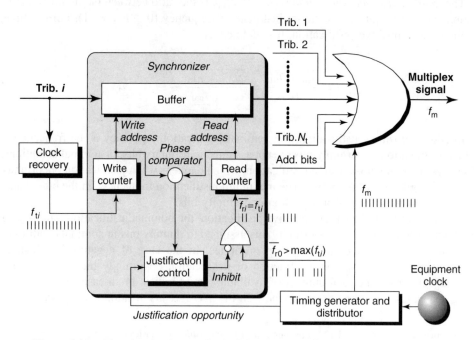

Figure 2.19. Functional scheme of principle of an asynchronous digital multiplexer

The bits read from the buffer (synchronizer output) are finally interleaved with the bits coming from other tributaries and with the additional bits, to form the output multiplex signal.

On the reception side, demultiplexing of an asynchronous multiplex signal is carried out in an analogous fashion as in synchronous demultiplexers. Once the additional bits and the possible stuffing bits have been extracted, the gapped-clock tributaries are separated and written into their own demultiplexing buffers. The write timing signal is gapped and very uneven, but the average write frequency results the same as the original tributary frequency, i.e.

$$\overline{f_{w_i}} = f_{ti} \tag{2.16}$$

The read frequency f_{ti} is not in fixed ratio with the multiplex frequency f_m, and thus cannot be synthesized directly from it like in synchronous demultiplexers. Moreover, f_{ti} may change its value dynamically, though staying within its tolerance range. For these reasons, f_{ti} is usually generated locally, by means of adaptive algorithms, based on the write gapped timing signal f_{wi}.

A PLL is usually adopted to smooth gaps in f_{wi} and then to return the regular clock f_{ti}. Its transfer function is low-pass: the jitter components beyond the cut frequency are damped, but the lower frequency components are still transferred to the output.

The functional scheme of principle of an asynchronous digital demultiplexer is depicted in Figure 2.20 (analogous to Figure 2.17), where only the ith tributary output port is detailed for the sake of simplicity. The tributary bits incoming from the multiplex input port are written into the buffer at the address selected by a write counter. The write frequency f_{wi} (gapped clock) is derived by inhibiting the multiplex clock f_m for the additional and stuffing bits and the other tributaries, so that f_{wi} is equal — *on the average* — to the tributary frequency f_{ti}. A PLL smoothes the gaps in f_{wi} and returns the regular clock f_{ti}.

The block performing the demapping of the tributary from the multiplex signal, thus comprising mainly the buffer together with the justification control, is usually called *desynchronizer* and fulfils the complementary function of the synchronizer of Figure 2.19:

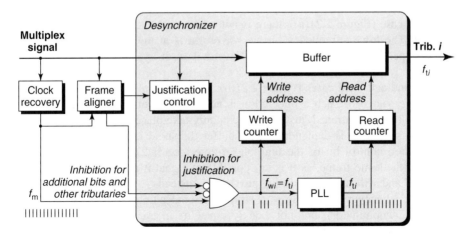

Figure 2.20. Functional scheme of principle of an asynchronous digital demultiplexer

it smoothes the gaps in the output tributary clock due to mapping thus returning a bit stream with regular clock (cf. the desynchronizer of Figure 2.17).

The jitter in the digital signal at the input of the PLL deserves a closer look. Four different components at different frequencies and their harmonics can be mainly distinguished in it:

- one component at the frame frequency due to the extraction of the alignment word and the miscellaneous overhead;
- one component at sector frequency ($2k + 1$ times the frame frequency) due to the extraction of the stuffing control bits;
- one component at the justification frequency due to the extraction of stuffing bits;
- one component at much lower frequencies called *waiting time jitter*.

These components are filtered by the PLL acting as a low-pass filter of the jitter. The first, second and third are easily damped out, but the fourth cannot be cancelled totally owing to its unlimited low frequency.

The waiting time jitter has been treated thoroughly by Duttweiler [2.15] and Chow [2.16]. Although not the highest in amplitude, it may be difficult to control in transmission chains, as its spectrum may approach the zero frequency and can therefore accumulate. Its name is due to the fact that justification does not take place exactly when it should (i.e., when the buffer threshold is trespassed), but only when a justification opportunity arrives in the multiplex frame.

Let us denote with $\Delta\varphi(t)$ the relative phase error between the buffer read and write clocks and consider only the effect of justification opportunity bits (we neglect thus the gaps due to additional bits). It is not difficult to demonstrate that it is quite possible for $\Delta\varphi(t)$ to contain significant low-frequency power. Figure 2.21 depicts the $\Delta\varphi(t)$ trend for three different cases (positive justification):

- $\rho = 1/2$ (i.e., justification occurs exactly every two frames);
- $\rho = 1/2^+$ (i.e., ρ is slightly higher than 1/2);
- $\rho = 1/2^-$ (i.e., ρ is slightly lower than 1/2).

In the first case (Figure 2.21(a)), there is not a low-frequency component in the phase error signal. The lowest frequency in this waveform is at the relatively high frequency of 1/2 cycle per justification opportunity. Nevertheless, it appears unrealistic to assume $\rho = 1/2$ exactly.

In the second and third cases (Figure 2.21(b) and (c)), there is a strong low-frequency envelope. The source of such a fluctuation is not self-evident indeed; it consists of the fact that, as already mentioned, justification can only take place in particular instants. For instance, if justification is demanded (i.e. a buffer threshold is trespassed) just after the first justification control bit of the multiplex frame (sector 2), then justification cannot take place in the same frame, as it is not possible to signal it anymore. Justification will be postponed and will take place in the next frame instead.

In general, whenever ρ is close to a simple rational number (i.e. a rational number with a small denominator), but not exactly equal to it, $\Delta\varphi(t)$ will have appreciable low-frequency power. This idea was developed by Kozuka *et al.* [2.26] [2.27]. Formulas are given for the peak-to-peak amplitude of the low-frequency envelope present on the

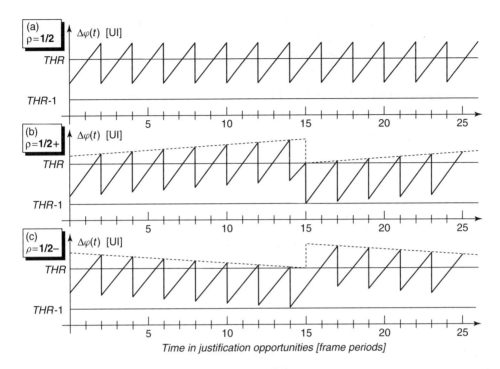

Figure 2.21. Relative phase error $\Delta\varphi(t)$ between the buffer read and write clocks for (a) $\rho = 1/2$, (b) $\rho = 1/2^+$, (c) $\rho = 1/2^-$. (Adapted from (2.15) by permission of Lucent Technologies Inc., ©1972, all rights reserved)

waveform $\Delta\varphi(t)$ when ρ is close, but not exactly equal, to a simple rational number. Questions arise, however, as to how close is close enough. An alternative approach, by Duttweiler [2.15], allows one to evaluate the power spectrum of the waiting time jitter and how it accumulates, but it is quite complex.

As a rule of thumb, let us approximate ρ as a rational number p/q (generating fraction), where p and q are integer, relatively prime numbers. Then, q expresses the interval (number of frames) along which $\Delta\varphi(t)$ takes again the initial value and p the number of frames with justification in the interval above ($p < q$). The greater is the number q, the lower is the waiting time jitter starting frequency. Theoretically, if ρ is an irrational number then the jitter power spectrum starts from the zero frequency. As far as the waiting time jitter peak amplitude is concerned, obviously it is not greater than one UI.

2.4 THE PLESIOCHRONOUS DIGITAL HIERARCHIES (PDH)

The *Plesiochronous Digital Hierarchy* (PDH) is a series of standard bit rates (hierarchical levels) defined by the CCITT (now ITU-T) for transmission in digital telephone networks [2.28]. It has been *the* standard for digital transmission systems since the 1970s worldwide. More recently, it has been complemented by the SDH standard working at higher rates, but it is expected that PDH interfaces will still be around in digital transmission systems for many years as well. Indeed, the SDH frames are designed to encapsulate every type of PDH signal to ensure full interworking of PDH and SDH networks.

Actually, two plesiochronous hierarchies, having different bit rates but following the same principle, have been defined by the ITU-T:

- the European PDH, based on the PCM Primary Multiplex signal E1 at 2.048 Mbit/s;
- the North American and Japanese PDHs, based on the PCM Primary Multiplex signal DS1 at 1.544 Mbit/s.

As expected, they are not compatible at all and have different bit rates, but they follow the same principle. Both are based on bit-interleaved asynchronous digital multiplexing, with positive bit justification. The multiplex signals of the ith level ($i > 1$) are made by multiplexing a given number of signals of the ($i - 1$)th level.

2.4.1   The European Plesiochronous Digital Hierarchy (Ei)

The signals of the various hierarchical levels are denoted with the short name Ei ($i = 1, 2, 3, 4, 5$).

The first level of the European hierarchy is the 2.048 Mbit/s PCM Primary Multiplex, carrying 30 telephone channels (see Section 2.2). The next levels are obtained by bit-interleaved asynchronous digital multiplexing (positive bit justification) of four signals of the lower level to produce one signal of the upper level, thus yielding the hierarchy summarized in Table 2.3. We notice, moreover, that sometimes the single PCM channel at 64 kbit/s is denoted as E0, for uniformity with the rest of the hierarchy.

The hierarchy is defined in the ITU-T Rec. G.702 [2.28]. The frame formats are specified in the ITU-T Recs. G.704 [2.19] and G.706 [2.20] for the E1 signal, in the G.741 [2.29] and G.742 [2.30] for the E2 signal and in the G.751 [2.32] for the E3 and E4 signals. The intra-office electrical interfaces are specified in the ITU-T Rec. G.703 [2.5]. The digital line systems are specified in the ITU-T Recs. G.952 [2.35], G.954 [2.37] and G.955 [2.38].

The fifth level (564.992 Mbit/s) has been defined at last by the ITU-T as a line system 4×139.264 Mbit/s in the Recs. G.954 [2.37] and G.955 [2.38]. This level is standard

Table 2.3   The European Plesiochronous Digital Hierarchy

Level	Nominal Bit Rate	Channels Carried	Notes
E1 First Level	2.048 Mbit/s	30	The PCM Primary Multiplex
E2 Second Level	8.448 Mbit/s	120	Obtained by asynchronous digital multiplexing of 4 E1 signals
E3 Third Level	34.368 Mbit/s	480	Obtained by asynchronous digital multiplexing of 4 E2 signals
E4 Fourth Level	139.264 Mbit/s	1920	Obtained by asynchronous digital multiplexing of 4 E3 signals
E5 Fifth Level	564.992 Mbit/s	7680	Obtained by asynchronous digital multiplexing of 4 E4 signals

as far as the frame format is concerned, but no intra-office electrical interface has been defined. For this reason, sometimes it is not recognized as part of the 'standard' PDH hierarchy, but it is considered as a sort of annex.

Moreover, it is worthwhile noticing that, while the frame formats and the intra-office electrical interfaces are standardized, no standards exist for the line interfaces in PDH systems. In commercial PDH digital line systems, therefore, the line interfaces are usually proprietary, i.e. both the line code and the line raw bit rate are designed differently by different suppliers. Line systems of different vendors are thus usually not compatible.

For ease of reference, finally, Table 2.4 summarizes the most relevant data of the European PDH signals.

2.4.2   The North American and Japanese Plesiochronous Digital Hierarchies (DSi/Ti)

The digital systems standardized in North America and in Japan base their hierarchies, unlike the European systems, on a PCM Primary Multiplex carrying 24 telephone channels and having the 1.544 Mbit/s bit rate. Moreover, even the coding and the quantization laws of the analogue speech signal are different: while the European PCM multiplexers follow the so-called *A-law*, North American equipment follows the so-called μ-*law*.

The upper hierarchical levels, as in European systems, are based on bit-interleaved asynchronous digital multiplexing, but different rules apply in building the hierarchy. In particular, the frame structures are more complicated than for the scheme depicted in Figure 2.18, which applies here only in principle. Multiframing is applied extensively in building the hierarchical signals.

The signals of the various hierarchical levels are denoted with the short name DSi (Digital Signal), often also called Ti ($i = 1, 2, 3, 4$)[18]. Only the first three levels in the North American hierarchy and the first four levels in the Japanese hierarchy (the first two being the same in both hierarchies) have been fully standardized by ITU-T. Also in this case, we notice that sometimes the single PCM channel at 64 kbit/s is denoted as DS0, for uniformity with the rest of the hierarchy.

The signals of the North American hierarchy are summarized in Table 2.5, while the signals of the Japanese hierarchy are summarized in Table 2.6.

These hierarchies are defined in the ITU-T Rec. G.702 [2.28]. The frame formats are specified in the ITU-T Recs. G.704 [2.19] and G.706 [2.20] for the DS1 signal, in the G. 741 [2.29] and G.743 [2.31] and in the G.752 [2.33] for the other standard signals. The intra-office electrical interfaces are specified in the ITU-T Rec. G.703 [2.5]. The digital line systems are specified in the ITU-T Recs. G.951 [2.34], G.953 [2.36] and G.955 [2.38].

Also in this case, as for the European PDH, it is worthwhile noticing that, while the frame formats and the intra-office electrical interfaces are standardized, no standards exist for the line interfaces. Therefore, line systems of different vendors are usually not compatible.

For ease of reference, finally, Table 2.7 summarizes the most relevant data of the North American and Japanese PDH signals. As far as frame alignment parameters are concerned, the relevant ITU-T Recs. specify only the alignment recovery and/or loss times.

---

[18] Properly speaking, DSi denotes the hierarchical signal, transmitted over a Ti line, which denotes instead the digital transmission system. Often, however, the two terms are used equivalently.

Table 2.4   Summary of characteristics of the European PDH signals

	2nd Level	3rd Level	4th Level	5th Level
Nominal bit rate	8.448 Mbit/s	34.368 Mbit/s	139.264 Mbit/s	564.992 Mbit/s
Frequency tolerance	±30 ppm	±20 ppm	±15 ppm	±15 ppm
Nominal tributary bit rate	2.048 Mbit/s	8.448 Mbit/s	34.368 Mbit/s	139.264 Mbit/s
Number of tributaries	4	4	4	4
Number of tel. channels carried	120	480	1920	7680
Frame alignment word	1111010000	1111010000	111110100000	111110100000
Overhead bits	AX	AX	AXXX	AXXX
Frame alignment strategy parameters[a]	$\alpha = 4$ $\delta = 2$ 4	$\alpha = 4$ $\delta = 2$ 4	$\alpha = 4$ $\delta = 2$ 6	$\alpha = 4$ $\delta = 2$ 7[b]
Number of sectors $2(k+1)$	4	4	6	
Frame length	848 bit	1536 bit	2928 bit	2688 bit
Number of bits per sector	212 bit	384 bit	488 bit	384 bit
Number of available tributary bits per tributary per frame[c]	206 bit	378 bit	723 bit	663 bit
Frame period	~100 μs	~44.7 μs	~21 μs	~4.8 μs
Number of justification opp. bits (per frame, per tributary)	1	1	1	1
Number of justification control bits $2k+1$ (per frame, per tributary)	3	3	5	5
Justification indication pattern (S bit is non-meaningful)	111	111	11111	11111
Non-justif. indication pattern (S bit is meaningful)	000	000	00000	00000
Nominal frame repetition rate (max. justification frequency)	9.962 kHz	22.375 kHz	47.563 kHz	210.190 kHz
Nominal justification ratio ρ_{nom}	0.42424	0.43575	0.41912	0.43906
	HDB3	HDB3	CMI	
Intra-office physical interface signal peak level	2.37 V	1 V	1 V	not defined

[a] According to what stated in the relevant ITU-T Recs., 'as it is not strictly necessary to specify the detailed frame alignment strategy, any suitable frame alignment strategy may be used provided the performance achieved is at least as efficient in all respects as that obtained by the above frame alignment strategy'.
[b] The 564.992 Mbit/s signal has a frame format slightly different from the basic scheme depicted in Figure 2.18. There are seven $(2k + 2 + 1)$ sectors in total: sectors II through VI carry the five $(2k + 1)$ justification control bits per each of the four tributaries, sector VII carries the justification opportunity bit per each tributary.
[c] Justification opportunity bits included.

Table 2.5 The North American Plesiochronous Digital Hierarchy

Level	Nominal Bit Rate	Channels Carried	Notes
DS1/T1 First Level	1.544 Mbit/s	24	The PCM Primary Multiplex
DS2/T2 Second Level	6.312 Mbit/s	96	Obtained by asynchronous digital multiplexing of four DS1 signals
DS3/T3 Third Level	44.736 Mbit/s	672	Obtained by asynchronous digital multiplexing of seven DS2 signals
DS4/T4 Fourth Level	274.176 Mbit/s	4032	Obtained by asynchronous digital multiplexing of six DS3 signals *(not an ITU-T standard)*

Table 2.6 The Japanese Plesiochronous Digital Hierarchy

Level	Nominal Bit Rate	Channels Carried	Notes
DS1/T1 First Level	1.544 Mbit/s	24	The PCM Primary Multiplex
DS2/T2 Second Level	6.312 Mbit/s	96	Obtained by asynchronous digital multiplexing of four DS1 signals
Third Level	32.064 Mbit/s	480	Obtained by asynchronous digital multiplexing of five DS2 signals
Fourth Level	97.728 Mbit/s	1440	Obtained by asynchronous digital multiplexing of three 3rd Level signals
Fifth Level	400.352 Mbit/s	5760	Obtained by asynchronous digital multiplexing of four 4th Level signals *(not an ITU-T standard)*

Exercise 2.1
Bit Justification in the PDH E4 Signal

The frame of the PDH E4 signal (nominal bit rate 139.264 Mbit/s) is divided into six sectors ($k = 2$), each having length 488 bit.

What is the maximum peak-to-peak deviation Δf of an E3 tributary (nominal bit rate 34.368 Mbit/s) that can be accommodated by the bit justification mechanism, at least in theory? (In other words, what is the difference between the maximum and the minimum frequency of the E3 tributary multiplexed in the E4?)

Solution

One frame is made of $2(k + 1) = 6$ sectors. Therefore, in each frame, there are $2k + 1 = 5$ justification control bits and one justification opportunity bit per tributary.

Table 2.7 Summary of characteristics of the standard North American and Japanese PDH signals

	2nd Level	3rd Level		4th Level
Nominal bit rate	6.312 Mbit/s[a]	32.064 Mbit/s	44.736Mbit/s[b]	97.728 Mbit/s
Frequency tolerance	±30 ppm	±10 ppm	±20 ppm	±10 ppm
Nominal tributary bit rate	1.544 Mbit/s	6.312 Mbit/s	6.312 Mbit/s	32.064 Mbit/s
Number of tributaries	4	5	7	3
Number of tel. channels carried	96	480	672	1440
Frame alignment word	01	11010/00101	1001	110/001
Multiframe alignment word	011x[c]	—	XXPP010[d]	—
Number of sectors	6	6	8	6
Frame length	294 bit	1920 bit	680 bit	1152 bit
Number of bits per sector	49 bit	320 bit	85 bit	192 bit
Multiframe length	1176 bit / 288 bit	— / 378 bit	4760 bit / 672 bit	— / 378 bit
Number of available tributary bits per tributary[e]	(per multifr.) 1	(per frame) 1	(per multifr.) 1	(per frame) 1
Number of justification opp. bits per tributary	(per multifr.) 3	(per frame) 3	(per multifr.) 3	(per frame) 3
Number of justification control bits per tributary	(per multifr.) 3	(per frame) 3	(per multifr.) 3	(per frame) 3
Justification indication pattern (S bit is non-meaningful)	111	111	111	111
Non-justif. indication pattern (S bit is meaningful)	000	000	000	000
Nominal (multi) frame repetition rate (max. justification frequency)	5.367 kHz	16.700 kHz	9.398 kHz	84.833 kHz
Nominal justif. ratio ρ_{nom}	0.33460	0.035928	0.39056	0.035363
Intra-office physical interface	B6ZS/B8ZS	scrambled AMI	B3ZS	scrambled AMI

[a]The 6.312 Mbit/s frame structure is more complicated than for the scheme depicted in Figure 2.18. Bits are organized in a multiframe made of four basic frames (one per tributary), each made of six sectors, three of which include the justification control bits, two include the frame alignment bits, one includes the multiframe alignment bit. For this reason, the scheme of Figure 2.18 applies only loosely.

[b]The 44.736 Mbit/s frame structure is more complicated than for the scheme depicted in Figure 2.18. Bits are organized in a multiframe made of seven basic frames (one per tributary), each made of eight sectors, three of which include the justification control bits, four include the frame alignment bits, one includes the multiframe alignment bit. For this reason, the scheme of Figure 2.18 applies only loosely.

[c]x may be used as an alarm service digit.
[d]X: bit assigned to service function; P: parity bit for the preceding multiframe.
[e]Justification opportunity bits included.

One frame has length 6×488 bit $= 2928$ bit. The nominal frame repetition rate is thus

$$\frac{139.264 \text{ Mbit/s}}{2928 \text{ bit}} \cong 47.563 \text{ kHz}$$

Since there is one justification opportunity bit per tributary and per frame, this is also the maximum justification frequency, i.e., in [bit/s], the optional amount of tributary bits that can be carried at the justification opportunity bit positions.

Therefore, $\Delta f = 47.563$ kHz. In fractional terms, this corresponds to a deviation from the nominal value

$$\frac{47.563 \text{ kHz}}{34.368 \text{ MHz}} \cong 1384 \text{ ppm}$$

This value may be compared to the E3 frequency tolerance ± 20 ppm (corresponding to 40 ppm peak-to-peak tolerance), specified by ITU-T Rec. G.703 [2.5].

2.5 THE SYNCHRONOUS DIGITAL HIERARCHY (SDH) AND SONET

Historically, pure synchronous digital multiplexing has been adopted quite seldom in digital transmission networks, until the introduction of SONET and SDH that follow the same principle but with some important innovations. This section outlines the basic characteristics of SDH and SONET, with special regard to the pointer justification technique designed purposely to cope with the issue of multiplexing asynchronous tributaries but maintaining a synchronous digital multiplexing structure. For further details, the reader is referred to the relevant ITU-T Recommendations (see Section 2.5.9) and to [2.39], one of the first books published on SDH and SONET.

2.5.1 A Bit of History

A first attempt to adopt synchronous digital multiplexing dates back to the 1970s with SYNTRAN (SYNchronous TRANsmission), developed in the United States. SYNTRAN was based on the synchronous digital multiplexing of 28 tributaries at 1.544 Mbit/s, with fixed allocation of octets within the frame, in order to allow direct access to the PCM channels. The multiplex signal bit rate was the same as that of the 3rd level signal DS3 of the North American PDH hierarchy (44.736 Mbit/s). The main problems of SYNTRAN were the complex frame structure, the scarce overhead for maintenance and service functions and, what's more, the impossibility of avoiding slips under lack of network synchronization. All these issues limited the diffusion of SYNTRAN to few experiments.

At the beginning of the 1980s, again in the United States, BELLCORE (BELL COmmunications REsearch) Laboratories designed a new synchronous digital hierarchy named Synchronous Optical NETwork (SONET). The great success of this new hierarchy was favoured by the modernization of the AT&T synchronization network (dating back to the 1970s) and by the innovative idea of the pointer justification technique.

In 1986, following a proposal of the United States of America, the CCITT (now ITU-T) body started the study activities aiming at the definition of a new world standard

Synchronous Digital Hierarchy (SDH), based on the USA standard SONET but with several extensions.

The SDH was defined by the CCITT exceptionally quickly and the early versions of the Recommendations on SDH were released after just 21 months of work (in 1988). The main differences of SDH compared to SONET can be summarized as follows:

- mapping schemes have been added in order to extend SDH capability to transport the signals of the European PDH as well as of the Japanese and North American PDH considered by SONET (the available standard schemes allow mapping of a very wide range of different signals, including PDH, ATM, FDDI and DQDB);

- the bit rate of the SDH first hierarchical level has been set to 155.520 Mbit/s, three times that of the SONET first level signal (51.840 Mbit/s), to be able to accommodate the E4 signal (139.263 Mbit/s), very common where European PDH is adopted;

- the SDH physical interface has been defined for three different transmission media, namely the optical fibre, the radio relay and the coaxial copper cable, while SONET — as the name itself says — uses solely the optical fibre.

2.5.2   Hierarchical Levels of SDH and SONET

The frames of the various standard hierarchical levels of the SDH [2.40] are named *Synchronous Transport Module* of level N (STM-N), for $N = 1, 4, 16, 64, \ldots$. The same name is used to refer extensively to the signals themselves. The bit rate of each level is given by simply multiplying the basic rate of the STM-1 (155.520 Mbit/s) by N.

The denomination of the signals of the SONET hierarchy, on the other hand, is different. Here, the signals are named *Synchronous Transport Signal* of level N (STS-N), for $N = 1, 3, 12, 48, 192, \ldots$, while the basic bit rate of the STS-1 signal (51.840 Mbit/s) is one third of the STM-1 rate. Also in SONET, the bit rates of the upper levels are given by multiplying the basic rate of the STS-1 by N. Moreover, the SONET optical signal is called *Optical Carrier* of level N (OC-N).

The highest bit rate of SDH/SONET commercial systems, at the date this has been written, is about 10 Gbit/s (STM-64/STS-192). Higher bit rates are being experimented: namely 40 Gbit/s (STM-256/STS-768). This seems really to be the physical border of practical realization of digital electronic circuits, achieved at highest cost.

To attain even higher transmission capacity on a single fibre, the optical magic of Dense Wavelength Division Multiplexing (DWDM) allows to multiplex in the optical-frequency domain even hundreds of STM-N/STS-N signals. Monster capacities in excess of 1 Tbit/s on a single fibre can be thus attained.

For reference, the standard hierarchical levels of SDH and SONET in use are summarized in Table 2.8, together with their bit rates.

2.5.3   SDH Frame Structure

The signals of all the hierarchical levels of SDH and SONET are organized in frames of same duration, equal to 125 µs (the sampling period of the telephone signal). In this

Table 2.8 Standard hierarchical levels of SDH and SONET

SDH Level	SONET Level	Bit Rate
(STM-0^a)	STS-1/OC-1	51.840 Mbit/s
STM-1	STS-3/OC-3	155.520 Mbit/s
STM-4	STS-12/OC-12	622.080 Mbit/s
STM-16	STS-48/OC-48	2.488 320 Gbit/s
STM-64	STS-192/OC-192	9.953 280 Gbit/s
STM-256	STS-768/OC-768	39.813 120 Gbit/s

^{a}The rate STM-0 (formerly known as *sub STM-1*), corresponding to the SONET STS-1, does not represent a level of SDH and is defined only for the special case of radio and satellite transmission medium.

way, each byte in a specified position within the frame can carry one telephone channel or equivalently a digital channel of capacity 64 kbit/s.

Indeed, the fact that the signals of all the hierarchical levels are organized in frames of duration equal to 125 µs, together with the synchronous multiplexing rules, allows in particular the synchronous transport of PCM telephone channels, i.e. the transmission of the samples of single PCM channels in octets placed in a fixed position within the SDH/SONET frames.

From now on, for the sake of brevity, we will focus mainly on SDH, which is the ITU-T standard designed as superset of the ANSI standard SONET, understanding that all principle considerations apply to SONET too, *mutatis mutandis*. Therefore, in this section the SDH frame structure and its synchronous multiplexing technique will be outlined. The frame structure of SONET is not different in principle, although different names are adopted.

However, this section and the following ones *do not* aim at providing an exhaustive treatise of SDH mapping rules, since they are very complex and they are well beyond the scope of this book. To probe further, the interested reader is referred to the bibliography.

The general scheme of the SDH STM-N frame is depicted in Figure 2.22. In this figure, one STM-N frame is represented as a matrix of bytes, made of $270 \times N$ columns and nine rows (2430 bytes or 19 440 bits in total in the case of the STM-1 frame). The bytes of the matrix are transmitted row by row, left to right, most significant bit first in each byte. The frame rate is 8 kHz, and therefore each byte of the frame carries one 64 kbit/s digital channel, as mentioned above. Moreover, the frame is scrambled before transmission, with the exception of the first nine bytes (first row of the RSOH submatrix).

The SDH frame is divided in two zones:

- an *OverHead* (OH), placed in the first left $9 \times N$ columns, further divided in
  - *Regenerator Section OverHead* (RSOH), placed in rows 1 through 3
  - *Multiplexer Section OverHead* (MSOH), placed in rows 5 through 9
  - N *AU pointers*, placed in row 4
- a *payload*, placed in the remaining $261 \times N$ right columns.

In the OH submatrix, a group of bytes are devoted to specific functions, viz. frame alignment, performance monitoring, alarming, transport of service management information, etc. For example, Figure 2.23 shows the layout of the STM-1 overhead. The STM-N

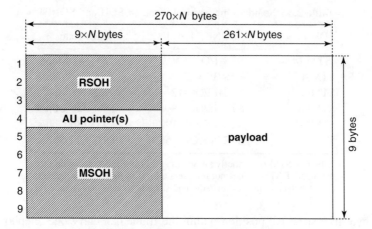

Figure 2.22. Scheme of the SDH STM-*N* frame. (Adapted from Figure 6-6/G.707 (2.40) by permission of ITU)

A1	A1	A1	A2	A2	A2	J0	✳	✳
B1	Δ	Δ	E1	Δ		F1	✕	✕
D1	Δ	Δ	D2	Δ		D3		

RSOH

AU pointer

B2	B2	B2	K1			K2		
D4			D5			D6		
D7			D8			D9		
D10			D11			D12		
S1					M1	E2	✕	✕

MSOH

✳ *Unscrambled reserved bytes*
✕ *Bytes reserved for national use*
Δ *Media dependent bytes*

Figure 2.23. Layout of the STM-1 overhead

OH layout for the upper levels is quite similar, more or less as if it was obtained from a byte-interleaving of *N* STM-1 OHs. The AU pointer, on the other hand, identifies the position within the payload submatrix where the payload itself is actually aligned (see Section 2.5.8).

In Table 2.9, the denomination and the functions of the OH bytes are briefly summarized. For further details, the reader is referred to the ITU-T Rec. G.707 [2.40].

The payload submatrix contains, mapped according to different synchronous multiplexing rules, the transported signals. These rules may be very complex, but they allow mapping of almost any kind of existing signals, including those of European, North American and Japanese PDH, ATM cells, etc. In the following sections, the synchronous multiplexing rules will be outlined.

Table 2.9   Summary of bytes of the STM-N overhead

RSOH		
A1, A2	$6 \times N$ bytes	Frame alignment bytes (A1 = 11110110, A2 = 00101000)
J0	N bytes	Regenerator Section Trace
B1	1 byte	Parity check on the previous frame, after scrambling, by means of Bit Interleaved Parity (BIP) code
E1	1 byte	RSOH orderwire channel for voice communication
F1	1 byte	User channel for voice service communication
D1–D3	3 bytes	192 kbit/s channel for the transmission of management messages, called Regenerator Section Data Communication Channel (RS-DCC)
MSOH		
B2	$3 \times N$ bytes	Parity check on the previous frame (BIP code), with the exception of the bytes of the RSOH, after scrambling
K1, K2 (bits 1-5)	2 bytes	Signalling protocol for the Automatic Protection Switching (APS) of the Multiplex Section
K2 (bits 6-8)		Multiplex Section Remote Defect Indication (MS-RDI)
E2	1 byte	MSOH orderwire channel for voice communication
D4-D12	9 bytes	576 kbit/s channel for the transmission of management messages, called Multiplexer Section Data Communication Channel (MS-DCC)
S1, bits 5-8	1 byte	Denotes the type of clock generating the equipment timing signal (synchronization status); standard Synchronization Status Messages (SSM) have been defined in order to allow automatic protection procedures of the network synchronization distribution (see Section 4.7)
M1	1 byte	Carries back, to the previous NE in the transmission chain, the number of errored blocks detected by means of the B2 byte (Multiplex Section Remote Error Indication, MS-REI)

2.5.4 Synchronous Multiplexing in SDH

Synchronous multiplexing of SDH and SONET is based on complex rules which combine several building blocks (*synchronous multiplexing elements*) in various ways. In particular, such synchronous multiplexing elements are structured, fixed-size sets of bytes, which are variedly byte-interleaved or mapped one into the other to eventually form STM-N frames.

Among them, the *Virtual Containers* (VCs), named Virtual Tributaries (VTs) in SONET, are the basic building blocks and are the conceptually most innovative elements compared to the traditional PDH technique. VCs have been defined in order to have a small set of subsignals, carried within STM-N frames along SDH transmission chains, which are processed in SDH network elements independently on their content (VC payload).

A VC is a structured set of bytes and maps (i.e. contains) a payload which can be a plesiochronous PDH signal as well as other synchronous multiplexing elements. VCs are *individually* and *independently* accessible within STM-N frames through a *pointer* information, directly associated with them on multiplexing. A pointer is located in a determined position within STM-N frames (e.g., the AU pointer is placed in the fourth row of the OH submatrix) or within VCs and identifies the position of the first byte of the pointed VC within the containing structure.

This means that it is then possible to access a single VC individually, without interfering with others, and makes easier inserting, extracting and cross-connecting tributary signals within the multiplex signal. A VC, once it is built in the source NE, is not disassembled nor modified until it reaches the destination node, and it travels as it is through the network, like a closed box. The only action that can be performed on a VC along that travel is a position shift (pointer action) within the STM-N frame or another multiplexing element containing it.

The multiplexing rules, combining the synchronous multiplexing elements to form a complete STM-N frame, are graphically described by the scheme of Figure 2.24. In this scheme, derived with few changes on the arrow styles from that depicted in ITU-T Rec. G.707 [2.40], three kinds of relationship among synchronous multiplexing elements (depicted as boxes and denoted by an acronym) are pointed out:

- *mapping* of a PDH signal into a synchronous Container (C-i); such mapping rules may include bit justification to accommodate a plesiochronous signal with variable frequency into a synchronous C-i having fixed size (asynchronous mapping, analogously to what discussed for asynchronous digital multiplexing in Section 2.3.4), but may not as well if the PDH tributary is locked to the same reference frequency as the STM signal (synchronous mapping);

- *synchronous multiplexing* of one or more multiplexing elements into another containing one; such synchronous multiplexing may be done by byte interleaving of some elements (multiplexing of more elements into the containing one) or by simply adding a fixed set of bytes to one element to yield the containing one; in both cases, the phase relationship between the contained elements and the containing one is fixed;

- *phase alignment* of one multiplexing element into the containing one; in this case, the phase relationship between the contained element and the containing one

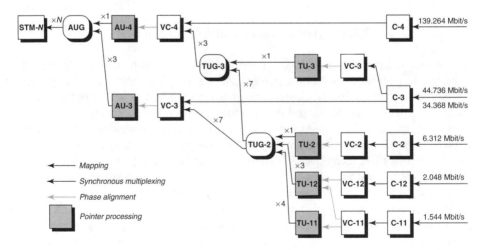

Figure 2.24. ITU-T SDH multiplexing structure. (Adapted from Figure 6-1/G.707 [2.40] by permission of ITU)

may vary and is coded by the pointer word, added to the former to yield the latter.

2.5.5   Synchronous Multiplexing Elements

The various synchronous multiplexing elements are listed below. For further details, the reader should refer to the ITU-T Rec. G.707 [2.40].

2.5.5.1   Container (C-i, for i = 11, 12, 2, 3, 4)

A Container is the basic synchronous element mapping the signal transported between the access points of the SDH network. Adaptation functions have been defined for many common network rates into a limited number of standard Containers, including those rates of the PDH defined by ITU-T Rec. G.702 [2.28]. Further adaptation functions have been defined, and others will be in the future, for new broadband rates such as ATM.

The mapping of PDH signals into a structured, fixed-size set of bytes C-i may be:

- *asynchronous*, where accommodation (synchronization) of a plesiochronous signal into the synchronous C-i is done through bit justification;
- *bit- or byte-synchronous*, where bit justification is not necessary because the mapped signal is synchronous with the timing signal of the C-i. An example of byte-synchronous mapping is the one defined for the 2.048 Mbit/s into the VC-12, where each PCM time slot of the mapped signal has been located in a fixed position within the VC-12.

A Container is made of tributary bits (true payload), fixed stuffing, justification opportunity and control bits, miscellaneous overhead.

2.5.5.2   Virtual Container (VC-i, for i = 11, 12, 2, 3, 4)

A Virtual Container consist of

- a Path OverHead (POH),
- a Container or other lower-order multiplexing elements (TUGs).

VCs are the basic building blocks of the SDH multiplexing structure. As mentioned above, a VC is not disassembled nor modified during its travel through the SDH network, whereas it may experience only phase shifts (pointer action).

VCs are of two kinds:

- lower order VCs, i.e. those mapped into TUs and pointed by a TU pointer, which is located in a determined position within the higher order VC that contains the lower order VC;
- higher order VCs, i.e. those mapped into AUs and pointed by an AU pointer, which is located in a determined position within the SOH of the STM-N frame.

2.5.5.3   Tributary Unit (TU-i, for i = 11, 12, 2, 3)

Tributary Units consist of

- a lower order VC-i ($i = 11, 12, 2, 3$),

- a TU-i (i = 11, 12, 2, 3) pointer indicating the offset of the pointed VC-i within the TU-i itself (the TU-i pointer is located in a fixed position within the higher order VC).

2.5.5.4   Tributary Unit Group (TUG-2, TUG-3)

Tributary Unit Groups have been defined in order to limit the large number of possible ways to combine lower order VCs into higher order VCs. They contain homogeneous groups of TUs, byte-interleaved and located in fixed positions.

- A TUG-2 may consist of:
  - — four TU-11,
  - — three TU-12,
  - — one TU-2.
- A TUG-3 may consist of:
  - — seven TUG-3,
  - — one TU-3.

2.5.5.5   Administrative Unit (AU-i, for i = 3, 4)

Administrative Units consist of

- a higher order VC-i (i = 3, 4),
- an AU-i (i = 3, 4) pointer indicating the offset of the pointed VC-i within the AU-i itself (the AU-i pointer is located in a fixed position within the SOH of the STM-N frame.).

2.5.5.6   Administrative Unit Group (AUG)

The purpose of Administrative Unit Groups is similar to that of TUGs. They contain an homogeneous group of AU-3 byte-interleaved or one AU-4, located in fixed positions. An AUG may consist of:

- three AU-3,
- one AU-4.

2.5.5.7   Synchronous Transport Module level N (STM-N)

A Synchronous Transport Module level N contains:

- N AUGs, byte-interleaved,
- one SOH.

2.5.6   Example of PDH Signal Transport on SDH: Asynchronous Mapping of the 139.264 Mbit/s

In order to provide a first, simple example of how the synchronous multiplexing rules outlined above apply in practice, and how the SDH frame structure is built starting from an input PDH signal, the case of the asynchronous mapping of the 139.264 Mbit/s into the VC-4 is presented in this section. This example is the simplest case to describe, but it is instructive indeed.

From the general scheme of Figure 2.24, by deleting the unwanted branches, it is possible to derive the simple multiplexing path depicted in Figure 2.25, showing the sequence of steps to take to build a STM-N frame from a PDH 139.264 Mbit/s signal.

The first step to take is blocking the asynchronous 139.264 Mbit/s bit stream (having a frequency tolerance of ±15 ppm [2.5]) into the synchronous Container C-4 (see Figure 2.26(a)). The C-4 has the fixed size of $260 \times 9 = 2340$ bytes, but can contain a variable number of tributary bits owing to nine justification opportunity bits. Positive bit justification has been adopted to accommodate different frequencies of the 139.264 Mbit/s stream.

The mapping of the 139.264 Mbit/s tributary bits into the C-4 is shown in Figure 2.27, which depicts only one of the nine rows of the C-4 structure. Each of the nine rows is partitioned into 20 blocks, consisting of 13 bytes each. In each row, 1934 information bits (I), one justification opportunity bit (S) and five justification control bits (C) are provided. The first byte of each block consists of either eight information bits (I), or eight fixed stuff bits (R), or one justification control bit (C) plus five fixed stuff bits (R) plus two overhead

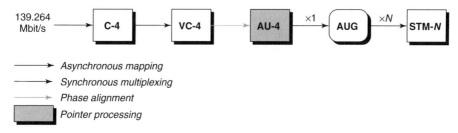

Figure 2.25. Steps in multiplexing a 139.264 Mbit/s signal into a SDH STM-N frame

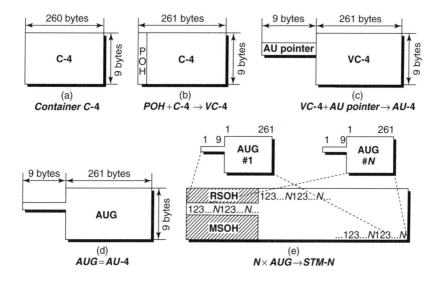

Figure 2.26. Elements in multiplexing a 139.264 Mbit/s signal into a SDH STM-N frame

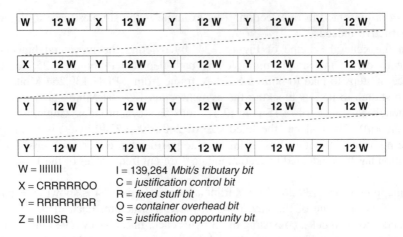

| W | 12 W | X | 12 W | Y | 12 W | Y | 12 W | Y | 12 W |

| X | 12 W | Y | 12 W | Y | 12 W | Y | 12 W | X | 12 W |

| Y | 12 W | Y | 12 W | Y | 12 W | X | 12 W | Y | 12 W |

| Y | 12 W | Y | 12 W | X | 12 W | Y | 12 W | Z | 12 W |

W = IIIIIIII
X = CRRRRROO
Y = RRRRRRRR
Z = IIIIIISR

I = 139,264 *Mbit/s tributary bit*
C = *justification control bit*
R = *fixed stuff bit*
O = *container overhead bit*
S = *justification opportunity bit*

Figure 2.27. Asynchronous mapping of the 139.264 Mbit/s tributary bits into the C-4 (one of the nine C-4 rows)

bits (O) reserved for generic overhead communication purposes, or six information bits (I) plus one justification opportunity bit (S) plus one fixed stuff bit (R). The last 12 bytes of each block consist of information bits (I).

The set of five justification control bits C in each row is used to control the corresponding opportunity bit S. The pattern CCCCC = 00 000 indicates that the S bit is an information bit, whereas CCCCC = 11 111 indicates that the S bit is a justification, dummy bit. Majority voting is used to make the justification decision, for protection against single and double bit errors in the C bits.

Resuming with the schemes of Figures 2.25 and 2.26, the next step is building the VC-4 by adding the higher order Path OverHead (POH) to the C-4 (see Figure 2.26(b)). The POH is one column of bytes and is the overhead associated with the VC on its creation. It travels with it along the SDH transmission chain, without being modified in any way by the intermediate NEs of the chain, whereas it is terminated by the last one, which opens the VC and extract the 139.264 Mbit/s stream.

Then, the VC-4 is phase aligned and the AU pointer value is determined to build the AU-4 (see Figure 2.26(c)). The meaning of *phase alignment* is that the position of the first byte of the VC-4 in the rectangle of 261 × 9 bytes at the right of the AU-4 can be whichever (actually, at multiples of three bytes). This position is coded by the AU pointer word (10 bits), which, multiplied by three, gives the offset of the position of the VC-4 first byte starting from the end of the pointer word (SOH fourth row). Thus, if the AU-4 pointer is 0, the VC-4 starts right after the SOH fourth row.

To clarify this concept, we point out that a VC-4 may — and usually does — cross the border between two STM-*N* frames as shown by Figure 2.28, where three consecutive frames are depicted and consecutive VC-4s are marked with alternate patterns. The VC-4 POH is also highlighted.

The last steps are building the AUG (Figure 2.26(d)), which in this case coincides with the AU-4 element, byte-interleaving *N* AUGs and adding the SOH to form the STM-*N* frame (Figure 2.26(e)).

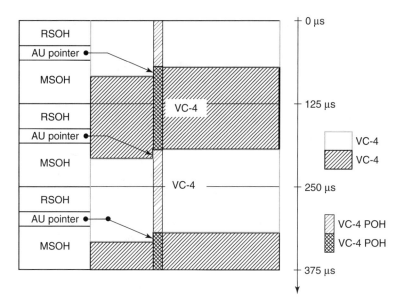

Figure 2.28. Location of VC-4s within STM-1 frames

Exercise 2.2
Bit Justification in the C-4 (Asynchronous Mapping of the E4 Signal)

In each C-4 row there are 1934 information bits I and one justification opportunity bit S. The nominal frequency of the E4 signal is 139.264 Mbit/s, while that of the STM-1 signal is 155.520 Mbit/s.

What is the nominal justification ratio ρ_{nom} (In other words, what is the fraction of S bits assigned to dummy bits in nominal conditions?)

What are the maximum positive and negative frequency deviations of the E4 tributary that can be accommodated by the bit justification mechanism, at least in theory?

Solution
In each VC-4 there are nine rows and thus there are nine S bits per frame. If all the nine S bits are dummy, the capacity available to the E4 signal in the VC-4 is the minimum value

$$f_{MIN} = 1934 \cdot 9 \cdot 8000 \text{ kbit/s} = 139.248 \text{ Mbit/s}.$$

Each single S bit filled with a tributary bit carry an additional capacity of 8 kbit/s. Therefore, the nominal E4 bit rate is carried by using 2 S bits per frame and the nominal justification ratio results

$$\rho_{nom} = \frac{7}{9}.$$

The maximum capacity available to the E4 signal in the VC-4 is (all the nine S bits filled)

$$f_{MAX} = (1934 + 1) \cdot 9 \cdot 8000 \text{ kbit/s} = 139.320 \text{ Mbit/s}.$$

Hence, the maximum positive and negative frequency deviations from the nominal value result, in fractional terms:

$$\frac{f_{MAX}}{f_{nom}} - 1 \cong +402 \text{ ppm} \qquad \frac{f_{MIN}}{f_{nom}} - 1 \cong -115 \text{ ppm}.$$

This value may be compared to the E4 frequency tolerance ± 15 ppm specified by ITU-T Rec. G.703 [2.5].

2.5.7 Example of PDH Signal Transport on SDH: Asynchronous Mapping of the 2.048 Mbit/s

The second mapping case studied here is the asynchronous mapping of the 2.048 Mbit/s PCM Primary Multiplex signal into the VC-12. The mapping of the 1.544 Mbit/s PCM Primary Multiplex signal into the VC-11 is not different in principle and follows similar rules[19].

Also in this case, from the general scheme of Figure 2.24, by deleting the unwanted branches, the simple multiplexing path depicted in Figure 2.29 has been derived. This figure shows the sequence of steps to take to build an STM-N frame from a PDH 2.048 Mbit/s signal.

The first step to take is blocking the asynchronous 2.048 Mbit/s bit stream (having a frequency tolerance of ± 50 ppm) into the synchronous Container C-12. The C-12 has the fixed size of 139 bytes, but can contain a variable number of tributary bits owing to two justification opportunity bits. Positive/negative bit justification has been adopted to accommodate different frequencies of the 2.048 Mbit/s stream.

Then, one lower order POH byte is added to build the VC-12. This POH, as it happens for the VC-4 POH, is the overhead associated to the VC-12 on its creation. It travels with it along the SDH transmission chain, without being modified in any way by the intermediate NEs of the chain, whereas it is terminated by the last one, which opens the VC and extracts the 2.048 Mbit/s stream.

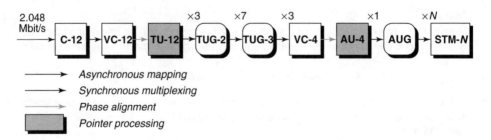

Figure 2.29. Steps in multiplexing a 2.048 Mbit/s signal into a SDH STM-N frame

---

[19] In addition, synchronous mappings of the 2.048 Mbit/s and 1.544 Mbit/s signals, into VC-12 and VC-11 respectively, have been defined [2.40]. The byte-synchronous mappings are especially remarkable, as the PCM channels are allocated in fixed positions into the VC-11/12, so that they are directly accessible after TU pointer addressing without further frame aligning on the PCM stream. With VC-11/12 byte-synchronous mapping, it would then be possible to feed digital switch input ports with SDH signals instead of PCM signals as usual until now.

The mapping of the 2.048 Mbit/s tributary bits into the VC-12 is shown in Figure 2.30. The VC-12 is made of 140 bytes, which are divided in four sectors, each made of 35 bytes. These sectors are mapped within four different, consecutive STM-N frames and thus VC-12 bytes span a period of 500 μs. This means that, to recover the 2.048 Mbit/s from the VC-12 sectors, multiframe alignment is needed. Standard parameters for the multiframe alignment strategy (see Section 1.1.3), assisted by the H4 byte within the VC-4 POH, have been specified [2.40].

In addition to the VC-12 POH, the VC-12 consists of 1023 information bits (I), six justification control bits (C1 and C2), two justification opportunity bits (S1 and S2), eight overhead bits (O) reserved for generic overhead communication purposes. The bytes J2, N2, K4 are reserved for other alarm and control functions. The remaining are fixed stuff bits (R).

Two sets (C1 and C2) of three justification control bits are used to control the two justification opportunities S1 and S2, respectively. The pattern $C_iC_iC_i = 000$ indicates that the controlled S_i bit is an information bit, whereas $C_iC_iC_i = 111$ indicates that the controlled S_i bit is a justification, dummy bit ($i = 1, 2$). Majority voting is used to make the justification decision for protection against single bit errors in the C bits.

The next step is phase aligning the VC-12 and determining the TU pointer value to build the TU-12 (144 bytes in total). The TU-12 single multiframe sector is made of 36 bytes and lasts 125 μs; it consists of a VC-12 sector (35 bytes) plus one byte (called V1, V2, V3, V4 according to the frame sequence number within the multiframe). The TU pointer word (10 bits) is written within the V1 and V2 bytes and gives the offset of the position of the VC-12 first byte, with single-byte increments, starting from the right of the pointer word.

The single multiframe sector of the TU-12 can be thus represented as a matrix of 4×9 bytes, as shown by Figure 2.31. If the TU-12 pointer is 0, the VC-12 starts right after the V2 byte, i.e. the VC5 byte (first byte of the VC-12) is the first following it at its right.

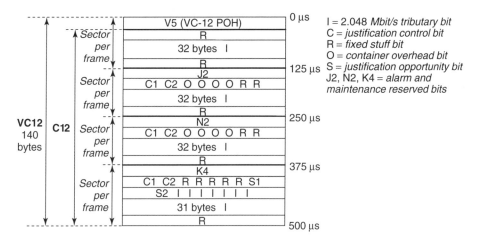

Figure 2.30.   Asynchronous mapping of the 2.048 Mbit/s tributary bits into the VC-12 (split in four segments over four frames). (Adapted from Figure 10-8/G.707 [2.40] by permission of ITU)

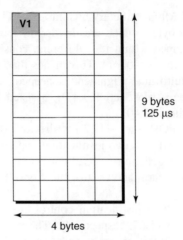

Figure 2.31.   Single multiframe sector of the TU-12 (VC-12 + TU pointer → TU-12)

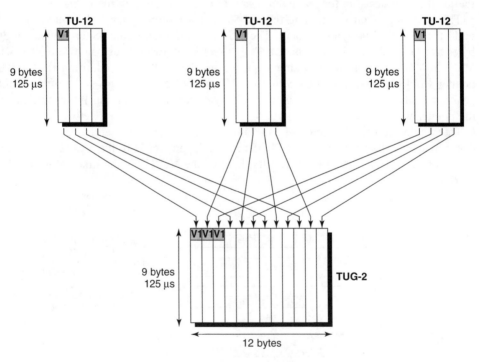

Figure 2.32.   Byte interleaving of three TU-12s to yield one TUG-2 (3 × TU-12 → TUG-2)

Then, three TU-12s (mapping three different E1 tributaries) are simply byte-interleaved to build one TUG-2, which can be thus represented as a matrix of 12 × 9 bytes, as shown in Figure 2.32. Let us notice that the multiframe alignment is the same for all TU-12s.

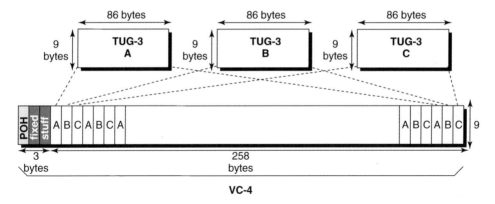

Figure 2.33.   Byte interleaving of three TUG-3s to yield one VC-4 (3 × TUG-3 + POH → VC-4)

Again, seven TUG-2s are byte-interleaved and two columns of fixed stuff are added to form one TUG-3, which can be represented as a matrix of 86 × 9 bytes.

Then, three TUG-3s are byte-interleaved and two columns of fixed stuff and the VC-4 POH (one column) are added to finally form the VC-4 (Figure 2.33), which can be represented as a matrix of 261 × 9 bytes. From this point, the same procedure followed in the previous section holds.

Exercise 2.3
Bit Justification in the C-12 (Asynchronous Mapping of the E1 Signal)

In each VC-12 there are 1023 information bits I and two justification opportunity bits S. The nominal frequency of the E1 signal is 2.048 Mbit/s, while that of the STM-1 signal is 155.520 Mbit/s.

What is the nominal justification ratio ρ_{nom}? (In other words, in nominal conditions what is the fraction of S bits filled by tributary bits?)

What are the maximum positive and negative frequency deviations of the E1 tributary that can be accommodated by the bit justification mechanism, at least in theory?

Solution

If all the S bits are dummy, the capacity available to the E1 signal in the VC-12 is the minimum value

$$f_{MIN} = 1023 \text{ bit} \cdot \frac{8000 \text{ frames/s}}{4 \text{ frames}} = 2.046 \text{ Mbit/s}$$

being the VC-12 transmitted in four frames. Similarly,

$$f_{MAX} = (1023 + 2) \text{ bit} \cdot \frac{8000 \text{ frames/s}}{4 \text{ frames}} = 2.050 \text{ Mbit/s.}$$

Each single S bit filled with a tributary bit carry an additional capacity of 2 kbit/s. Hence, the nominal E1 bit rate is carried by using one of the two S bits and the nominal justification ratio results

$$\rho_{nom} = \frac{1}{2}.$$

The maximum positive and negative frequency deviations from the nominal value result, in fractional terms:

$$\frac{f_{MAX}}{f_{nom}} - 1 \cong +976 \text{ ppm} \qquad \frac{f_{MIN}}{f_{nom}} - 1 \cong -976 \text{ ppm}.$$

This value may be compared to the E1 frequency tolerance ±50 ppm specified by ITU-T Rec. G.703 [2.5].

2.5.8   The Pointer Justification Mechanism: Synchronous Multiplexing in Asynchronous Networks

The pointer justification mechanism deserves a closer look. While different kinds of pointer processors and all the related timing issues are treated in Chapter 3, this section details the pointer justification technique, aiming at explaining how pointers may change value and VCs may shift in the STM-N frame in order to compensate a frequency offset between the clock of the VC-originating NE and the clock of the NE where the pointer is processed.

Figure 2.34 (example of VC-4 containing several lower-order VCs) summarizes the role of pointers in SDH frames.

- The AU-n pointer indicates the position of the first byte of the corresponding higher order VC-n ($n = 3, 4$), free to float within the payload submatrix of the STM-N frame. Therefore, the AU-n pointer provides a method for allowing dynamic alignment of the higher order VC-n within the AU-n element.

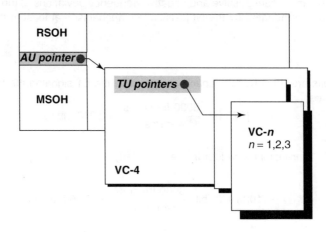

Figure 2.34.   Role of AU and TU pointers

- The TU-n pointer indicates the position of the first byte of the corresponding lower order VC-n ($n = 11$, 12, 2, 3), free to float within the containing multiplexing element. Therefore, the TU-n pointer provides a method for allowing dynamic alignment of the lower order VC-n within the TU-n element.

The pointer justification mechanism allows to compensate in an NE frequency offsets between the input VCs, timed by the clock of the NE originating them in the transmission chain, and the output STM-N frames, timed by the local clock.

In asynchronous multiplexing (PDH), frequency offsets among tributaries and multiplex signals are compensated by bit justification. Justification opportunity bits, by accommodating a variable amount of tributary bits, make the tributary frame alignment to float within the multiplex frame. Therefore, in an asynchronous multiplex frame structure, it is not possible to know what is the phase relationship between the multiplex frame and the contained tributary frames.

In SDH synchronous multiplexing, on the other hand, frequency offsets among VCs and multiplex signals are compensated by pointer justification. As it will be shown, justification bytes make the VC alignment free to float within the multiplex frame or the higher order VCs. In this case, contrary to PDH, this alignment information is coded in the pointer words, which track the VC position.

The need of adjusting pointer values is pointed out in the example of Figure 2.35, showing a situation in which AU-4 pointer action takes place along a chain of SDH NEs where VC-4 are not open, but retransmitted in the output STM-N frames (this is the case, for example, of Digital Cross-Connects 4/4). The SDH NE 1, timed by Clock 1, generates the VC-4 mapping a PDH 139.264 Mbit/s signal and the STM-N signal that carries it. Both are timed by the same clock (Clock 1) and thus there is no need to accommodate excess or lack of VC-4 bits in the output STM-N frames. Conversely, in the SDH NE 2, while at the input the VC-4 is timed by Clock 1, at the output the STM-N signal is generated being timed by Clock 2. The same situation holds for the SDH NE 3 (at the input the VC-4 is timed by Clock 1, but at the output the STM-N signal is generated being timed by Clock 3) and for any further NE in the SDH transmission chain.

The transport of VC-4s timed by the clock of the originating NE, over STM-N signals timed by the different clocks along a chain of NEs, is possible only by allowing the relative shift of VC-4s within the carrying frames, in order to compensate their frequency offsets.

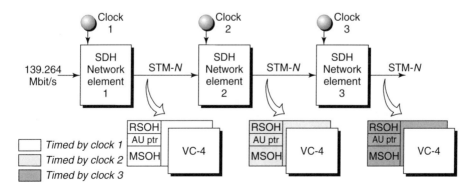

Figure 2.35. Example of SDH transmission chain where AU-4 pointer action takes place

To give a deeper insight into the pointer mechanism, the case of the AU-4 pointer positive and negative justification is now expounded in particular. The AU-4 pointer is located in the bytes H1, H2 of the fourth row of the SOH, as shown by Figure 2.36(a). The pointer contained in H1 and H2 designates the location of the byte where the VC-4 begins: multiplied by three, it indicates how many bytes there are between the end of the SOH fourth row and the first byte of the VC-4, counting only bytes of the payload submatrix.

The two bytes H1 and H2 allocated to the pointer function can be viewed as one word, as shown by Figure 2.36(b). The last 10 bits of this word carry the pointer value. The AU-4 pointer value is a binary number with a range of 0 to 782 which indicates the offset, in three-byte increments, between the pointer itself and the first byte of the VC-4.

If there is a frequency offset between the frame rate of the AUG and that of the incoming VC-4, then the pointer value is incremented or decremented as needed, accompanied by a corresponding positive or negative byte justification. Consecutive pointer adjustments are separated by at least three frames in which the pointer value remains constant (i.e., there shall not be more than one pointer justification every four frames).

If the frame rate of the incoming VC-4 is slower compared to that of the AUG (i.e., referring to NE2 in Figure 2.35, Clock 2 is faster than Clock 1), then the alignment of the VC-4 is periodically delayed (seen from Clock 2) and the pointer value is incremented by one (*positive pointer justification*). This operation is indicated by inverting odd bits 7, 9, 11, 13 and 15 (I-bits) of the pointer word to allow 5-bit majority voting at the receiver. Three positive justification bytes appear immediately after the last H3 byte in the AU-4 frame containing inverted I-bits. Subsequent pointers will contain the new offset. This is illustrated by Figure 2.37.

If the frame rate of the VC-4 is faster compared to that of the AUG (i.e., referring to NE2 in Figure 2.35, Clock 2 is slower than Clock 1), then the alignment of the VC-4 is periodically advanced in time (seen from Clock 2) and the pointer value is decremented by one

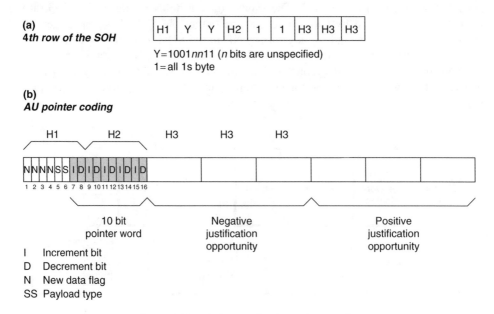

Figure 2.36. Scheme of AU-4 pointer coding

(*negative pointer justification*). This operation is indicated by inverting even bits 8, 10, 12, 14 and 16 (D-bits) of the pointer word to allow 5-bit majority voting at the receiver. Three negative justification bytes appear in the H3 bytes in the AU-4 frame containing inverted D-bits. Subsequent pointers will contain the new offset. This is illustrated by Figure 2.38.

Moreover, bits 1-4 (N-bits) of the pointer word carry a New Data Flag (NDF), which allows an arbitrary change of the pointer value, for example due to a change in the payload. Normal operation is indicated by the 0110 code in the N-bits. NDF is indicated by inversion of the N-bits to 1001. The new alignment is indicated by the pointer value accompanying the NDF and takes effect at the offset indicated.

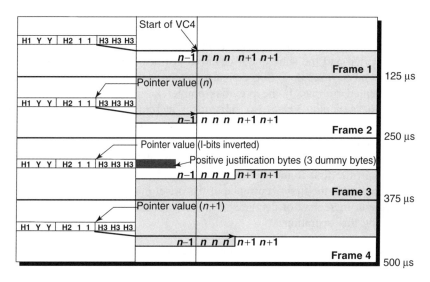

Figure 2.37.   AU-4 pointer positive justification

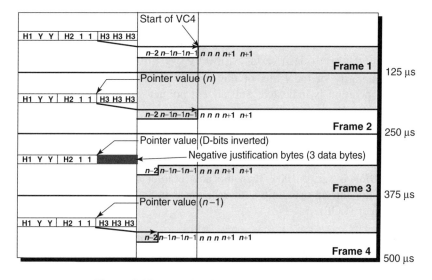

Figure 2.38.   AU-4 pointer negative justification

The rules for AU pointer generation at the transmitter are summarized as follows (ITU-T Rec. G.707 [2.40]):

(1) During normal operation, the pointer locates the start of the VC-n within the AU-n frame. The NDF is set to 0110.

(2) The pointer value can only be changed by operation 3, 4 or 5.

(3) If a positive justification is required (see Figure 2.37), the current pointer value is sent with the I-bits inverted and the subsequent positive justification opportunity is filled with dummy bits. Subsequent pointers contain the previous pointer value incremented by one. If the previous pointer is at its maximum value, the subsequent pointer is set to zero. No subsequent increment or decrement operation is allowed for at least three frames following this operation.

(4) If a negative justification is required, the current pointer value is sent with the D-bits inverted and the subsequent negative justification opportunity is overwritten with actual payload data. Subsequent pointers contain the previous pointer value decremented by one. If the previous pointer is zero, the subsequent pointer is set to its maximum value. No subsequent increment or decrement operation is allowed for at least three frames following this operation.

(5) If the alignment of the VC-n changes for any reason other than rules (3) or (4), the new pointer value is sent accompanied by NDF set to 1001. The NDF only appears in the first frame that contains the new value. The new location of the VC-n begins at the first occurrence of the offset indicated by the new pointer. No subsequent increment or decrement operation is allowed for at least three frames following this operation.

The rules for AU pointer interpretation at the receiver are summarized as follows (ITU-T Rec. G.707 [2.40]):

(1) During normal operation, the pointer locates the start of the VC-n within the AU-n frame.

(2) Any variation from the current pointer value is ignored unless a consistent new value is received three times consecutively or it is preceded by that stated by one of the rules (3), (4) or (5). Any consistent new value received three times consecutively overrides (i.e. takes priority over) rules (3) or (4).

(3) If the majority of the I-bits of the pointer word is inverted, a positive justification operation is indicated. Subsequent pointer values are then expected to be incremented by one.

(4) If the majority of the D-bits of the pointer word is inverted, a negative justification operation is indicated. Subsequent pointer values are then expected to be decremented by one.

(5) If the NDF is set to 1001, then the new pointer value of the same frame shall replace the current one, unless the receiver is in a state that corresponds to a Loss Of Pointer (LOP).

AU-4 pointers are adjusted by increments or decrements of three bytes, as described above. AU-3 pointers work in an analogous way, but are adjusted by increments or decrements of one byte. Three AU-3 pointers, each made of two bytes H1, H2 and one justification opportunity byte H3, are located in the fourth row of SOH, byte-interleaved.

Also TU pointers works in an analogous way. The role of the H1, H2, H3 bytes is taken by the V1, V2, V3 bytes of the TU element. However, the TU pointer generation and interpretation rules are substantially the same as for the AU pointer. TU pointers are adjusted by increments or decrements of one byte. Moreover, being the TU-i ($i = 11, 12, 2$) transmitted over a 500-μs multiframe, the maximum justification frequency is 500 Hz (not more than one justification over four multiframes).

A detailed treatise of all pointer codes and rules is beyond the scope of this book. The interested reader is thus referred to the relevant ITU-T Rec. G.707 [2.40] for exact specifications.

Exercise 2.4
Example of AU-4 Pointer Negative Justification

Draw the scheme of four consecutive STM-1 frames where an AU-4 pointer negative justification is taking place, with the pointer value updated from the value $n = 87$ to $n = 86$. Specify the decimal value of the 10-bits pointer word in all frames and highlight the position of the VC-4 with its POH.

Solution

The AU-4 pointer value $n = 87$ indicates that the VC-4 is aligned with offset $87 \times 3 = 261$ bytes (i.e., right one payload row) from the end of the SOH fourth row. The AU-4 pointer negative justification requested is thus illustrated by the following figure.

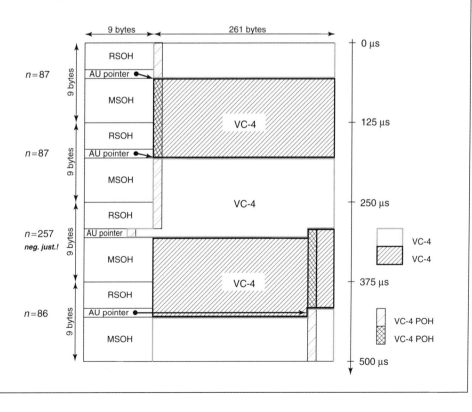

Exercise 2.5
AU-4 Pointer Justification Rate

The maximum AU-4 pointer justification rate is 2000 justifications per second. Referring to Figure 2.35, what is the maximum fractional frequency offset $\Delta f = |CK_1 - CK_2|$ that can be accommodated by AU-4 pointer justification taking place in the NE2, at least in theory?

Solution

One single AU-4 pointer justification moves the VC-4 back or forward by 3 bytes and therefore accommodates a phase error between CK_1 and CK_2 equal to 24 bits of the incoming VC-4.

Therefore, a maximum variation of ± 48 kbit/s in the arrival rate of VC-4 bits can be accommodated (i.e., a maximum absolute frequency offset between CK_1 and CK_2 equal to ± 48 kHz).

Being the nominal VC-4 bit rate (within the carrying STM-N signal)

$$f_{VC-4} = N \cdot 155.520 \text{ Mbit/s} \cdot \frac{261}{N \cdot 270} = 150.336 \text{ Mbit/s}$$

in fractional terms the requested theoretical maximum frequency offset results

$$\Delta f = |CK_1 - CK_2| = \frac{\pm 48 \text{ kbits/s}}{150.336 \text{ Mbit/s}} = \pm 319 \text{ ppm.}$$

This value may be compared to the SDH clock frequency tolerance ± 4.6 ppm specified by ITU-T Rec. G.813 [2.56].

Exercise 2.6
TU-12 Pointer Justification Rate

The maximum TU-12 pointer justification rate is 500 justifications per second (not more than one TU-12 pointer justification every four multiframes). One VC-12 is made of 140 bytes, transmitted in four segments over four frames.

Referring to a scheme analogous to that of Figure 2.35, in which the input tributary is a 2.048 Mbit/s signal and VC-12 floating within VC-4 is considered instead, what is the maximum fractional frequency offset $\Delta f = |CK_1 - CK_2|$ that can be accommodated by TU-12 pointer justification taking place in the NE2, at least in theory?

Solution

One single TU-12 pointer justification moves the VC-12 back or forward by 1 byte and therefore accommodates a phase error between CK_1 and CK_2 equal to 8 bits of the incoming VC-12.

Therefore, a maximum variation of ± 4000 bit/s in the arrival rate of VC-12 bits can be accommodated (i.e., a maximum absolute frequency offset between CK_1 and CK_2 equal to ± 4000 Hz).

Being the nominal VC-12 bit rate (within the carrying STM-N signal)

$$f_{VC-12} = N \cdot 155.520 \text{ Mbit/s} \cdot \frac{140/4}{N \cdot 270 \cdot 9} = 2.240 \text{ Mbit/s}$$

in fractional terms the requested theoretical maximum frequency offset results

$$\Delta f = |CK_1 - CK_2| = \frac{\pm 4000 \text{ b/s}}{2.240 \text{ Mb/s}} = \pm 1785 \text{ ppm.}$$

Also in this case, this value may be compared to the SDH clock frequency tolerance ± 4.6 ppm specified by ITU-T Rec. G.813 [2.56].

Exercise 2.7
Time Step Associated to One AU-4/TU-12 Pointer Adjustment

The VC-4 size is $261 \times 9 = 2349$ byte, transmitted over 1 STM-N frame. The VC-12 size is 140 byte, transmitted in four segments over four STM-N frames.
    What is the phase step associated to one AU-4/TU-12 pointer adjustment, expressed in seconds (time step)?

Solution
The phase step experienced by a VC-4 under a AU-4 pointer adjustment is 24 bit, at the VC-4 rate. This corresponds to a time step

$$\frac{24 \text{ bit}}{155.520 \text{ Mbit/s} \cdot \dfrac{261}{270}} \cong 160 \text{ ns}$$

The phase step experienced by a VC-12 under a TU-12 pointer adjustment is 8 bit, at the VC-12 rate. This corresponds to a time step

$$\frac{8 \text{ bit}}{155.520 \text{ Mbit/s} \cdot \dfrac{35}{270 \cdot 9}} \cong 3.57 \text{ } \mu\text{s}$$

2.5.9   SDH Equipment

The various types of SDH equipment are defined and specified in functional terms by the ITU-T Rec. G.783 [2.43]. According to the scheme of Figure 2.39, mainly three kinds of SDH equipment are distinguished:

- *regenerators*, to regenerate the STM-N signals along very long transmission lines;
- *multiplexers*, to assemble several PDH or SDH tributaries into a multiplex SDH signal;
- *Digital Cross-Connects (DXC)*, to cross-connect SDH/PDH signals from input to output ports.

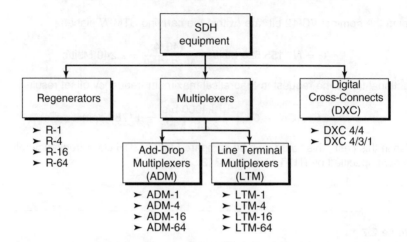

Figure 2.39.   Taxonomy of SDH equipment

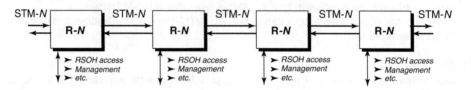

Figure 2.40.   Chain of SDH STM-N regenerators

Regenerators for SDH are more complex equipment than traditional digital line regenerators. They do not only perform simple bit decision and then retransmission of a 'clean' digital signal, but they also process the RSOH bytes to simple maintenance and management purposes. SDH regenerators can be classified according to the hierarchical level N (1, 4, 16, 64) of their line interfaces and may be shortly referred to as R-N equipment. Regenerators are cascaded to build very long transmission lines (as in Figure 2.40), whereas the range spanned by optically amplified lines is usually not longer than a few hundred kilometres.

SDH multiplexers can be classified according to their functions. Thus, they result commonly divided in *Line Terminal Multiplexers* (LTM) and *Add-Drop Multiplexers* (ADM). On the other hand, SDH multiplexers can be also classified according to the hierarchical level N (1, 4, 16, 64) of their line interfaces. Then, they are shortly referred to as LTM-N and ADM-N equipment.

A Line Terminal Multiplexer of level N (LTM-N) is used at the termination of an SDH transmission chain to multiplex/demultiplex several tributaries on one STM-N line[20] (see Figure 2.41). Tributaries may be PDH signals of any kind and hierarchical level as well as SDH STM-M signals ($M \leq N$). Two examples of commercial LTM equipment are the following:

---

[20] With *line*, we mean a bi-directional link (i.e., for example a couple of fibres for reception and transmission). Moreover, an LTM can be equipped with two lines for protection.

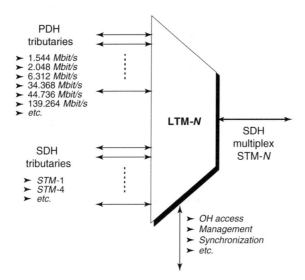

Figure 2.41.   SDH line terminal multiplexer of level STM-*N* (LTM-*N*)

- LTM-16, multiplexing for instance up to 16 × STM-1 or 16 × E4 signals in one STM-16 signal;

- LTM-1, multiplexing for instance up to 63 × E1 or 84 × DS1 signals in one STM-1 signal; mixed payloads (such as 1 × E3 + 42 × E1) are also possible;

Add-Drop Multiplexers of level *N* (ADM-*N*) are used in intermediate nodes of SDH transmission chains to insert or drop tributaries from the STM-*N* line in transit (see Figure 2.42). Also in this case, tributaries may be PDH signals of any kind and hierarchical level, as well as SDH STM-*M* signals ($M \leq N$). Commercial ADM equipment can have on the tributary side any mix of PDH and SDH interfaces up to the available capacity of the STM-*N* line (e.g., for the ADM-1, up to 63 × E1 or 84 × DS1 signals, or 1 × E3 + 42 × E1, etc.).

Moreover, an interesting feature of multiplexers is the flexible configurability of the cross-connection matrices. For example, in the case of ADMs, this allows to choose freely what VCs terminate from the STM-*N* line to the tributary interfaces and what to retransmit unchanged in the output line.

A Digital Cross-Connect (commonly abbreviated as DXC in Europe or DCS in the USA) takes at the input ports both PDH and SDH signals (see Figure 2.43). It allows flexible cross-connection of the signals and of the VCs from any input port *i* to any output port *j*, according to a cross-connection matrix which can be set by the management system. Mainly two types of SDH DXCs are commercially available:

- the DXC 4/4, which accepts at the input/output interfaces
  - PDH E4 signals,
  - SDH STM-*N* signals

  and which allows to cross-connect the VC-4s (the input PDH E4 signals are mapped into VC-4s);

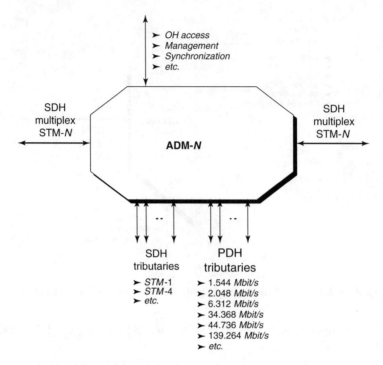

Figure 2.42. SDH add-drop multiplexer of level STM-*N* (ADM-*N*)

- the DXC 4/3/1, which accepts at the input/output interfaces
  — PDH signals of any hierarchical level,
  — SDH STM-*N* signals

  and which allows to cross-connect the VCs of any level, in particular VC-11s, VC-12s, VC-3s and VC-4s (the input PDH signals are mapped into the corresponding VCs).

For both DXC types, the number of ports depends on the supplier and may be very large (the DXC 4/4 with 256 STM-1 ports is widely deployed).

The DXC 1/0 is not a SDH equipment. It accepts at the input/output interfaces E1 or DS1 signals and allows to cross-connect the single 64 kbit/s channels, as well as any subrate signal therein multiplexed (e.g., 14.4 kbit/s, 28.8 kbit/s, etc.). DXC 1/0 equipment is used to make data-circuit networks, for provisioning leased lines.

SDH multiplexers and digital cross-connect systems process the RSOH and MSOH bytes to maintenance and management purposes and process the AU/TU pointers.

If a PDH signal is at the input port, the SDH equipment maps it into the corresponding VC and adds the POH. Then, if the VC is higher order, it generates the AU pointer. If the VC is lower order, instead, it generates the TU pointer, builds the TUGs and the higher order VC and then generates the AU pointer. Finally, it adds the MSOH and RSOH. The opposite process happens in the reverse direction.

If an SDH signal is at the input port, we have to distinguish whether the SDH equipment allows flexible cross-connection only of higher order VCs or of both higher order and lower order VCs. In the former case, the SDH equipment terminates the RSOH and

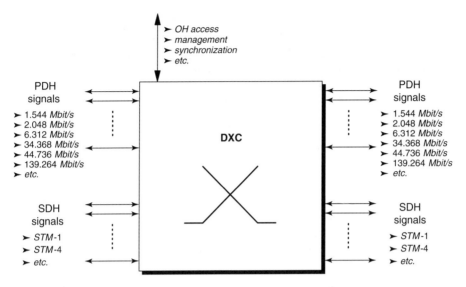

Figure 2.43.   SDH digital cross-connect (DXC)

MSOH bytes, interprets the AU pointers, cross-connects the higher order VCs, generates the new AU pointers and adds the RSOH and MSOH bytes, then rebuilding the STM-N frame at output. In the latter case, the SDH equipment terminates the RSOH and MSOH bytes, interprets the AU pointers, open the higher order VCs, interprets the TU pointers, cross-connects the lower order VCs, generates the new TU pointers, rebuilds the higher order VCs, cross-connects the higher order VCs, generates the new AU pointers and adds the RSOH and MSOH bytes, then rebuilding the STM-N frame at output.

2.5.10   ITU-T Standards on SDH

The main ITU-T standards on SDH or relevant to SDH are summarized as follows, not following the official ITU numbering order:

- Rec. G.707 *Network Node Interface for the Synchronous Digital Hierarchy (SDH)* [2.40] defines the SDH and the frame format;

- Rec. G.780 *Vocabulary of Terms for Synchronous Digital Hierarchy (SDH) Networks and Equipment* [2.42] provides the vocabulary of terms for SDH;

- Rec. G.783 *Characteristics of Synchronous Digital Hierarchy (SDH) Equipment Functional Blocks* [2.43] provide the functional specification of SDH equipment;

- Recs. G.774 *Synchronous Digital Hierarchy (SDH) Management Information Model for the Network Element View* [2.41] (and related Recs. G.774.01, G.774.02, G.774.03, G.774.04, G.774.05, G.774.06, G.774.07, G.774.08 and G.774.01), G.784 *Synchronous Digital Hierarchy (SDH) Management* [2.44] and G.831 *Management Capabilities of Transport Networks Based on the Synchronous Digital Hierarchy (SDH)* [2.47] deal with the management of SDH equipment and networks;

- Rec. G.803 *Architectures of Transport Networks Based on the Synchronous Digital Hierarchy (SDH)* [2.45] deals with the SDH network architecture;

- Rec. G.826 *Error Performance Parameters and Objectives for International, Constant Bit Rate Digital Paths at or Above the Primary Rate* [2.46] specifies the error performance objectives;

- Rec. G.832 *Transport of SDH Elements on PDH Networks: Frame and Multiplexing Structures* [2.48] deals with PDH–SDH interworking;

- Rec. G.841 *Types and Characteristics of SDH Network Protection Architectures* [2.49] and Rec. G.842 *Interworking of SDH Network Protection Architectures* [2.50] deal with SDH network protection;

- Rec. G.703 *Physical/Electrical Characteristics of Hierarchical Digital Interfaces* [2.5], G.957 *Optical Interfaces for Equipments and Systems Relating to the Synchronous Digital Hierarchy* [2.51] and G.958 *Digital Line Systems Based on the Synchronous Digital Hierarchy for Use on Optical Fibre Cables* [2.52] specify SDH physical interfaces and line systems;

- New Recs. G.810 *Definitions and Terminology for Synchronisation Networks* [2.53], G.811 *Timing Characteristics of Primary Reference Clocks* [2.54], G.812 *Timing Requirements of Slave Clocks Suitable for Use as Node Clocks in Synchronization Networks* [2.55] and G.813 *Timing Characteristics of SDH Equipment Slave Clocks (SEC)* [2.56] specify the characteristics of clocks of synchronization networks in a SDH environment.

- Rec. G.825 *The Control of Jitter and Wander within Digital Networks which are Based on the Synchronous Digital Hierarchy* [2.6] specifies the jitter limits at SDH equipment and network interfaces.

It is worthwhile noticing that the management of SDH equipment and networks is a topic addressed by ITU-T far more widely than mentioned above. However, TMN standards are well beyond the scope of this book.

Moreover, we notice that the new Recs. G.810, G.811, G.812 and G813 are a coherent set of Recommendations specifying the characteristics of clocks suitable for operation in digital telecommunications networks (including SDH). While Rec. G.813 has been specifically written for SDH equipment clocks, the others have replaced the older Recs. G.810, G.811, G.812 [2.57] (published in the Blue Book, 1988) written for PDH networks.

2.5.11   Summary of Characteristics of the SDH Signals

For ease of reference, finally, Tables 2.10 through 2.15 summarize some of the most relevant characteristics of the SDH signals, with special regard to the two asynchronous mappings of the 139.264 Mbit/s (via VC-4) and of the 2.048 Mbit/s (via VC-12). For these two mappings and for all the pointer types, the limit values of the frequency deviation range, which can be accommodated by bit or pointer justification, have been evaluated (as done in Exercises 2.2, 2.3 and 2.5, 2.6) and reported. Such limit values are theoretical and take into account simply the ratio between the payload size and the number of available justification opportunity bits, but do not consider other issues such as the buffer size, etc.

Table 2.10 Summary of basic characteristics of the STM-*N* signals

	STM-1	STM-4	STM-16	STM-64
Nominal bit rate	155.520 Mbit/s	622.080 Mbit/s	2488.320 Mbit/s	9953.280 Mbit/s
Frequency tolerance	±4.6 ppm	±4.6 ppm	±4.6 ppm	±4.6 ppm
No. of tel. channels carried via VC-12	1890	7560	30240	120960
No. of tel. channels carried via VC-11	2016	8064	32256	129024
Frame length	19440 bit	77760 bit	311040 bit	1244160 bit
Frame period	125 μs	125 μs	125 μs	125 μs
Frame alignment word (hex)	3 × F6, 3 × 28	12 × F6, 12 × 28	48 × F6, 48 × 28	192 × F6, 192 × 28
Frame alignment strategy parameters	(see Section 1.1.3)	(see Section 1.1.3)	(see Section 1.1.3)	(see Section 1.1.3)

Table 2.11   Summary of characteristics relevant to the asynchronous mapping of the 139.264 Mbit/s into STM-N via VC-4

Number of tributaries	N
VC-4 nominal bit rate	150.336 Mbit/s
No. of justification opportunity bits S (per VC-4)	9
Nominal justification ratio ρ_{nom}	7/9
Limit value of the frequency deviation range of the 139.264 Mbit/s which can be accommodated	$-115/+402$ ppm
Number of justification control bits C (per each justification opportunity bit)	5
Justification indication pattern	11111
Non-justification indication pattern	00000

Table 2.12   Summary of characteristics relevant to the asynchronous mapping of the 2.048 Mbit/s into STM-N via VC-12

Number of tributaries	$N \times 63$
VC-12 nominal bit rate	2.240 Mbit/s
No. of justification opportunity bits S (per VC-12)	2
Nominal justification ratio ρ_{nom}	1/2
Limit value of the frequency deviation range of the 2.048 Mbit/s which can be accommodated	±976 ppm
No. of justification control bits C (per each justification opportunity bit)	3
Justification indication pattern	111
Non-justification indication pattern	000

Table 2.13   Summary of characteristics relevant to AU pointer action

AU-4 Pointer	
Pointer legal values	$0 \div 782$
No. of justification bytes	3
Maximum justification frequency	2 kHz
VC-4 size	$261 \cdot 9 = 2349$ bytes (over 1 STM-N frame)
Limit value of the frequency deviation range of the incoming STM-N which can be accommodated	±319 ppm
Phase/time step of one AU-4 pointer adjustment	24 UI (at VC-4 rate) $\cong$ 160 ns

AU-3 Pointer	
Pointer legal values	$0 \div 782$
No. of justification bytes	1
Maximum justification frequency	2 kHz
VC-3 size	$85 \cdot 9 = 765$ bytes (over 1 STM-N frame)
Limit value of the frequency deviation range of the incoming STM-N which can be accommodated	±327 ppm
Phase/time step of one AU-3 pointer adjustment	8 UI (at VC-3 rate) $\cong$ 163 ns

Table 2.14 Summary of characteristics relevant to TU-3 and TU-2 pointer action

TU-3 Pointer	
Pointer legal values	$0 \div 764$
No. of justification bytes	1
Maximum justification frequency	2 kHz
VC-3 size	$85 \cdot 9 = 765$ bytes (over 1 STM-N frame)
Limit value of the frequency deviation range of the incoming STM-N which can be accommodated	± 327 ppm
Phase/time step of one TU-3 pointer adjustment	8 UI (at VC-3 rate) $\cong 163$ ns
TU-2 Pointer	
Pointer legal values	$0 \div 427$
No. of justification bytes	1
Maximum justification frequency	500 Hz
VC-2 size	428 bytes (over 4 STM-N frames)
Limit value of the frequency deviation range of the incoming STM-N which can be accommodated	± 584 ppm
Phase/time step of one TU-2 pointer adjustment	8 UI (at VC-2 rate) $\cong 1.17$ μs

Table 2.15 Summary of characteristics relevant to TU-12 and TU-11 pointer action

TU-12 Pointer	
Pointer legal values	$0 \div 139$
No. of justification bytes	1
Maximum justification frequency	500 Hz
VC-12 size	140 bytes (over 4 STM-N frames)
Limit value of the frequency deviation range of the incoming STM-N which can be accommodated	± 1785 ppm
Phase/time step of one TU-12 pointer adjustment	8 UI (at VC-12 rate) $\cong 3.57$ μs
TU-11 Pointer	
Pointer legal values	$0 \div 103$
No. of justification bytes	1
Maximum justification frequency	500 Hz
VC-11 size	104 bytes (over 4 STM-N frames)
Limit value of the frequency deviation range of the incoming STM-N which can be accommodated	± 2404 ppm
Phase/time step of one TU-11 pointer adjustment	8 UI (at VC-11 rate) $\cong 4.81$ μs

2.6 ASYNCHRONOUS VS SYNCHRONOUS MULTIPLEX FRAME STRUCTURES

As a concluding remark, we point out in this section the distinguishing features of asynchronous vs synchronous multiplex frame structures, due to their different multiplexing methods based on bit justification or not.

In a *synchronous multiplex frame structure*, given the frame alignment information, it is immediately possible to access single tributaries, because the bits of the different tributaries are allocated to positions within the multiplex frame that are known, fixed or coded by a pointer word.

The phase relationship between the multiplex frame and the lower order tributary frames is determined (synchronous structure). Either it is fixed, or it may change but it is controlled by the pointer, which tracks the phase shifts of the lower order tributary structures.

A first practical example of synchronous multiplex frame structure is the PCM Primary Multiplex. Given the frame start, each frame byte is allocated to a determined channel (tributary). Thus, each channel can be extracted or switched in a very straightforward way. Moreover, any signal built by synchronous digital multiplexing, as described in Section 2.3.1, features a synchronous multiplex frame structure. In particular, SDH/SONET signals feature a synchronous multiplex frame structure and any contained multiplexing element can be accessed through at the most two (the AU and TU pointers) deferred addresses.

On the other hand, in an *asynchronous multiplex frame structure*, it is not possible to know in any way what the phase relationship is between the multiplex frame and the contained lower order tributary frames. Given the multiplex frame start, the frame starts of the multiplexed tributaries may be anywhere; what's more, they may even shift according to any variation of the instantaneous frequency of the tributaries.

Any signal built by asynchronous digital multiplexing as described by Section 2.3.4 (e.g. PDH signals) features such a structure. Therefore, for example, to extract a single 2.048 Mbit/s E1 signal from a 139.264 Mbit/s E4 signal, it is necessary to fully demultiplex the E4 signal in all its tributaries by taking all the following steps in sequence:

- to frame-align onto the 139.264 Mbit/s E4 signal and then to distribute its bits among the output desynchronizers in order to demultiplex four 34.368 Mbit/s E3 signals;

- then, to frame-align onto the desired 34.368 Mbit/s E3 signal and to distribute its bits among the output desynchronizers in order to demultiplex four 8.448 Mbit/s E2 signals;

- then, to frame-align onto the desired 8.448 Mbit/s E2 signal and to distribute its bits among the output desynchronizers in order to demultiplex four 2.048 Mbit/s E1 signals.

This is the price to pay if adopting bit justification to cope with the issue of digital multiplexing asynchronous tributaries. Conversely, in SDH/SONET it is possible to access single VCs individually, without interfering with others, at the most through two deferred addresses.

2.7   SUMMARY

Digital transmission deals with transferring digital signals from one point in a network to another far one. *Digital multiplexing* consists of transmitting several digital signals on a single shared transmission channel.

A *timing signal* may be defined as a periodic or pseudo-periodic signal used to control the timing of actions (e.g. the symbol decision at reception) on digital signals, by triggering events at the so-called significant instants.

According to their phase and frequency relationship, isochronous digital signals (such as all the standard digital signals) can be classified in synchronous and asynchronous signals. The latter ones are further divided in mesochronous, plesiochronous and heterochronous.

Jitter denotes extensively any kind of phase noise of the digital signal symbols. Low-frequency jitter is often called the specific name *wander*. The frequency border between jitter and wander is purely conventional and is located around 10 Hz. Jitter and wander are commonly measured and expressed in terms of [UI]. The primary source of jitter and wander in digital transmission systems are regenerators, multiplexers and the transmission lines themselves.

The two fundamental information transfer modes are the *Asynchronous Transfer Mode* (ATM), where the sources send information in independent instants with interval dependent on source demand, and the *Synchronous Transfer Mode* (STM), where the sources can start sending their information units only in preassigned time-slots.

The various Time Division Multiplexing (TDM) techniques are outlined as follows.

- TDM may apply to analogue *Pulse Amplitude Modulation* (PAM) signals, yielding PAM–TDM signals which are analogue signals where N PAM signals are multiplexed in TDM. The PAM–TDM technique is not suited for transmission, due to the issue of dealing with analogue sampled signals.

- *PCM multiplexing* consists of TDM multiplexing N analogue telephone signals, having converted them to PCM digital form.

- *Digital multiplexing* (in the proper sense of the word) means specifically to TDM-multiplex N digital signals (tributaries). The distinctive feature of digital multiplexing, compared to PCM multiplexing, is that the tributary signals to multiplex are already in digital form. Digital multiplexing is further divided into:
  - *synchronous digital multiplexing*, where the tributary signals are synchronous;
  - *asynchronous digital multiplexing*, where the tributary signals are asynchronous (usually plesiochronous) and the bit justification technique is adopted to synchronize tributaries into the multiplex signal. Bit justification consists of transmitting either tributary or dummy bits at certain bit positions within the multiplex signal (*justification opportunity bits*), following the input frequency of the tributaries.

The PDH is a series of standard bit rates (hierarchical levels) defined by the CCITT (now ITU-T) for transmission in digital telephone networks. Two plesiochronous hierarchies, having different bit rates but following the same principle, are defined by the ITU-T: the European and the North American (with the Japanese variant) hierarchies. Both are based on bit-interleaved asynchronous digital multiplexing.

Two standard PCM multiplex signals are the first levels of PDH digital hierarchies: the European E1 signal, with rate 2.048 Mbit/s and carrying 30 telephone channels, and the North American DS1 signal, with rate 1.544 Mbit/s and carrying 24 telephone channels.

SONET and SDH transmission hierarchies follow the principle of the synchronous digital multiplexing but with some important innovations, in particular the pointer justification technique, designed purposely to cope with the issue of multiplexing asynchronous tributaries but maintaining a synchronous digital multiplexing structure.

Synchronous multiplexing of SDH and SONET is based on complex rules which combine several building blocks (synchronous multiplexing elements) in various ways. In particular, such synchronous multiplexing elements are structured, fixed-size sets of bytes, which are variedly byte-interleaved or mapped one into the other to eventually form STM-N (in the case of SDH) frames. Among them, the *Virtual Containers* (VCs), named *Virtual Tributaries* (VT) in SONET, are the basic building blocks and are the conceptually most innovative elements compared to the traditional PDH technique. They map a payload and are individually and independently accessible through a pointer information, directly associated with them on multiplexing.

Mainly three kinds of SDH equipment can be distinguished: *regenerators, multiplexers* and *Digital Cross-Connects*. Regenerators regenerate the STM-N signals along very long transmission lines. Multiplexers assemble several PDH or SDH tributaries into a multiplex SDH signal. Digital Cross-Connects cross-connect SDH/PDH signals from the input to the output ports.

2.8   REFERENCES

[2.1] S. Benedetto, E. Biglieri, V. Castellani. *Digital Transmission Theory*. Englewood Cliffs, NJ: Prentice Hall Inc., 1987.

[2.2] J. G. Proakis. *Digital Communications*. New York: McGraw–Hill, 1989.

[2.3] S. Haykin. *An Introduction to Analog and Digital Communications*. New York: John Wiley & Sons, 1989.

[2.4] A. J. Jerri. The Shannon sampling theorem — its various extensions and applications: a tutorial review. *Proceedings of the IEEE* 1977; **65**: 1565–1596.

[2.5] ITU-T Rec. G.703 *Physical/Electrical Characteristics of Hierarchical Digital Interfaces*. Geneva, Oct. 1998.

[2.6] ITU-T Recs. G.823 *The Control of Jitter and Wander within Digital Networks which are Based on the 2048 kbit/s Hierarchy;* G.824 *The Control of Jitter and Wander within Digital Networks which are Based on the 1544 kbit/s Hierarchy*; G.825 *The Control of Jitter and Wander within Digital Networks which are Based on the Synchronous Digital Hierarchy*. Geneva, March 1993.

[2.7] A. J. Viterbi. *Principles of Coherent Communications*. New York: McGraw–Hill, 1966.

[2.8] W. C. Lindsey. *Synchronization Systems in Communications and Control*. Englewood Cliffs, NJ: Prentice Hall Inc., 1972.

[2.9] A. Blanchard. *Phase-Locked Loops*. New York: John Wiley & Sons, 1976.

[2.10] F. M. Gardner. *Phaselock Techniques*. New York: John Wiley & Sons, 1979.

[2.11] R. E. Best. *Phase Locked Loops*. New York: McGraw–Hill Book Company, 1984.

[2.12] H. Meyr, G. Ascheid. *Synchronization in Digital Communications. Vol. 1: Phase-, Frequency-Locked Loops, and Amplitude Control*. New York: John Wiley & Sons, 1990.

[2.13] Y. Takasaki. *Digital Transmission Design and Jitter Analysis*. Norwood, MA: Artech House, 1991.

[2.14] ITU-T Rec. G.701 *Vocabulary of Digital Transmission and Multiplexing, and Pulse Code Modulation (PCM) Terms*. Geneva, March 1993.

[2.15] D. L. Duttwailer. Waiting time jitter. *Bell System Technical Journal*. 1972; **51**: 165–207.

[2.16] P. E. K. Chow. Jitter due to pulse stuffing synchronization. *IEEE Transactions on Communications* 1973; **COM-21**(7): 854–859.

[2.17] P. R. Trischitta, E. L. Varma. *Jitter in Digital Transmission Systems*. Norwood, MA: Artech House, 1989.

[2.18] ITU-T Rec. G.211 *Make Up of a Carrier Link*. Blue Book, Geneva 1988.

[2.19] ITU-T Rec. G.704 *Synchronous Frame Structures Used at 1544, 6312, 2048, 8488 and 44 736 kbit/s Hierarchical Levels*. Geneva, Oct. 1998.

[2.20] ITU-T Rec. G.706 *Frame Alignment and Cyclic Redundancy Check (CRC) Procedures relating to Basic Frame Structures Defined in Rec. G.704*. Geneva, April 1991.

[2.21] ITU-T Rec. G.731 *Primary PCM Multiplex Equipment for Voice Frequencies*. Blue Book, Geneva, 1988.

[2.22] ITU-T Rec. G.732 *Characteristics of Primary PCM Multiplex Equipment Operating at 2048 Kbit/s*. Blue Book, Geneva, 1988.

[2.23] ITU-T Rec. G.733 *Characteristics of Primary PCM Multiplex Equipment Operating at 1544 Kbit/s*. Blue Book, Geneva, 1988.

[2.24] ITU-T Rec. G.734 *Characteristics of Synchronous Digital Multiplex Equipment Operating at 1544 Kbit/s*. Blue Book, Geneva, 1988.

[2.25] ITU-T Rec. G.736 *Characteristics of Synchronous Digital Multiplex Equipment Operating at 2048 Kbit/s*. Geneva, April 1993.

[2.26] S. Kozuka. Phase controlled oscillator for pulse stuffing synchronization system. *Review of the Electrical Communication Laboratory* 1969; **17**(5–6): 376–387.

[2.27] Y. Matsuura, S. Kozuka, K. Yuki. Jitter characteristics of pulse stuffing synchronization. *Proc. of IEEE ICC '68*, June 1968: 259–264.

[2.28] ITU-T Rec. G.702 *Digital Hierarchy Bit Rates*. Blue Book, Geneva, 1988.

[2.29] ITU-T Rec. G.741 *General Considerations on Second Order Multiplex Equipments*. Blue Book, Geneva, 1988.

[2.30] ITU-T Rec. G.742 *Second Order Digital Multiplex Equipment Operating at 8448 kbit/s and Using Positive Justification*. Blue Book, Geneva, 1988.

[2.31] ITU-T Rec. G.743 *Second Order Digital Multiplex Equipment Operating at 6312 kbit/s and Using Positive Justification*. Blue Book, Geneva, 1988.

[2.32] ITU-T Rec. G.751 *Digital Multiplex Equipments Operating at the Third Order Bit Rate of 34 368 kbit/s and the Fourth Order Bit Rate of 139 264 kbit/s and Using Positive Justification*. Blue Book, Geneva, 1988.

[2.33] ITU-T Rec. G.752 *Digital Multiplex Equipments Based on a Second Order Bit Rate of 6312 kbit/s and Using Positive Justification*. Blue Book, Geneva, 1988.

[2.34] ITU-T Rec. G.951 *Digital Line Systems Based on the 1544 kbit/s Hierarchy on Symmetric Pair Cables*. Blue Book, Geneva, 1988.

[2.35] ITU-T Rec. G.952 *Digital Line Systems Based on the 2048 kbit/s Hierarchy on Symmetric Pair Cables*. Blue Book, Geneva, 1988.

[2.36] ITU-T Rec. G.953 *Digital Line Systems Based on the 1544 kbit/s Hierarchy on Coaxial Pair Cables*. Blue Book, Geneva, 1988.

[2.37] ITU-T Rec. G.954 *Digital Line Systems Based on the 2048 kbit/s Hierarchy on Coaxial Pair Cables*. Blue Book, Geneva, 1988.

[2.38] ITU-T Rec. G.955 *Digital Line Systems Based on the 1544 kbit/s and the 2048 kbit/s Hierarchy on Optical Fibre Cables*. Geneva, Nov. 1996.

[2.39] M. Sexton, A. Reid. *Broadband Networking: ATM, SDH and SONET*. Norwood, MA: Artech House, 1997.

[2.40] ITU-T Rec. G.707 *Network Node Interface for the Synchronous Digital Hierarchy (SDH)*. Geneva, March 1996.

[2.41] ITU-T Rec. G.774 *Synchronous Digital Hierarchy (SDH) Management Information Model for the Network Element View*; Rec. G.774.01 *Synchronous Digital Hierarchy (SDH) Performance Monitoring for the Network Element View*; Rec. G.774.02 *Synchronous Digital Hierarchy (SDH) Configuration of the Payload Structure for the Network Element View*; Rec. G.774.03 *Synchronous Digital Hierarchy (SDH) Management of Multiplex-Section Protection for the Network Element View*; Rec. G.774.04 *Synchronous Digital Hierarchy (SDH) Management of the Subnetwork Connection Protection for the Network Element View*; Rec. G.774.05 *Synchronous Digital Hierarchy (SDH) Management of*

Connection Supervision Functionality (HCS/LCS) for the Network Element View. Geneva, 1994/1998.

[2.42] ITU-T Rec. G.780 *Vocabulary of Terms for Synchronous Digital Hierarchy (SDH) Networks and Equipment*. Geneva, June 1999.

[2.43] ITU-T Rec. G.783 *Characteristics of Synchronous Digital Hierarchy (SDH) Equipment Functional Blocks*. Geneva, April 1997.

[2.44] ITU-T Rec. G.784 *Synchronous Digital Hierarchy (SDH) Management*. Geneva, June 1999.

[2.45] ITU-T Rec. G.803 *Architectures of Transport Networks Based on the Synchronous Digital Hierarchy (SDH)*. Geneva, June 1997.

[2.46] ITU-T Rec. G.826 *Error Performance Parameters and Objectives for International, Constant Bit Rate Digital Paths at or Above the Primary Rate*. Geneva, Feb. 1999.

[2.47] ITU-T Rec. G.831 *Management Capabilities of Transport Networks Based on the Synchronous Digital Hierarchy (SDH)*. Geneva, March 1993.

[2.48] ITU-T Rec. G.832 *Transport of SDH Elements on PDH Networks: Frame and Multiplexing Structures*. Geneva, Oct. 1998.

[2.49] ITU-T Rec. G.841 *Types and Characteristics of SDH Network Protection Architectures*. Geneva, Oct. 1998.

[2.50] ITU-T Rec. G.842 *Interworking of SDH Network Protection Architectures*. Geneva, April 1997.

[2.51] ITU-T Rec. G.957 *Optical Interfaces for Equipments and Systems Relating to the Synchronous Digital Hierarchy*. Geneva, June 1999.

[2.52] ITU-T Rec. G.958 *Digital Line Systems Based on the Synchronisation Digital Hierarchy for Use on Optical Fibre Cables*. Geneva, Nov. 1994.

[2.53] ITU-T Rec. G.810 *Definitions and Terminology for Synchronisation Networks*. Geneva, Aug. 1996.

[2.54] ITU-T Rec. G.811 *Timing Characteristics of Primary Reference Clocks*. Geneva, Sept. 1997.

[2.55] ITU-T Rec. G.812 *Timing Requirements of Slave Clocks Suitable for Use as Node Clocks in Synchronization Networks*. Geneva, June 1998.

[2.56] ITU-T Rec. G.813 *Timing Characteristics of SDH Equipment Slave Clocks (SEC)*. Geneva, August 1996.

[2.57] ITU-T Rec. G.810 *Considerations on Timing and Synchronization Issues*; G.811 *Timing Requirements at the Outputs of Primary Reference Clocks Suitable for Plesiochronous Operation of International Digital Links*; G.812 *Timing Requirements at the Outputs of Slave Clocks Suitable for Plesiochronous Operation of International Digital Links*. Blue Book, 1988.

3

TIMING ASPECTS IN SDH NETWORKS

The SDH technique poses peculiar timing issues that deserve a closer look. Historically, the timing aspects of telecommunications networks have often been reduced to a set of pragmatic parameters and limits to verify, while their underlying rationale has not been normally considered part of the education of telecommunications engineers. The introduction of SDH has given a new perspective on timing and has yielded the need for a deeper insight, changing for example many of the original assumptions used in deriving such parameters and limits.

Moreover, new timing issues have been raised by SDH, which implies new phenomena generating jitter and wander. SDH further developed the concepts of asynchronous and synchronous digital transmission, as explained in the previous chapter. Hence the need for a closer look on the timing aspects of SDH.

This chapter deals with some of the key topics related to timing in SDH networks. First, the different causes of jitter and wander in an SDH transmission chain are reviewed. Then, the blocks where digital signals with asynchronous bit rates are arranged in SDH equipment are treated: the synchronizer, the desynchronizer and the pointer processor. Finally, the SDH equipment clock is described in terms of functional blocks.

3.1 CAUSES OF JITTER AND WANDER IN AN SDH TRANSMISSION CHAIN

In spite of the numerous advantages offered by SDH compared to PDH, SDH made the issue of controlling the jitter and wander of the transported tributaries more delicate than in traditional PDH networks. While all the well known and studied causes of jitter and wander in a PDH transmission chain are still present, in fact, there are some further phenomena in SDH which may produce additional jitter and wander in the tributaries transported along an SDH transmission chain.

Some phenomena affect the phase of the whole SDH signals transmitted along the chain and hence the phase of the tributary signals mapped into them. Some others, instead, generate jitter only on the mapped tributaries, without affecting the phase of the SDH carrier signals. For example, the wander due to variations of the environmental conditions (temperature) and the jitter generated in digital regenerators belong to the first kind, as they affect the whole SDH signals.

SDH multiplexers, on the other hand, feature processes which generate jitter and wander solely on the tributary signals mapped and in no way on the SDH carrier signals. Such processes include:

- insertion of additional bits (e.g. miscellaneous overhead and fixed stuff) in the mapping of tributaries into the STM frames;
- bit justification, to map the asynchronous tributaries within the synchronous VCs;
- pointer justification, which allows VCs to shift within other VCs or STM frames.

This section reviews the main or more interesting causes of jitter and wander in a SDH transmission chain, by outlining the different phenomena affecting the phase of the tributaries transported.

3.1.1   Environmental Conditions (Temperature)

Variations of temperature along the day and the year are a well known cause of wander (*diurnal* and *annual wander*), since the first long-haul digital transmission systems have been set up. While this kind of wander was very important on long copper cable lines, its amplitude is far lower in optical fibre systems. Though, it may still reach several UI of amplitude if the bit rate of the transmission system is high.

Such wander cannot be reduced, as it is the result of uncontrollable changes in the cable environment. The most that can be done to reduce the effects of temperature variation is to bury the cable deep. This reduces the daily temperature variation to a fraction of a degree, although larger temperature swings will be seen annually.

Moreover, diurnal and annual wander cannot be filtered out, owing to their extremely low frequency. The only way to cope with them is making larger the size of elastic stores, in order to avoid slips due to them or to reduce their rate.

The main reason of such wander is that the propagation delay of the light in an optical fibre depends on the refractive index of the fibre core and on the fibre length, which both depend on the temperature. Hence, the digital signals exhibit different propagation delays during the day and the night, as well as during the winter and the summer. This phenomenon yields a pseudo-periodical variation in the phase of the digital signal received, having a period of about one day or one year (diurnal and annual wander).

In more detail, the propagation delay of an optical signal transmitted through an optical fibre of length l is given by [3.1][3.2]

$$\tau = \frac{l n_c}{c} \tag{3.1}$$

where n_c is the group index of refraction of the fibre core and c is the light speed in the vacuum. It is evident that a slow change in l or n_c results in wander. Both optical fibre temperature changes and laser wavelength changes will change n_c, while an optical fibre temperature change also results in a slight variation in the length of the optical fibre.

Variations in the optical fibre temperature are the most significant source of diurnal and annual wander. Even a slight variation in temperature of the optical fibre can cause a significant amount of wander over a long distance, since both the group index of refraction and the length of the fibre are temperature-dependent.

Fluctuations in the laser transmitter wavelength are another source of wander in fibre optic systems. Wavelength variations cause wander because the group refractive index of the optical fibre is wavelength-dependent. Again, the drift in laser wavelength is mainly due to changes in laser temperature.

The Table 3.1 lists some practical values of dependence of the quantities mentioned above on temperature [3.1]. Based on such values, for example, the effect of a 1°C fibre temperature change along a 1000 km transmission path containing 20 sections of 50 km each can be evaluated as about 40 ns of accumulated peak-to-peak diurnal wander, assuming that the temperature change is systematic along the entire route. In the case of an SDH STM-16 transmission system, having a bit rate approximately equal to 2.5 Gb/s, this wander is equivalent to 100 UI.

3.1.2 Overhead and Fixed Stuffing in the Mapping Structure

On designing the mapping of tributary signals into the STM-N frame, much care was taken in dispersing the tributary bits along the frame. For example, the Section OverHead (SOH) is not transmitted all at once, but it is divided into nine segments per frame (the STM-N frame of Figure 2.22 is transmitted row by row).

Nevertheless, the bits of the tributary extracted do not arrive at a uniform rate. The arrival rate of the bits of a certain tributary is given by a gapped-clock signal, obtained from the regular clock associated with the incoming STM-N multiplex signal, by inhibiting the pulses corresponding to the unwanted bits (e.g. to miscellaneous overhead, fixed stuff, justification bits, etc.).

The gapped clock associated with the mapped tributary is highly discontinuous. Its peak frequency is that of the regular STM-N clock, but, owing to its numerous gaps, its average frequency is that of the mapped tributary. The desynchronizer in the demultiplexing equipment smoothes its gaps thus returning a bit stream with regular clock at its average frequency.

The jitter of the mapped-tributary gapped clock has very wide peak-to-peak amplitude, but it has most of its power at very high frequencies. Therefore, desynchronizers properly

Table 3.1 Typical values of optical fibre and laser characteristic quantities (3.1)

$\dfrac{\partial n_c}{\partial \theta}$	$1.2 \times 10^{-5}/°C$ [3.3]
$\dfrac{1}{l}\dfrac{\partial l}{\partial \theta}$	$8.0 \times 10^{-7}/°C$ [3.4]
$\dfrac{1}{c}\dfrac{\partial n_c}{\partial \lambda}$	$17\dfrac{ps}{nm\ km}$ (at $\lambda = 1.5$ μm)
$\dfrac{\partial \lambda}{\partial \theta}$	0.1 nm/°C

θ fibre temperature
θ_l laser temperature
λ signal wavelength
l fibre length
n_c group index of refraction

designed can cancel most of it. The residual jitter is very low and can be usually neglected compared to the jitter coming from other sources.

3.1.3   Bit Justification

The frequency offsets of plesiochronous tributaries are accommodated, on mapping into the synchronous Containers, by using the justification opportunity bits, as explained in Sections 2.5.6 and 2.5.7.

As broadly discussed in Section 2.3.5, the process of bit justification yields *waiting time jitter*. This jitter, though of very low amplitude, is impossible to be filtered out completely by the output desynchronizer, owing to its unlimited low frequency, and can therefore accumulate along a chain of PDH-to-SDH and SDH-to-PDH mappings and demappings.

The overall jitter due to bit justification and to overhead and fixed stuff in the mapping structure, addressed in the previous section, is often referred to as *mapping jitter*.

3.1.4   Pointer Justification

As discussed in Section 2.5.8, the pointer justification mechanism allows to compensate in an NE the phase fluctuations between the input VCs, timed by the clock of the NE where the VC is originated, and the output STM-N frames, timed by the local clock.

Each pointer adjustment shifts the VC back or forward in the carrier STM-N frame: a VC-4 is moved by three bytes, all other VCs are moved by one byte for each Pointer Justification Event (PJE). Therefore, viewing the VC bits extracted from the incoming frame, one notices a sudden pause or acceleration in the bit flow after each pointer adjustment.

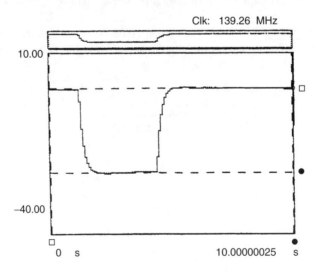

Figure 3.1. Phase error [UI], measured on a SDH piece of equipment, between the demapped 139.264 Mbit/s signal and the carrying STM-1 signal with two alternate AU-4 pointer adjustments. (Reproduced from [3.6] by permission of IEEE)

Said in a more formal way, the phase diagram of the VC digital signal exhibits a positive or negative step on each positive or negative pointer adjustment. The amplitude of the phase step corresponds to the number of bits justified by one PJE. In the case of the AU-4 pointer, one PJE justifies 24 bits of the VC-4, corresponding to about 160 ns.

The bit flow of the demapped tributary exhibits a proportional phase step, corresponding to the average number of tributary bits in the word justified. Again in the case of the E4 mapped into the VC-4, the step amplitude corresponds to about 22 bits at the bit rate of 139.264 Mbit/s.

This phase step in the demapped tributary digital signal is low-pass filtered by the desynchronizer, which smoothes it as a negative exponential curve, at least under the simple assumption of single-pole low-pass phase filter. To give an idea of the residual phase hit, Figure 3.1 reports the phase error (expressed in UI units) measured with a time counter on an SDH piece of equipment, between the demapped 139.264 Mbit/s signal and the carrying STM-1 signal under two alternate AU-4 pointer adjustments [3.6]. The step amplitude is something more than 22 UI and the time constant of the exponential transients is less than one second.

3.2   SYNCHRONIZATION PROCESSES ALONG AN SDH TRANSMISSION CHAIN

In this section, the different synchronization processes taking place in an SDH transmission chain are summarized. Figure 3.2 shows a typical PDH-SDH-PDH path: a PDH tributary is mapped in the SDH frames and transported along an SDH transmission chain until it is extracted by the terminating demultiplexer.

Basically, three synchronization processes take place in the SDH equipment of the scheme depicted in Figure 3.2:

- the mapping of the PDH tributary into SDH frames in the first multiplexer,

- the re-synchronization of the VCs in intermediate nodes along the SDH transmission chain[1],

- the demapping of the PDH tributary from SDH frames in the terminating demultiplexer.

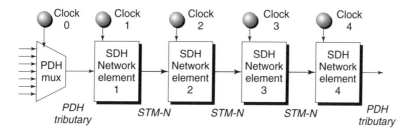

Figure 3.2.   Transport of a PDH tributary along an SDH transmission chain

---

[1] We note that, in normal conditions, all the nodes of an SDH network are synchronized directly or not by a suitable network synchronization system. However, the pointer justification mechanism allows re-synchronization of the VCs along a transmission chain where the nodes are asynchronous.

3.2.1   Mapping of the PDH Tributary into SDH Frames

The PDH tributary signals at the input of SDH multiplexers are typically asynchronous (see Chapter 2). The adaptation of the bit rate of the PDH tributary (Clock 0) to the bit rate of the mapping SDH VC (Clock 1) is performed, as in PDH multiplexers, by means of *bit justification*.

As explained in the examples of Sections 2.5.6 and 2.5.7, the VCs based on asynchronous mapping are made of tributary bits (true payload), miscellaneous overhead, fixed stuffing, justification opportunity and control bits. The functional block carrying out such adaptation, and thus proportioning the justification opportunity bits according to the variable frequency offset between Clock 0 and Clock 1, is called *synchronizer*.

The operation of an SDH synchronizer is perfectly analogous to that of PDH synchronizers. Section 3.3 details the functional blocks and the operation of SDH synchronizers.

3.2.2   Re-Synchronization of the VCs in SDH Intermediate Nodes

Let us consider, for example, the case of a DXC or an ADM, where the VCs of the incoming STM-N are cross-connected. Since, in general, input signals are timed by different clocks, the node must re-synchronize, according to the local equipment clock, all the VCs before cross-connecting and retransmitting them in the output STM-N signals.

Referring to Figure 3.2, the VC mapping the PDH signal, timed by Clock 1, is transported along the SDH chain by STM-N frames timed by the different clocks of the chain (Clocks 2 and 3). In each node, the VC timed by Clock 1 is re-synchronized to fit into the STM-N frames, timed by a different clock, by means of *pointer justification*.

As explained in Section 2.5.8, the pointer justification mechanism allows to compensate in an NE the phase fluctuations between the input VCs, carried in STM-N frames timed by the clock of the previous NE in the transmission chain, and the output STM-N frames, timed by the local clock.

Let us focus for example on NE3. At its input, the STM-N signal is timed by Clock 2. Inside that STM-N signal, the VC signal is still timed by Clock 1, which generated it. Pointer justification is the magic that allows the VC flowing at its original rate inside the bearer signal STM-N, which has an independent rate. At the output of NE3, the STM-N signal is timed by Clock 3. Inside that STM-N signal, the VC signal still keeps its original timing (Clock 1).

Re-synchronization of VCs is accomplished by means of an elastic store, in which the VC bytes are written according to the timing signal extracted by the incoming STM-N signal and are read according to the local equipment clock. Pointer justification allows to compensate the unavoidable differences between the input and output frequencies. This block based on an elastic store is called *pointer processor*. Section 3.4 details the functional blocks and the operation of SDH pointer processors.

3.2.3   Demapping of the PDH Tributary from SDH Frames

As discussed in Sections 3.1.2 through 3.1.4, the bits of the tributary extracted from the SDH frames do not arrive at a uniform rate, but according to a gapped clock.

Therefore, demapping involves smoothing those gaps in order to return a regular bit stream.

The block performing the demapping of the tributary from its VC is called *desynchronizer* and fulfils the complementary function of the synchronizer of Section 3.2.1. Its task is to reduce all the jitter components mentioned in Section 3.1.

The operation of an SDH desynchronizer is perfectly analogous to that of PDH desynchronizers. Section 3.3 describes in detail the functional blocks and the operation of SDH desynchronizers.

3.2.4 The Average Frequency of Tributary Signals Is Transferred along the Chain

The average frequency of the tributary returned at the end of the chain is, as obvious, the *same* as the original Clock 0, except for some residual jitter. In fact, regardless of what pointer action may happen along the chain, no bits are added or deleted to or from the tributary bit stream along its path! This concept is well expressed by what we humorously like to call the

Fundamental Law of Digital Transmission Chains

ALL BITS THAT GET IN ON ONE SIDE MUST GET OUT FROM THE OTHER

In the same way, the average frequency of the VC keeps the same along the chain, regardless of what the frequencies of the various bearer STM-N signals may be. If the VC frequency at the output of NE2 was that of Clock 2, it would mean that some bits would have been added or deleted to or from the VC stream, originally with rate driven by Clock 1. Pointer adjustments allow to write the bytes of the VC, incoming with rate driven by Clock 1, on a stream that is generated with rate driven by Clock 2, while retaining their original *average* rate.

Exercise 3.1
Synchronization Processes along a PDH–SDH Transmission Chain

Let us consider the PDH–SDH transmission chain in the figure below, where a PDH E4 (timed by Clock 0, CK$_0$) is mapped in its VC-4 and transported along a chain of SDH multiplexers and DXCs until it is demultiplexed at the end of the chain. NEs are timed by different clocks (numbered #0 through #5), characterized by different frequency offsets Δf from the nominal values. Nominal bit rates are 139.264 Mbit/s for the E4 signal and $N \times 155.520$ Mbit/s for the STM-N signal.

What is the average frequency of the E4 signal output by the last multiplexer at point E?

What is the AU pointer justification rate (as [justifications/s], with sign) at points A, B, C and D?

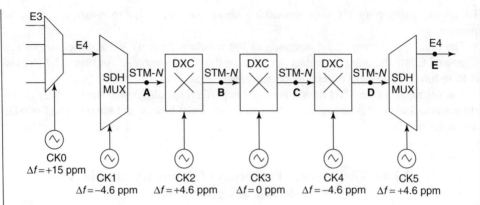

Solution

The average frequency of the E4 signal at point E is simply that of CK_0, i.e. 139.264 Mbit/s $\cdot (1 + 15 \cdot 10^{-6}) \cong 139.266$ Mbit/s (cf. the *Fundamental Law of Digital Transmission Chains*).

At point A, there is no pointer action and the AU pointer justification rate is thus 0 justifications/s, because the STM-N frame and the VC-4 are both timed by CK_1.

At point B, the frequency offset between the STM-N frame and the VC-4 is $CK_2 - CK_1 = +9.2$ ppm. Being the nominal VC-4 bit rate (within the bearer STM-N signal)

$$f_{VC-4} = N \cdot 155.520 \text{ Mbit/s} \cdot \frac{261}{N \cdot 270} = 150.336 \text{ Mbit/s}$$

at the pointer processor buffer of the DXC before point B there is a *deficit* of

$$150.336 \text{ Mbit/s} \cdot 9.2 \cdot 10^{-6} = 1383 \text{ bit/s}$$

which is made up through pointer action. Since one AU-4 pointer justification recovers a phase offset of 24 bits, the AU pointer justification rate at point B results then in

$$\frac{1383 \text{ bit/s}}{24 \text{ bit}} \cong 57.6 \text{ justifications/s.}$$

Similarly, at point C, the frequency offset between the STM-N frame and the VC-4 is $CK_3 - CK_1 = +4.6$ ppm. Therefore, the AU pointer justification rate at point C results in one half of that at point A and thus 28.8 justifications/s.

At point D, the frequency offset between the STM-N frame and the VC-4 is $CK_4 - CK_1 = 0$ ppm. Therefore, the AU pointer justification rate at point D results in 0 justifications/s (provided that the phase noise of $CK_4 - CK_1$ is not large enough to exceed the thresholds in the buffer of the pointer processor that determine pointer action).

3.3   SDH SYNCHRONIZER AND DESYNCHRONIZER

The SDH multiplexing structure comprises both the concepts of asynchronous and synchronous digital multiplexing: asynchronous mappings are defined to map asynchronous tributaries into VCs, by use of bit justification, while the VCs are mapped into STM-N frames by synchronous multiplexing, although allowing dynamic phase aligning by use of pointer justification.

Therefore, SDH synchronizers and desynchronizers must cope with severe issues of jitter reduction, having to deal with a wide set of mappings. All the causes of jitter and wander in an SDH transmission chain discussed in Section 3.1 must be properly counteracted. Even multiple PDH-to-SDH and SDH-to-PDH conversions must be taken into account. Hence, in the design of SDH synchronizers and desynchronizers, the most advanced techniques developed for jitter reduction are applied.

In this section, after having outlined the basic principles of operation of SDH synchronizers and desynchronizers, we will provide some further details on their implementation. First, the PLLs used in desynchronizers for jitter reduction will be addressed. Then, methods based on stuff threshold modulation for an enhanced design of synchronizers and methods based on bit leaking for an enhanced design of desynchronizers will be described.

3.3.1 Principle of Operation of an SDH Synchronizer

The functional scheme of principle of the mapping of an asynchronous tributary into an SDH VC (*synchronizer*) is depicted in Figure 3.3, where only the ith tributary input port is detailed for the sake of simplicity. The case of positive justification, as in the mapping of the 139.264 Mbit/s into the VC-4, is considered for example.

The bits incoming from the tributary input port are written into the buffer with their arrival rate f_{ti} (regular clock) at the address selected by a write counter. These bits are read from the buffer at instants triggered by a VC timing generator and distributor, driven

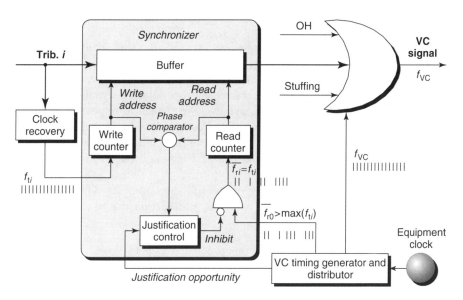

Figure 3.3. Functional scheme of principle of the mapping of an asynchronous tributary into an SDH VC (synchronizer)

by the local equipment clock, which synthesizes the regular clock^2 signal f_{VC} (frequency of the VC generated in that node) and generates a buffer read timing signal f_{r0} by properly gapping the f_{VC} signal. The average frequency $\overline{f_{r0}}$ is slightly greater than the maximum value tolerated for the write frequency f_{ti}.

The buffer write and read addresses are compared so that when the buffer gets empty, reading is inhibited (AND gate at the read counter input) at justification opportunity positions. Therefore, the actual buffer read frequency f_{ri} results equal — *on the average* — to the write frequency f_{ti}.

The block performing the mapping of the tributary into the VC signal, thus comprising mainly the buffer together with the justification control, is called *synchronizer*, as it accomplishes the synchronization of the tributary bits into the VC available positions (cf. the PDH synchronizer of Section 2.3.5).

The bits read from the buffer (synchronizer output) are finally interleaved with the OH bytes and the stuffing additional bits, to form the output VC signal.

3.3.2   Principle of Operation of an SDH Desynchronizer

The functional scheme of principle of the demapping of an asynchronous tributary from an SDH signal (*desynchronizer*) is depicted in Figure 3.4, where only the ith tributary output port is detailed for the sake of simplicity.

The STM-N signal is processed: starting from the frame and pointer alignment and the justification control information, it is possible to enable writing of the tributary bits into the demultiplexing buffer at the address selected by a write counter. The write timing signal is gapped and very uneven, owing to fixed overhead gaps and to variable gaps due to pointer adjustments and to the bit justification inside the VC, but the average write frequency results in the same as the original tributary frequency, i.e.

$$\overline{f_{wi}} = f_{ti} \tag{3.2}$$

As in the asynchronous demultiplexer of Section 2.3.5, the read frequency f_{ti} is not in fixed ratio with the multiplex frequency f_m and thus cannot be synthesized directly from it like in synchronous demultiplexers. For this reason, f_{ti} is usually generated locally, by means of adaptive algorithms, based on the write gapped timing signal f_{wi}.

A PLL is usually adopted to smooth gaps in f_{wi} and then to return the regular clock f_{ti}. Its transfer function is low-pass: the jitter components beyond the cut frequency are damped, but the lower frequency components are still transferred to the output. A simple second-order PLL with a 3 dB cut-off bandwidth on the order of 10^2 Hz is mostly used.

The block performing the demapping of the tributary from the STM-N signal, thus comprising mainly the buffer together with the justification control, is usually called

---

^{2} Actually, it has been neglected, for the sake of simplicity, that also the clock signal of the VC mapped into the output STM-N multiplex signal (in case through TUGs) is gapped, owing to the SDH multiplexing structure. However, this does not impact the substance of what is explained: the scheme might be modified by making the VC timing generator and distributor to generate directly the VC clock signal gapped according to the STM-N frame structure, or also, alternatively, the same scheme may be left unchanged, considering that the adaptation of the VC to the TU or AU (i.e., its phase aligning and its pointer generation) follows, in the functional block scheme specified by the ITU-T Rec. G.783 [3.5], the adaptation of the tributary signal to the VC (i.e., the generation of the VC itself).

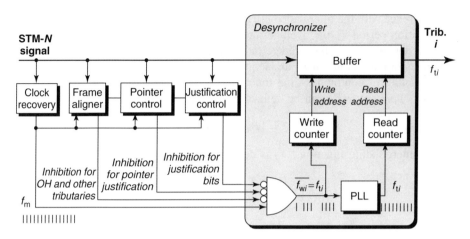

Figure 3.4. Functional scheme of principle of the demapping of an asynchronous tributary from an SDH signal (desynchronizer)

desynchronizer and fulfils the complementary function of the synchronizer of Figure 3.3: it smoothes the gaps in the output tributary clock due to mapping thus returning a bit stream with regular-clock timing (cf. the desynchronizer of Figure 2.3.2).

3.3.3 Phase-Locked Loops Used in Desynchronizers for Jitter Reduction

As stated above, a PLL is mostly used in desynchronizers to smooth gaps in the incoming jittered clock signal to derive a regular-clock signal for the output data. Its transfer function is low-pass: the jitter components beyond its cut-off frequency are damped. The PLL must be properly designed so that the time deviations of the recovered clock from an ideal (i.e., jitter free) clock signal are within the allowable limits.

The performance of PLLs in filtering continuous-time signals is well known and studied in several works on PLL basics [3.7]–[3.12] and on synchronization and jitter analysis [3.13][3.14]. As an approximation, it is often customary to extend this continuous-time signal analysis to discrete-time systems, such as systems associated with digital clock signals. However, the details of such approximation are hard to find in the literature. In this section, we follow the approach proposed by Abeysekera and Cantoni in their paper [3.15], which the reader is referred to for the mathematical details.

Recalling that stated in Section 2.1.3, the *true jitter* $h[n]$ of a write clock signal is defined as the difference between the occurrence time B_n of the nth clock pulse (significant instant) and the occurrence time of the nth pulse of an ideal clock at the same average frequency. Therefore,

$$B_n = nT_w + h[n] \tag{3.3}$$

where T_w is the average interval between the occurrences of the incoming clock, i.e.

$$T_w = \lim_{n \to \infty} (B_n/n) \tag{3.4}$$

The difference with the jitter called $e[nT]$ in Section 2.1.3 is that, instead of the nominal pulse repetition period T, the average interval T_w is used here to point out that the true jitter is the deviation from the average rate of a clock, rather than from the nominal rate. Moreover, note that B_n and $h[n]$ are sequences of time variables and do not correspond to sample values of continuous-time functions.

A *'phase' function* $\alpha(\Gamma)$ can be defined by the linear interpolating rule[3]

$$\alpha(\Gamma) = [B_n \delta(\Gamma - n)] * l(\Gamma) \tag{3.5}$$

where the interpolating function $l(x)$ is defined as

$$l(x) = \begin{cases} 1 + x & -1 < x < 0 \\ 1 - x & 0 < x < 1 \\ 0 & x \leq -1, \quad x \geq 1 \end{cases} \tag{3.6}$$

$\Gamma = t/T_w$ is the buffer address variable ($\Gamma \in R^+$) and $n = \lfloor \Gamma \rfloor$[4] is the buffer write address ($n \in N$). Therefore, $B_n = \alpha(n)$. The role of the variables and functions t, n, Γ, $\alpha(\Gamma)$, $h[n]$, B_n and the relationships among them are clarified by Figure 3.5, which depicts the buffer write address corresponding to the incoming jittered clock.

A smoothed clock can be obtained by feeding the sequence B_n into a digital low-pass filter. Alternatively, working with analogue signals, a smoothed clock can be achieved by low-pass filtering the function to obtain $\tilde{\alpha}(\Gamma)$. The nth occurrence time of the smoothed clock is thus $\tilde{B}_n = \tilde{\alpha}(n)$.

A direct implementation of such digital or analogue low-pass filtering, in order to smooth the jittered clock B_n, is not feasible, as the function $\alpha(\Gamma)$ and the sequence

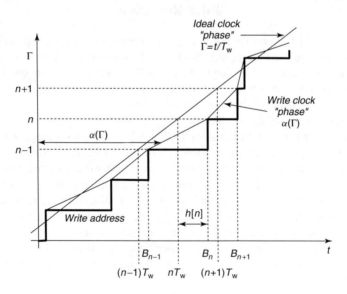

Figure 3.5. Buffer write address and write clock 'phase' $\alpha(\Gamma)$. (Adapted from [3.15] by permission of Kluwer Academic Publishers)

---

[3] The operator * denotes the convolution of two signals: $x(t) * y(t) = \int_{-\infty}^{\infty} x(\tau) y(t - \tau) d\tau$.

[4] The operator $\lfloor x \rfloor$ denotes *'the largest integer less than or equal to x'*.

B_n are unbounded; furthermore, B_n is not a sequence defined at uniform time intervals. Nevertheless, the jitter filtering circuits can be designed by evaluating the difference between the occurrences of the incoming clock and those of the outgoing clock through a feedback approach. The difference can then be low-pass filtered to drive either a plain VCO or a digitally controlled oscillator (DCO). The former approach yields an Analogue PLL (APLL), the latter a Digital PLL (DPLL).

3.3.3.1 *Implementation Using a VCO (Analogue PLL)*

The VCO approach (APLL) is shown in Figure 3.6. A flip-flop (FF), set by the incoming clock B_n and toggled by the filtered clock $\tilde{B}_n$ output by the VCO, is used to obtain a continuous-time signal $g(t)$. Such a flip-flop is conventionally identified as a saw-tooth phase detector [3.16]. The low-pass filtered signal $g_F(t)$ is then used to drive the VCO, to obtain the filtered clock $\tilde{B}_n$.

Let $h[n]$ and $\tilde{h}[n]$ be the jitter associated with B_n and $\tilde{B}_n$ respectively, i.e.

$$B_n = nT_w + h[n]$$

$$\tilde{B}_n = nT_w + \tilde{h}[n] + \Delta \tag{3.7}$$

where Δ is a constant delay whose value is about $T_w/2$. The output signal of the flip-flop $g(t)$ can be expressed as the difference of two pulse duration modulation signals modulated by B_n and $\tilde{B}_n$ respectively, i.e. $g(t) = g_1(t) - g_2(t)$, where

$$g_1(t) = V_0 \sum_{n=-\infty}^{+\infty} [U(t - nT_w) - U(t - nT_w - \tilde{B}_n)]$$

$$g_2(t) = V_0 \sum_{n=-\infty}^{+\infty} [U(t - nT_w) - U(t - nT_w - B_n)] \tag{3.8}$$

and $U(t)$ is the Heaviside step function. A sample plot of $g(t)$ is depicted in Figure 3.7.

Denoting with $X_h(e^{j\omega T_w})$ and $X_{\tilde{h}}(e^{j\omega T_w})$ the discrete Fourier transforms of the sequences $h[n]$ and $h[n]$ respectively, the analysis carried out by Abeysekera and Cantoni in [3.15] leads to the following result

$$X_{\tilde{h}}\left(e^{j\omega T_w}\right) = \frac{V_0 K_0 \left[\sum_{m=-\infty}^{+\infty} \frac{H(\omega + 2\pi m/T_w)}{(\omega + 2\pi m/T_w)} \right]}{1 + V_0 K_0 \left[\sum_{m=-\infty}^{+\infty} \frac{H(\omega + 2\pi m/T_w)}{(\omega + 2\pi m/T_w)} \right]} X_h\left(e^{j\omega T_w}\right) \tag{3.9}$$

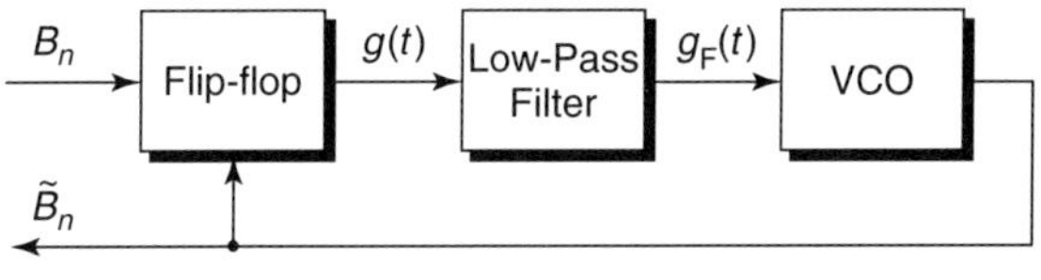

Figure 3.6. Jitter filtering through an analogue PLL. (Adapted from [3.15] by permission of Kluwer Academic Publishers)

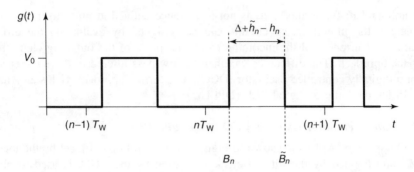

Figure 3.7. Sample plot of the pulse duration modulation signal $g(f)$. (Adapted from [3.15] by permission of Kluwer Academic Publisher)

where K_0 is the VCO gain constant, V_0 is the $g(t)$ peak amplitude and $H(\omega)$ is the low-pass filter transfer function. By inspection of Equation (3.9), it is evident that the sequence $\tilde{h}[n]$ is a low-pass filtered version of the sequence $h[n]$.

3.3.3.2   Implementation Using a DCO (Digital PLL)

The DCO approach (DPLL) is shown in Figure 3.8. Again, a flip-flop (FF), set by the incoming clock B_n and toggled by the filtered clock $\tilde{B}_n$, is used to obtain the continuous time signal $g(t)$. The low-pass filtered signal $g_F(t)$ is then used to drive the DCO, to obtain clock periods τ_p. The occurrence of the nth smoothed clock pulse $\tilde{B}_n$ is then given by the accumulation of τ_p.

Using a similar approach as done for the APLL, it is possible to obtain the following relationship between the original and the DPLL-filtered jitter sequences:

$$X_{\tilde{h}}\left(e^{j\omega T_W}\right) = \frac{T_W V_0 K_0 \overline{K}\left[\displaystyle\sum_{m=-\infty}^{+\infty} H(\omega + 2\pi m/T_W)\right]}{\sin(\omega T_W) + T_W V_0 K_0 \overline{K}\left[\displaystyle\sum_{m=-\infty}^{+\infty} H(\omega + 2\pi m/T_W)\right]} X_h\left(e^{j\omega T_W}\right) \qquad (3.10)$$

where, as before, K_0 is the VCO gain constant, V_0 is the $g(t)$ peak amplitude and $H(\omega)$ is the low-pass filter transfer function. The constant $\overline{K}$ $(0 \le \overline{K} \le 1)$ depends on the DCO model.

It is worthwhile noticing that the circuit of Figure 3.8 is not a fully digital implementation, owing to the analogue filtering of $g(t)$. A fully digital implementation can be done by replacing the flip-flop by a counter, set by $h[n]$ and reset by $\tilde{h}[n]$, to measure the jitter

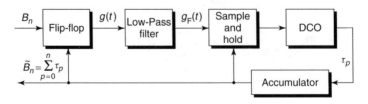

Figure 3.8. Jitter filtering through a digital PLL. (Adapted from [3.15] by permission of Kluwer Academic Publishers)

difference $h[n] - \tilde{h}[n]$. The counter is then followed by a digital low-pass filter, clocked by the smoothed clock $\tilde{B}_n$. The analysis carried out for this full-digital circuit yields, instead of the relationship of Equation (3.10),

$$X_{\tilde{h}}\left(e^{j\omega T_W}\right) = \frac{T_W V_0 K_0 \overline{K} \cdot \overline{H}_D\left(e^{j\omega T_W}\right)}{\sin(\omega T_W) + T_W V_0 K_0 \overline{K} \cdot \overline{H}_D\left(e^{j\omega T_W}\right)} X_h\left(e^{j\omega T_W}\right) \qquad (3.11)$$

where $\overline{H}_D\left(e^{j\omega T_W}\right)$ denotes the transfer function of the digital low-pass filter. Again, by inspection, it is evident that both Equations (3.10) and (3.11) express low-pass filtering.

3.3.4   Enhanced Design of Synchronizers: Stuff Threshold Modulation

As widely discussed in the previous sections, bit justification allows to multiplex together several lower-speed asynchronous digital signals into a unique higher-speed stream. This allows, moreover, providing timing transparency for tributaries multiplexed at different levels. Unfortunately, no matter how bit justification is implemented, it has the potential to give rise to waiting time jitter in the recovered payload clock, in addition to the jitter caused by the mapping and demapping process (the payload clock signal is always gapped according to the multiplex frame structure).

This waiting time jitter has been the subject of several papers and books which analysed it in depth [3.17]–[3.19]. It may be annoying owing to its unlimited low frequency which makes it troublesome to filter it out (see Section 2.3.5).

In designing SDH synchronizers and desynchronizers, one of the main concerns is to effectively reduce the waiting time jitter due to bit justification and the jitter due to the pointer justification process. The latter problem will be specifically addressed in Sections 3.3.5 and 3.4, since an enhanced design of the desynchronizer and of the pointer processor can effectively reduce the jitter produced by pointer action. The focus of this section, on the other hand, is on an enhanced design of synchronizers, coping in particular with the waiting time jitter due to bit justification.

To overcome this problem, methods based on *stuff threshold modulation* have gained widespread attention since the late 1980s. Their objective is to force the waiting time jitter to fall outside the bandwidth of the PLL filter in the desynchronizer on the receiver side, in order to reduce the amount of residual jitter at its output.

These methods were initially proposed by Grover *et al.* [3.20]–[3.22], who suggested a modified version of the bit justification technique with two buffer thresholds modulated by a sawtooth waveform. A mathematical analysis of this method was provided by Kusyk *et al.* [3.23] in the frequency domain and by Pierobon and Valussi [3.24] in the time domain. Further analysis was carried out by Abeysekera and Cantoni [3.25]–[3.27], who discussed also the issue of selecting the most suitable dithering sequence for modulating the buffer thresholds.

In this section, we provide some details on the methods based on stuff threshold modulation, following the analysis of Abeysekera, Cantoni and Sreeram in [3.25].

3.3.4.1   Simplified Model of Synchronizer with Positive/Negative Justification

For the purpose of this section, we simplify the scheme of the synchronizer previously shown in Figure 3.3, but introducing the concept of positive/negative justification to let

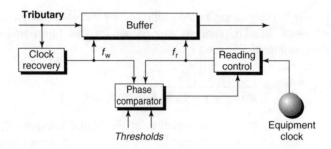

Figure 3.9. Simplified model of the synchronizer with double justification threshold

the discussion be more general. The new scheme [3.24][3.25] is depicted in Figure 3.9. The incoming data stream is written into the elastic store using the recovered clock f_w, but is read using the local equipment clock f_r, purposely inhibited for stuffing and for the inclusion of various overhead bits. The 'reading control' block merges the functions of the 'justification control', the AND and the 'timing generator and distributor' blocks of Figure 3.3.

The *positive/negative justification* is based on comparing the write and read phase difference with two buffer thresholds (such thresholds may be also time-varying, in general, as in the case of stuff threshold modulation which is going to be analysed). There are two justification opportunity bits per multiplex frame and per tributary: one devoted to positive justification, one to negative justification (cf. the positive justification treated in Section 2.3.5 where only one justification opportunity bit per frame per tributary is available).

The phase comparator examines the difference between the input and output phases at a fixed time in each frame period (the justification decision instant, in correspondence with the justification opportunity bits in the multiplex frame). If the phase difference is less than or equal to the lower justification threshold, then in the next frame the reading from the memory is inhibited in correspondence with the positive justification opportunity bit, so that this carries a dummy information instead of a tributary bit. Similarly, if the phase difference is greater than the upper threshold, the reading, usually inhibited during the negative justification opportunity unit carrying a dummy bit, is allowed.

For example, the PDH multiplex signals use positive justification (see Section 2.4). The mapping of the 139.264 Mbit/s into the C-4 of the SDH use positive justification as well (see Section 2.5.6). On the other hand, an example of positive/negative justification is the mapping of the 2.048 Mbit/s into the C-12 of the SDH (see Section 2.5.7).

3.3.4.2   *Waiting Time Jitter*

In the phase comparator, the write and read addresses are compared to take a justification decision. By sampling the synchronizer phase difference $\Phi_s(t) = \Phi_r(t) - \Phi_w(t)$, where $\Phi_w(t) = f_w t$ is the buffer write phase and $\Phi_r(t)$ is the buffer read phase defined as the linear interpolation of buffer addresses, we get the *waiting time jitter* sequence $h[n]$[5].

---

[5] Properly speaking, and according to the tradition, the phase difference $\Phi_s(t)$ (which has been called $\Delta\varphi(t)$ in Section 2.3.5) in the synchronizer buffer should be considered the sum of two jitter waveforms: the former being *pulse stuffing jitter* and the latter *waiting time jitter*. According to the definition of Duttweiler [3.17], pulse stuffing jitter is defined as the jitter that would be present if justification could occur timely on demand

The jitter sequence $h[n]$ can be expressed as

$$h[n] = \sum_{n=-\infty}^{+\infty} \left(\rho n - \sum_{m=-\infty}^{n-1} y[m] \right) \tag{3.12}$$

The justification ratio ρ, according to Equation (2.15), is defined as

$$\rho = \frac{f_w - f_{r0}}{f_j} \tag{3.13}$$

where f_{r0} denotes the nominal read clock frequency and f_j denotes the maximum justification frequency, i.e. the maximum rate at which justification can take place. In the case of standard PDH signals, for example, f_j is the frame rate (there is one justification opportunity per frame, per tributary). The sequence $y[m]$ provides the stuff decisions: a sample $y[m]$ is equal to 0 if no justification takes place at instant m, to 1 if a positive justification takes place, to -1 if a negative justification takes place.

Therefore, as far as the jitter generated is concerned, the synchronizer can be modelled by the block diagram shown in Figure 3.10. The block Δ represents a unitary delay. A three-level (positive, null, negative) justification thresholder is used to decide justifications. The function $\mu(x)$ describes the thresholder characteristic as

$$\mu(x) = \begin{cases} \text{sgn}(x) & |x| \geq 0.5 \\ 0 & |x| < 0.5 \end{cases} \tag{3.14}$$

According to Duttweiler's analysis [3.17], the total jitter power resulting from the synchronizer model of Figure 3.10 and low-pass filtered by the desynchronizer can be expressed as [3.26]

$$B_0^H(\rho) = \sum_{k=1}^{\infty} \frac{1}{2\pi^2 k^2} \sum_{m=-\infty}^{+\infty} H(m - \rho k) \tag{3.15}$$

where $H(\overline{f})$ is the power transfer function of the analogue PLL filter in the desynchronizer, related to the power transfer function of a digital PLL by the following relationship

$$H_D(f) = \sum_{r=-\infty}^{+\infty} H(\overline{f} - r) \tag{3.16}$$

where $f = \overline{f}/f_j$.

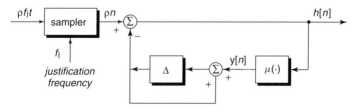

Figure 3.10.   Model of the jitter generated by the two-thresholds synchronizer. (Adapted from [3.25], © 1998 IEEE, by permission of IEEE)

(i.e., the saw-toothed high-frequency component of $\Delta\varphi(t)$ as drawn in Figure 2.21), while waiting time jitter is defined as the low-frequency jitter in $\Delta\varphi(t)$ (i.e., the envelope of $\Delta\varphi(t)$ as drawn in Figure 2.21) because in actuality there is a 'waiting time' between stuff demand and stuff execution. For the sake of simplicity, here we follow Duttweiler [3.17] and Abeysekera *et al.* [3.25] to neglect to make such distinctions and we shall simply call $\Phi_s(t)$ and its sampled version $h[n]$ the *waiting time jitter* waveform.

3.3.4.3 Stuff Threshold Modulation

The principle of the *stuff threshold modulation* technique for waiting time jitter reduction is based on modulating the buffer thresholds with a suitable dithering waveform, in order to shift the waiting time jitter out of the bandwidth of the PLL filter in the desynchronizer on the receiver side.

From the mathematical modelling point of view, modulating the thresholds is equivalent to keeping them fixed and adding the dithering waveform to the buffer phase difference. It is worthwhile noticing that a similar technique can be used in analogue-to-digital converters to reduce the quantization noise [3.28]: by adding a dithering noise of known characteristics to the analogue signal to convert, it is possible to enhance the resolution of the quantizer forcing a dithering of the least significant digits.

Therefore, the jitter generated by a synchronizer with stuff threshold modulation can be modelled by the block diagram of Figure 3.11 [3.20]. A dither sequence $\varphi[n]$ is added to the jitter sequence $h[n]$ modelled by Equation (3.12) (i.e., the write–read phase difference at the synchronizer buffer). Obviously, this is equivalent to shifting the thresholds by the same dither sequence.

From the model of Figure 3.11, the new jitter sequence with the added dithering noise results

$$\xi[n] = h[n] + \varphi[n] \tag{3.17}$$

By moving the dithering sequence $\varphi[n]$ to the input, we get a scheme analogous to that of Figure 3.10 but with input $\rho n + \varphi[n]$. It is then possible to apply Duttweiler's analysis, analogously to what was done for getting $B_0^H(\rho)$ of Equation (3.15), to calculate the low-pass filtered jitter power $B_\varphi^H(\rho)$ resulting from the synchronizer with stuff thresholds modulated by a dithering sequence $\varphi[n]$.

Actually, Duttweiler assumed the threshold function $\mu(x)$ to be a multi-level quantizer of the form

$$\mu(x) = \lfloor x + 0.5 \rfloor \tag{3.18}$$

generating justification decisions which result in an integer level of stuffing. To apply Duttweiler's analysis in a system using a three-level thresholder, it must be ensured that multiple justification decisions do not occur. This is true if

$$|\rho| \le 1 - \max |\varphi[n] - \varphi[n-1]| \quad \forall n \in Z \tag{3.19}$$

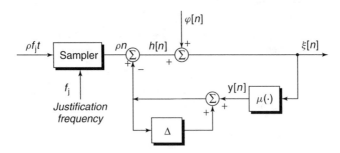

Figure 3.11. Model of the jitter generated by a synchronizer with stuff threshold modulation. (Adapted from [3.25], © 1998 IEEE, by permission of IEEE)

Under the assumption (3.19), Abeysekera *et al.* derived the following result (see [3.25] for the mathematical details):

$$B_\varphi^H(\rho) = \sum_{k=1}^{+\infty} \left(\frac{1}{2\pi^2 k^2}\right) \int_{-0.5}^{+0.5} H_D(k\rho - f) Z_k(f) \, df + \int_{-0.5}^{+0.5} S_\varphi(f) H_D(f) \, df \quad (3.20)$$

where $H_D(f)$ is the power transfer function (3.16) of the digital PLL in the desynchronizer, $S_\varphi(f)$ is the power spectral density of the dither sequence $\varphi[n]$ and $Z_k(f)$ is the discrete-time Fourier transform of the kth order autocorrelation sequence $z_k[n]$ of $e^{-j2\pi\varphi[n]}$, i.e.

$$Z_k(f) \leftrightarrow z_k[n] \equiv \lim_{M\to\infty} \frac{1}{M} \sum_{m=0}^{M-1} e^{-j2\pi k(\varphi[n+m]-\varphi[m])} \quad (3.21)$$

3.3.4.4   Periodic Dither Sequences

Most dither sequences studied in the literature are periodic. A periodic dither sequence with period N is such that

$$\varphi[n] = \frac{p\left(n - \left\lfloor \frac{n}{N} \right\rfloor N\right)}{N} \quad \forall n \quad (3.22)$$

where $p[k]$ (for $k = 1, 2, \ldots, N - 1$) takes all the values from the integer set $\{0, 1, \ldots, N - 1\}$.

Being the dither sequence periodic with N samples per period, the Fourier transform of the sequence $z_k[n]$ becomes a discrete Fourier transform

$$Z_k(f) = \sum_{m=-\infty}^{+\infty} A_{k,m}\delta\left(f - \frac{m}{N}\right) \quad (3.23)$$

with coefficients

$$A_{k,m} = \left| \frac{1}{N} \sum_{n=0}^{N-1} e^{-j2\pi k\varphi[n]} e^{-j2\pi m \frac{n}{N}} \right|^2 \quad (3.24)$$

Therefore, using the above coefficients (3.24), for periodic dither sequences with N samples per period, Equation (3.20) results

$$B_\varphi^H(\rho) = \sum_{k=1}^{\infty} \left(\frac{1}{2\pi^2 k^2}\right) \sum_{m=-\infty}^{+\infty} A_{k,m} H\left(k\rho - \frac{m}{N}\right) + \int_{-0.5}^{+0.5} S_\varphi(f) H_D(f) \, df \quad (3.25)$$

The double summation in the first term of the right hand side of Equation (3.25) can be made more understandable when depicted as shown in Figure 3.12. Each line represents one term of the summation over k, while along each kth line the replicas of the spectrum $H(f)$, translated by m/N and shrunk by the factor k, are the terms of the summation over m. This graph has been drawn under the assumption that the bandwidth B_{PLL} (normalized to the maximum justification frequency f_j) of the desynchronizer PLL is such that $B_{PLL} < 1/N$, so that the contributions from each coefficient $A_{k,m}$ do not overlap.

The expression for $B_\varphi^H(\rho)$ can be simplified in some remarkable cases.

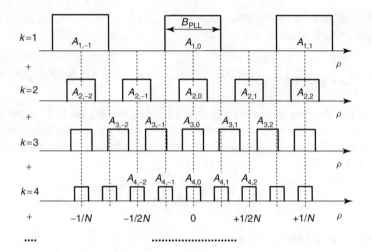

Figure 3.12. Graphic description of the double summation in the first term of $B_\varphi^H(\rho)$ (Equation (3.25)). (Adapted from [3.25], © 1998 IEEE, by permission of IEEE)

For *small justification ratios and small PLL bandwidths* (i.e., $\rho \cong 0$ and $B_{\mathrm{PLL}} < 1/N$), the contributions from each coefficient $A_{k,m}$ do not overlap and thus only the summation term with $m = 0$ contributes to the first term of Equation (3.25), to yield

$$B_\varphi^H(\rho \cong 0) = \sum_{k=1}^{\infty} \left(\frac{1}{2\pi^2 k^2}\right) A_{k,0} + \int_{-0.5}^{+0.5} S_\varphi(f) H_D(f)\, df \qquad (3.26)$$

If the periodic dither sequence is a *saw-tooth sequence*, Equation (3.22) becomes (being $p[k] = k$)

$$\varphi[n] = \rho_0 n - \lfloor \rho_0 n \rfloor \quad \forall n \qquad (3.27)$$

where $\rho_0 = 1/N$ is the slope of the saw-tooth waveform. Therefore

$$Z_k(f) = \sum_{m=-\infty}^{+\infty} \delta(f + k\rho_0 - m) \leftrightarrow z_k[n] = e^{-j2\pi k(\rho_0 n - \lfloor \rho_0 n \rfloor)} \qquad (3.28)$$

(where $z_k[n]$ is periodic of period N as the underlying dither sequence) and Equation (3.20) can be simplified as

$$B_{\mathrm{ST}}^H(\rho) = \sum_{k=1}^{\infty} \left(\frac{1}{2\pi^2 k^2}\right) \sum_{m=-\infty}^{+\infty} H(k\rho + k\rho_0 - m) + \int_{-0.5}^{+0.5} S_\varphi(f) H_D(f)\, df$$

$$= B_0^H(\rho + \rho_0) + \int_{-0.5}^{+0.5} S_\varphi(f) H_D(f)\, df \qquad (3.29)$$

This last expression is equivalent to the expression derived by Pierobon and Valussi by time-domain analysis [3.24]. As noted there, if the integral term is neglected, the jitter power observed for a given justification ratio ρ with saw-toothed justification threshold modulation is the same as that observed without justification threshold modulation for a justification ratio translated by the slope of the saw-toothed waveform.

3.3.4.5   Random Dither Sequences

Random dither sequences for justification threshold modulation were first proposed by Abeysekera in [3.26]. Let $\varphi[n]$ be a random sequence of independent samples (i.e., a sequence with white spectrum), which are identically distributed with an arbitrary probability density function $P_\varphi(\Phi)$. Therefore,

$$z_k[n] = E\left\{e^{-j2\pi k(\varphi[n+m]-\varphi[m])}\right\} \tag{3.30}$$

where $E\{\cdot\}$ denotes the *'expected value'* operator. Being the sequence of independent samples, we have

$$\begin{cases} z_k[n] = 1 & n = 0 \\ z_k[n] = E\left\{e^{-j2\pi k\varphi[n+m]}\right\} \cdot E\left\{e^{+j2\pi k\varphi[m]}\right\} & n \neq 0 \end{cases} \tag{3.31}$$

Also with random dither sequences, the expression for $B_\varphi^H(\rho)$ can be simplified in some remarkable cases. If $\varphi[n]$ is *uniformly distributed over the interval (0,1]*, then

$$\begin{cases} z_k[n] = 1 & n = 0 \\ z_k[n] = 0 & n \neq 0 \end{cases} \tag{3.32}$$

and

$$S_\varphi(f) = \frac{1}{12}, \quad Z_k(f) = 1 \tag{3.33}$$

Equation (3.20) results finally

$$B_{\mathrm{RD}}^H(\rho) \cong \frac{B_{\mathrm{PLL}}}{6} \tag{3.34}$$

3.3.4.6   Guidelines for the Selection of the Optimal Dither Sequence

The designer is interested in selecting a suitable dither sequence $\varphi[n]$ for modulating the thresholds of the synchronizer, aiming at minimizing the residual jitter output by the desynchronizer. This means aiming at minimizing the functional $B_\varphi^H(\rho)$.

To summarize, the filtered jitter power $B_\varphi^H(\rho)$ depends on the following parameters:

- $H(f)$, i.e. the desynchronizer PLL power transfer function;
- $S_\varphi(f)$, i.e. the power spectral density of the dither sequence $\varphi[n]$;
- $A_{k,m}$, i.e. the coefficients of the discrete Fourier transform of $z_k[n]$, the kth order autocorrelation sequence of $e^{-j2\pi\varphi[n]}$.

Therefore, comparing the performance of different dither sequences, given the desynchronizer PLL power transfer function, requires the knowledge of the justification ratio ρ at which to compare the values of $B_\varphi^H(\rho)$ for different dither sequences $\varphi[n]$. Nevertheless, in many applications ρ is known only within some operating range. Therefore, it may be ambiguous to decide what is the 'best' dither sequence, because a justification modulation scheme may perform better at a certain justification ratio, but worse at another.

One possibility could be to evaluate some mean of $B_\varphi^H(\rho)$ over the range of ρ of interest (e.g., the root mean square). On the other hand, a more meaningful figure of merit may be to take the maximum value of the jitter power over the whole range of ρ of interest. This means to compare the worst cases, i.e. the jitter power at the values of

ρ where the justification modulation schemes work worst. The figure of merit suggested by Abeysekera *et al.* in [3.25] is thus the maximum filtered jitter power

$$J_{\text{filt}} = \max_{|\rho| \le \rho_{\max}} \left[B_{\varphi}^{H}(\rho) \right] \tag{3.35}$$

Based on this figure of merit, in [3.25] different justification modulation schemes are evaluated. The main conclusions are summarized as follows.

- *If the design of the PLL filter is unrestricted*, the best dither sequence for justification threshold modulation is a *random sequence*. The filtered jitter power is then given by Equation (3.34) as $B_{\text{PLL}}/6$ (i.e., the slower PLL, the less jitter power is returned).

- *If the design of the PLL filter is constrained* instead, for example by implementation issues or by time delay requirements, then the choice of the optimal dither sequence is among the *random sequence* and the so-called *modified-wave sequence*, based on pseudo-random codes and proposed in [3.23] as a superior periodic dither sequence for justification threshold modulation. The choice depends on the values of $\rho_{\max}$ and B_{PLL}: while for the random sequence the jitter power is given by $B_{\text{PLL}}/6$, for the modified-wave sequence the following approximated limit for the maximum filtered jitter power has been estimated [3.25]

$$J_{\text{filt}} \cong \frac{\rho_{\max}}{12} + \frac{1}{12N^2} + \frac{B_{\text{PLL}}^2}{20} \tag{3.36}$$

It is worthwhile noticing that the minimum jitter power attainable through a periodic modified-wave sequence is given by $\rho_{\max}/12$.

3.3.5   Enhanced Design of Desynchronizers: Reducing Pointer Adjustment Jitter by Bit Leaking

As stated previously, an enhanced design of the desynchronizer can effectively reduce the jitter produced by pointer action. In this section, we summarize the review provided by Sari and Karam [3.29] on bit-leaking techniques often adopted to cancel pointer-adjustment jitter.

Cancelling pointer-adjustment jitter may be extremely difficult by using a single PLL, due to the very low frequency of pointer-adjustment jitter. This would require a PLL with cut-off frequency so narrow, to make unfeasible a practical implementation with analogue circuitry. Moreover, a digital PLL would require an internal oscillator whose frequency is by several orders of magnitude larger than the STM-1 clock.

It may be worthwhile recalling, at this point, that pointer-adjustment jitter in SDH systems takes the form of phase steps of substantial amplitude. For example, AU-4 pointer adjustments yield phase steps with a good 24-bit[6] magnitude at the VC-4 rate, equivalent to about 160 ns in absolute terms. TU pointer adjustments, on the other hand, yield even larger steps: 8 bits at the VC-12/11/2 rate, equivalent for example to a time step of about 3.57 µs in the case of the VC-12. For the sake of comparison, bit justification

---

[6] Often, in technical papers, phase steps and differences are measured in [bit] units, obviously equivalent to [UI] units.

in PDH multiplexers yields jitter with only 1-bit magnitude, at the tributary rate. In the following, for the sake of simplicity, we will focus in particular on the case of AU-4 pointer justification.

3.3.5.1 The Bit-Leaking Technique

To cancel effectively pointer-adjustment jitter, the desynchronizer is often designed as cascade of two elements. The first has the task of spreading each of the incoming 24-bit phase steps into 24 1-bit phase steps in some way. This task is often referred to as *bit leaking* or even *phase spreading*. The second element may be a conventional second-order PLL that filters the 1-bit phase steps at the output of the first stage. The bit-leaking circuit can be fixed or adaptive.

Fixed bit leaking assumes that the time interval separating two consecutive leaked bits (one-bit phase steps) is fixed once for all and is maintained in both the normal operation mode (i.e., when highly stable synchronization reference signals are available to all the SDH network elements) and in the degraded mode (i.e., when some synchronization reference signal fails and equipment clocks enter the free-run mode).

This bit-leaking speed should be set in accordance with the maximum frequency offset between SDH NEs that must be handled in the degraded mode. For example, if a fractional frequency offset on the order of 10^{-5} is handled, the bit-leaking speed must be set to at least 1 bit every 5 STM-N frames[7]. In normal mode, such a speed would be excessive and the jitter specifications would hardly be met. To meet jitter specifications in normal mode, the bit-leaking speed must be set smaller, but a smaller speed would not be sufficient in degraded mode and thus be leading to buffer overflow. To summarize, fixed bit leaking does not allow one to meet performance objectives in both normal and degraded modes.

In *adaptive bit leaking*, the bit-leaking speed is somehow adjusted following the pointer adjustment rate. The ideal goal would be to have the 1-bit phase steps evenly spread in time, thus maximizing the jitter filtering action of the second-stage PLL.

3.3.5.2 A Basic Adaptive Bit-Leaking Circuit

Most proposed adaptive bit leaking circuits are based on digital PLL operation. As an example, in Figure 3.13 we depict the scheme of the algorithm reported in [3.29], whose

---

[7] A fractional frequency offset $\Delta f = 10^{-5}$ between the clocks of two SDH NEs corresponds to a absolute difference between the write and read frequencies in the AU-4/TU-12 pointer processor equal to (cf. Exercises 2.5, 2.6 and 3.1)

$$10^{-5} \cdot 155.520 \text{ Mbit/s} \cdot \frac{261}{270} = 1503.36 \text{ bit/s} \qquad \text{(AU-4 pointer processor)}$$

or

$$10^{-5} \cdot 155.520 \text{ Mbit/s} \cdot \frac{35}{270 \cdot 9} = 22.4 \text{ bit/s} \qquad \text{(TU-12 pointer processor).}$$

These frequency differences are compensated by pointer justification. That is, on the average,

$$1 \text{ bit every } \frac{8000}{1503.36} = 5.32 \text{ frames} \qquad \text{(AU-4 pointer processor)}$$

or

$$1 \text{ bit every } \frac{8000}{22.4} = 357 \text{ frames} \qquad \text{(TU-12 pointer processor)}$$

is justified.

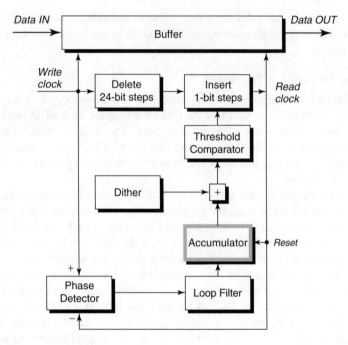

Figure 3.13. Block diagram of a PLL-type bit-leaking circuit. (Adapted from [3.29], © 1994 IEEE, by permission of IEEE)

first version (without the accumulator drawn in grey thick line) was originally proposed by British Telecom in contributions to the ETSI and ANSI standard bodies [3.30]. Such former version will be now described first.

The phase difference between the read and write clocks of the elastic buffer is filtered by a low-pass single-pole digital filter (loop filter). The resulting signal is then passed to a control circuit after having added a dithering noise. The 'insert 1-bit steps' block leaks a bit every time the signal at its input exceeds a fixed threshold. Thus, the read clock is obtained by suppressing the 24-bit steps in the write clock and inserting the 1-bit steps provided by the control circuit.

In such a digital PLL system, the output phase is governed by the difference equations

$$\varphi_{n+1} = \varphi_n - u_n \qquad (3.37)$$

and

$$u_n = \sum_{k=0}^{\infty} f_k \varepsilon_{n-k} \qquad (3.38)$$

where φ_n is the output phase at the time instant t_n, ε_n is the phase detector output and the $\{f_k\}$ sequence is the impulse response samples of the digital loop filter. The sum on the right-hand side of Equation (3.38) is the convolution of the $\{\varepsilon_k\}$ sequence with the loop filter impulse response sequence. In the first version of the circuit proposed by [3.30], the loop filter is a simple first-order filter, with a single pole and no zeroes. Nevertheless, as it will be shown in the following, a more appropriate choice of the loop filter allows controlling better the loop transient and steady-state properties.

In an analogous fashion to the stuff threshold modulation technique described in Section 3.3.4, the additive dithering noise (with uniform probability density) in the loop modulates the low-frequency components on the phase jitter and up-converts them to higher frequencies, which may be more easily filtered out by the PLL at the second stage of the desynchronizer.

In this first version of the digital circuit proposed by [3.30], the output phase is updated (i.e., a 1-bit step is inserted) when the instantaneous control signal u_n exceeds a threshold (say 0.5 UI) in absolute value. This may be expressed mathematically by replacing u_n by its three-level quantized version

$$q\,(u_n) = \begin{cases} \text{sgn}\,(u_n) & \text{if } |u_n| > 0.5 \text{ UI} \\ 0 & \text{otherwise} \end{cases} \tag{3.39}$$

where $sgn(\cdot)$ denotes the sign function. This quantization is coarse and the properties of the resulting loop differ substantially from those of the original loop with unquantized u_n, like in an analogue PLL.

3.3.5.3 *Improving the Basic Bit-Leaking Circuit*

The bit-leaking technique as described above proves to have several weaknesses and limitations. Mainly, being the loop filter a simple first-order filter with a single pole and no zeroes, such a PLL system does not allow to set independently the basic loop parameters: gain, natural frequency and damping factor. Therefore, both small jitter in normal operation mode and small static phase error in degraded mode cannot be achieved.

A first change, aiming at improving the basic bit-leaking circuit originally proposed by [3.30], consists of replacing the first-order low-pass digital loop filter with a more appropriate one. A choice suggested by [3.29] is given by

$$u_n = \alpha\varepsilon_n + \beta\sum_{k=0}^{n}\varepsilon_k \tag{3.40}$$

where α and β are two constants that determine the loop transient and steady-state properties. Note that this digital loop filter is the equivalent of an active analogue filter of the form

$$F(s) = -\frac{s\tau_2 + 1}{s\tau_1} \tag{3.41}$$

shown to be the best filter for second-order PLLs in [3.8].

Next, to better approximate the operation of conventional PLLs with finer phase quantization than that of Equation (3.39), the control signal u_n may be accumulated before driving the threshold comparator (a three-level quantizer), which determines whether φ_n has to be updated or not. Therefore, a further improved version of the basic circuit includes the accumulator depicted in the thick grey line in Figure 3.13. A bit-leaking circuit based on this concept is governed by the set of equations

$$\begin{cases} \varphi_{n+i} = \varphi_n & \text{for } i < K \\ \varphi_{n+K} = \varphi_n - \text{sgn}\left(\displaystyle\sum_{i=0}^{K-1} u_{n+i}\right) \end{cases} \tag{3.42}$$

where K is the smallest integer for which

$$\left|\sum_{i=0}^{K-1} u_{n+i}\right| > 0.5 \text{ UI} \tag{3.43}$$

A phase update is thus made every time the absolute value of the accumulated control signal exceeds 0.5 UI. Moreover, phase updating is accompanied by an accumulator reset, which consists of subtracting 1 UI from its output if the output is positive and adding 1 UI if it is negative.

Although the quantization function is still given by Equation (3.39), the fact that this operator is applied to the accumulated control signal makes the loop much less affected by quantization.

In normal mode of operation, this improved bit-leaking circuit modified as above allows to obtain arbitrarily small phase jitter in the demapped tributary. The graph of Figure 3.14, again reproduced from [3.29], shows the output jitter [bit] on the E4 demapped tributary versus time (in STM-N frame period units, T), evaluated in response to a single 24-bit phase step, when:

- the α and β parameters of the loop filter in the first stage of the desynchronizer are set so as to have noise equivalent bandwidth $B_L = 2$ Hz and damping factor $\zeta = 0.7$;

- the second stage of the desynchronizer is a PLL acting as a two-poles low-pass phase filter whose poles are at 100 Hz and 1000 Hz;

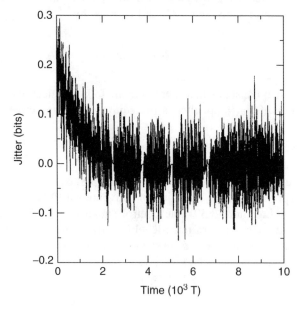

Figure 3.14. Output jitter of the improved PLL-type desynchronizer evaluated in response to a single 24-bit phase step. (Reproduced from [3.29], © 1994 IEEE, by permission of IEEE)

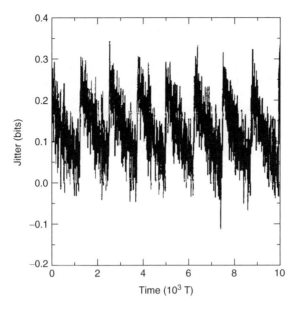

Figure 3.15. Output jitter of the improved PLL-type desynchronizer evaluated in response to a sequence of pointer adjustments (1 ppm frequency offset between the clocks of the SDH NEs). (Reproduced from [3.29], © 1994 IEEE, by permission of IEEE)

- the output jitter is observed in the bandwidth of a single-pole high-pass filter with pass frequency f_0 =200 Hz, according to ITU-T Rec. G.823 [3.31] for the PDH E4 signal[8].

The resulting peak-to-peak amplitude of this high-pass filtered output jitter is on the order of 0.3 UI.

Also in degraded mode of operation, this improved bit-leaking circuit proves effective in cancelling the phase jitter in the demapped tributary. The graph of Figure 3.15, again reproduced from [3.29], shows the output jitter [bit] on the E4 demapped tributary versus time (in STM-N frame period units, T), evaluated in response to a sequence of pointer adjustments corresponding to 1-ppm frequency offset between the clocks of the SDH NEs. The resulting peak-to-peak jitter is on the order of 0.4 UI. The main period of the jitter waveform is the time interval between two consecutive pointer adjustments.

3.4 SDH POINTER PROCESSOR

The pointer justification mechanism was detailed in Section 2.5.8. This justification technique was designed to cope with the issue of multiplexing asynchronous tributaries but retaining a synchronous digital multiplexing structure.

---

[8] The authors of [3.29] remark that observing the output jitter through some high-pass filter is justified by the fact that low-frequency jitter is tracked by clock recovery circuits of downstream regenerators and thus should not be a problem, at least under the point of view of bit decision at input interfaces. Obviously, this assumption depends on the context. It is reasonable for an E4 signal, but it would be questionable for an E7 signal, which may be used to vary synchronization.

Re-synchronization of VCs in an NE, where the pointer is processed, is accomplished by means of an elastic store, in which the VC bytes are written according to the timing signal extracted by the incoming STM-N signal and are read according to the local equipment clock. Pointer justification allows to compensate for the unavoidable differences between the input and output frequencies. This block, based on an elastic store, is called *pointer processor*.

The aim of this section is to describe the principle of operation of basic and enhanced pointer processors, the latter ones designed to limit jitter generation due to pointer action even under lack of network synchronization, by outlining their functional blocks.

3.4.1   Basic Pointer Processor

The principle of operation of a basic pointer processor is outlined in Figure 3.16. The incoming VC data are written into the elastic buffer according to their gapped clock signal, which is extracted by the incoming STM-N stream by inhibiting the pulses corresponding to all the bits not belonging to that VC. As remarked in Section 3.2, the average clock of the incoming VC is that of the first NE in the transmission chain, that one generating the VC.

The VC data are read from the elastic buffer according to a gapped clock signal generated from the local equipment clock, by inhibiting the pulses corresponding to all bits not belonging to that VC in the outgoing STM-N frames.

Such two gapped clock signals differ mainly due to three different reasons. First, the phases of the incoming and the outgoing STM-N frames are independent and in general not aligned. Second, the regular clock signals from which the gapped ones are derived (i.e., the clock timing the incoming STM-N stream and the clock timing the outgoing STM-N stream) have in general independent phases and frequencies and may be both affected by jitter and wander. Third, the pointer adjustments in the incoming STM-N stream yield additional phase hits in the incoming VC gapped clock signal, while making its average frequency equal to that of the clock of the NE originating the VC.

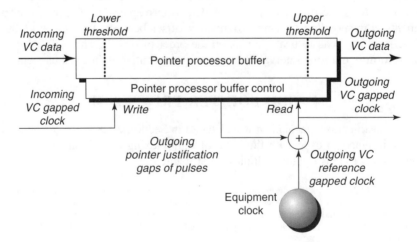

Figure 3.16.   Scheme of principle of a basic pointer processor

Therefore, the buffer fill level varies according to the phase difference between the two incoming and outgoing VC gapped clock signals. If the incoming and outgoing VC gapped clock signals differed only for the first of the three reasons above, i.e. for the not-aligned STM-N frame phases, then the buffer fill level would always follow a deterministic trend and would never overflow. More in general, even a large buffer size cannot avoid buffer overflows. Pointer justification, instead, can control the buffer fill level, by adjusting the outgoing VC phase by means of timely pointer increments or decrements.

The buffer has two thresholds, limiting the fill level range not yielding pointer action. When the lower threshold is trespassed (i.e., when the incoming VC data arrive at a lower average rate than that of the buffer read clock), a positive pointer justification takes place: VC data reading is inhibited in correspondence to one (for TU and AU-3 pointers) or three (for the AU-4 pointer) bytes immediately following the pointer in the outgoing STM-N frame (cf. Section 2.5.8). When the upper threshold is trespassed (i.e., when the incoming VC data arrive at a higher average rate than that of the buffer read clock), a negative pointer justification takes place: VC data reading is activated in correspondence to one (TU and AU-3) or three (AU-4) H3 bytes at the end of the pointer word in the outgoing STM-N frame (cf. Section 2.5.8). In such a way, clock pulses are gapped from or added to the outgoing reference gapped clock signal to produce the actual outgoing VC gapped clock signal.

In order to avoid that small phase variations between the two clocks generate unnecessary pointer action, a dead-interval is specified between the two thresholds. ITU-T Rec. G.783 [3.5] recommends that such threshold spacing should be at least 12 bytes for AU-4, at least 4 bytes for AU-3, at least 4 bytes for TU-3 and at least 2 bytes for TU-1 and TU-2.

As a final remark, it may be worthwhile to notice that this pointer processor scheme may be regarded as a non-linear phase-locked loop. Similar forms of control loops are used in other control applications.

The number and the statistics of pointer justification events depend on the buffer threshold spacing and on the statistics of the phase error between the buffer read and write clocks. Actually, the pointer processor has the effect of *quantizing* the phase error between the incoming VC gapped clock and the outgoing VC gapped clock and of *encoding* it in the pointer value. Therefore, the VC timed by the first clock of the chain can be transported by an STM-N frame timed by a different clock, retaining its *average* frequency.

In the example plot shown in Figure 3.17, a random phase error between the buffer read and write clocks yields some pointer action. When the peak-to-peak amplitude of the random phase error is sufficiently small (i.e., smaller than the buffer threshold spacing), it is absorbed by the elastic store. Otherwise, when the random phase noise exhibits higher deviations, pointer justifications take place. Pointer justifications are rendered in Figure 3.17 as threshold shifts, although actually thresholds stay at fixed places while it is the buffer fill level that exhibits steps, whenever a pointer justification occurs.

The phase of the outgoing VC signal follows the trend of these threshold shifts, i.e. the input phase error quantized by the pointer mechanism. Obviously, these phase steps cannot be accepted on the output signal. While the desynchronizer at the output port is able to substantially smooth such phase hits, an enhanced design of the pointer processor itself allows to limit jitter generation due to pointer action even at this stage.

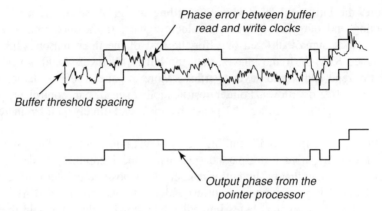

Figure 3.17. Response of a pointer processor to a random phase error between the buffer read and write clocks

3.4.2 Advanced Pointer Processors

In the case of the basic pointer processor of Figure 3.16, special care must be taken in order to limit wander generation and accumulation along a transmission chain. First, all equipment clocks have to be synchronized and the overall network synchronization performance has to be controlled. In the ideal case, this would avoid any pointer action. Nevertheless, residual synchronization impairments, such as the internal clock noise and the wander on long timing links, still pose additional requirements on the threshold hysteresis spacing and make it very difficult or costly to control wander accumulation.

Therefore, advanced pointer processor schemes have been devised which feature enhanced performance with respect to the wander generated, even under poor or no network synchronization. In this section, we summarize the ideas of Urbansky and Klein [3.32]–[3.36], who were among the first ones to study such advanced pointer processors, coping with some deficiency of the basic pointer processor of Figure 3.16.

To introduce the ideas underlying the design of advanced pointer processors, first we redraw with more detail the scheme of a basic pointer processor in Figure 3.18. The buffer fill level varies according to the phase error between the read and write clocks. The phase detector recovers such phase error information from the buffer fill level. The decision circuit (quantizer) monitors the phase and generates pointer justification requests by comparison to thresholds.

Assuming that clock and network synchronization instabilities and drifts can be neglected, this implementation approach may solve the issue of eliminating pointer action, if the buffer threshold spacing is wide enough to accommodate the various overhead gaps in the read and write clocks, due to frame phase offset.

Therefore, a good synchronization of the network elements to a common master clock and the use of equipment clocks with a good short-term stability, in conjunction with a certain threshold spacing, may be adequate to solve the problem of pointer action. Nevertheless, as it will be shown in the following, an enhanced design of the pointer

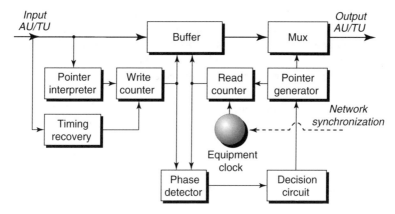

Figure 3.18. Block diagram of basic pointer processor. (Adapter from [3.35], © 1994 IEEE, by permission of IEEE)

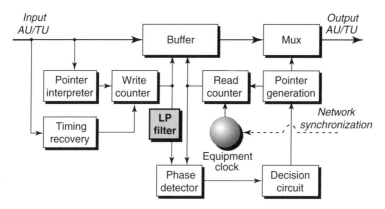

Figure 3.19. Block diagram of a improved pointer processor (with phase-smoothing filter). (Adapter from [3.35], © 1994 IEEE, by permission of IEEE)

processor can do more: reduce the wander generated by the elastic store in the pointer processor with this straightforward implementation[9].

A first step to improve the basic pointer processor of Figure 3.18 is to insert a smoothing low-pass filter on the write phase signal at the input of the phase detector, as shown in Figure 3.19. For the sake of simplicity, the filter function is here indicated neglecting the modulo effect of the write counter. Digital signal processing allows a very efficient and cost-effective digital implementation of this filter function.

This filter may reduce the pointer justification rate, as the phase detector does not see anymore fast fluctuations occurring in the write phase. Nevertheless, the issue of limiting the wander generated, thus ensuring a good timing transparency of the carried tributary,

---

[9] The wander, generated by the elastic store in the pointer processor, will be smoothed by the output desynchronizer at the next stage, as discussed in the previous section. Nevertheless, reducing the wander since its generation at this stage improves the overall performance.

remains. Even if the threshold spacing may be reduced, pointer justification may create very low frequency jitter components with amplitudes of about a pointer step size.

Therefore, in an analogous way to what was done in the synchronizer/desynchronizer design, methods based on threshold modulation have been proposed, which allow to shift these low frequency components to higher frequencies, where they can be more effectively filtered out.

Threshold modulation can be done by adding a *fixed* modulation waveform, such as a saw-tooth signal, to the phase error signal at the input of the quantizer. On the other hand, instead of a fixed waveform, an adaptive modulation waveform, derived from the quantization error itself (i.e., the difference between the quantizer input and output), can be added. This is called *adaptive threshold modulation*.

The two different approaches are shown in Figure 3.20, which illustrates the decision circuit of Figure 3.19 in the two cases of fixed and adaptive threshold modulation. The decision circuit implementing adaptive threshold modulation represents a first-order sigma-delta modulator, which may be considered as a numerical controlled oscillator. In conjunction with the pointer processor feed-back loop described above (phase detector, decision circuit, pointer generator, read counter) a second-order sigma-delta modulator characteristic is achieved, providing improved spectral characteristics of the quantization error compared to the fixed threshold modulation methods [3.36]. Further details on adaptive threshold modulation strategies for advanced pointer processors and their performance are described by [3.37][3.38].

This advanced pointer processor with threshold modulation does not work limiting the pointer adjustment rate. On the contrary, pointers justify frantically! The point is that this advanced pointer processor generates fast alternating pointer justification events. The corresponding mean value represents the phase of the VC that provides timing transparency. These high-frequency pointer justification events generated by the advanced pointer processor allow the implementation of desynchronizers with a shorter time constant (higher cut-off frequency), or alternatively a much higher jitter reduction in a desynchronizer with the same bandwidth.

Moreover, such an advanced pointer processor improves the dynamic behaviour of a chain of NEs, due to its reduced time constant. In an extreme case, if a chain is made *exclusively* of NEs equipped with excellent pointer processors and desynchronizers adopting carefully designed threshold modulation, it may be possible to not synchronize the network, still not having to face high wander on the carried tributary. The provision of synchronization links to other network clients (e.g., telephone exchanges) via primary rate signals, mapped in the VCs carried along the chain, would then be made

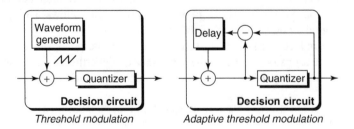

Threshold modulation   Adaptive threshold modulation

Figure 3.20. Decision circuit of an advanced pointer processor with fixed and adaptive threshold modulation. (Adapted from [3.35], © 1994 IEEE, by permission of IEEE)

possible by the good timing transparency provided by these types of pointer processors and desynchronizers.

3.5 THE SDH EQUIPMENT CLOCK

A considerable effort has been devoted in standard bodies to the specification of the SDH Equipment Clock (SEC). Standards, for less recent types of equipment clock, did not specify much more than the external physical interface and some simple accuracy requirements. On the contrary, SEC characteristics are addressed in several ETSI and ITU-T Recommendations, which provide thorough specifications of its accuracy and stability and its detailed functional description. In particular, the SEC incorporates specific functions of timing generation, filtering and extraction from incoming signals.

Interesting design aspects of SECs are treated in [3.39].

3.5.1 ITU-T Functional Description

The ITU-T Rec. G.783 [3.5] specifies the SDH equipment by means of a functional approach. Neither hardware blocks nor implementation details are addressed. Instead, equipment is described by decomposition in functional blocks, which represent abstractions of the functions carried out by the equipment. This approach allows to specify *what* the equipment is required to do, but not *how* it should implement its tasks, thus not impeding the designers in any way from finding the most efficient solutions.

In the functional specifications of ITU-T Rec. G.783, followed also by ETSI, the SEC is represented by the Synchronous Equipment Timing Source (SETS) functional block. The information flow associated with this block is shown in Figure 3.21.

The synchronization source of the SETS may be selected from any of three reference points:

- T1, reference derived from a STM-N input signal;
- T2, reference derived from a PDH input signal;
- T3, reference derived from an external synchronization signal (through the Synchronous Equipment Timing Physical Interface, SETPI).

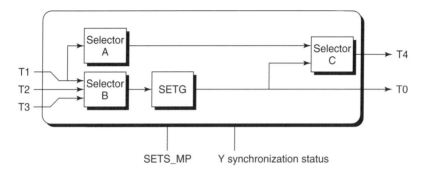

Figure 3.21. Synchronous Equipment Timing Source (SETS) functional block as defined by ITU-T and ETSI. (Adapted from Figure 9-1/G.783 [3.5] and from Figure 7/EG 201 793 [3.40]. © ETSI 2000, by permission of ITU and ETSI)

In addition, the SETS may take its synchronization source from an internal oscillator (contained in the SETG block, described in the following).

On the output side, the SETS supplies synchronization:

- to all the other functional blocks modelling the SDH equipment itself, through the reference point T0;
- to the external synchronization port, through the reference point T4 between the SETS and the SETPI.

Loss of all incoming timing references is reported to the Synchronous Equipment Management Function (SEMF) at the reference point SETS_MP (SETS Management Point). At point Y, the synchronization status is reported.

The reference points above are conceptual points, where information (in this case, synchronization) is exchanged, and may thus represent several physical signals. Selectors A, B and C represent the function of picking one signal among those available. At T1 point, Selector A selects one reference synchronization source from signals derived from input STM-N signals. At T1, T2 and T3 points, Selector B selects one reference synchronization source from signals derived from input STM-N, PDH and external reference signals.

The Synchronous Equipment Timing Generator (SETG) function contains a clock of characteristics specified by either ITU-T Rec. G.812 or Rec. G.813 (but G.813 in most common cases). It may operate in the following modes, as defined in ITU-T Rec. G.810 (further details will be provided in Chapter 5):

- *locked* to the input reference source selected by Selector B;
- *hold-over* mode;
- *free-run* mode.

In free-run mode, the frequency accuracy specified by ITU-T Rec. G.813 is ±4.6 ppm. Moreover, the SETG provides filtering functions to ensure compliance with the stability requirements specified by ITU-T Recs. G.812 or G.813. It is worthwhile noticing that the SONET equipment clock is specified in less detail and with less strict requirements. In free-run mode, its frequency accuracy specified by ANSI is ±20 ppm.

Selector C is activated by an operator command and may select either T0 or T1, as selected by Selector A. This means that the synchronization supplied by the SEC to the external synchronization port, through the reference point T4, may be derived from either the output of the SETG (*filtered* timing from an external reference or the internal oscillator) or from a *non-filtered* reference derived directly from an input STM-N signal. The relevance of this fact will be clear in Section 4.1.6, when the architecture for timing transfer on SDH links will be presented.

Moreover, a squelching function at Selector C is provisioned by the operator to disable T4 when necessary, under particular network synchronization architectures.

The SETPI function provides the interface between the external physical synchronization signal and the SETS in both directions. Therefore, the SETPI function takes timing at reference point T4 from the SETS to form the output synchronization signal. The SETPI passes the timing information to the synchronization interface transparently. On the reverse direction, the SETPI function extracts timing from the received synchronization signal and, after decoding, passes timing information to the SETS.

3.5.2   Timing Modes

The SETS can be synchronized and distributes its timing according to several options (*timing modes*), depending on the network synchronization distribution architecture.

In all modes, the outgoing STM-N signals from all line and tributary ports (included the selected reference port) are timed from the SETG. Outgoing PDH tributary signals are normally not retimed from the SETG, but they retain their timing from the upstream originating node.

In the following, we will consider an ADM-N network element, i.e. a SETS with two STM-N bidirectional lines (T1), say East and West lines, some PDH or STM-N tributaries (T1, T2) and one external synchronization reference input (T3) and output (T4). The ETSI document EG 201 793 [3.40] outlines the following four timing modes.

- *Line timing* (see Figure 3.22(a)). The reference signal for the SETG is derived from either the west- or the east-line signal (T1). This is the common timing mode in chains or rings.
- *Tributary timing* (see Figure 3.22(b)). The reference signal for the SETG is derived from a tributary port, which may be an STM-N signal (T1) or a PDH signal (T2).
- *External timing* (see Figure 3.22(c)). The network element is synchronized from a dedicated external reference signal (T3). This timing mode applies for example when the network element is synchronized from a synchronization-network clock.
- *Internal timing* (see Figure 3.22(d)). The clock of the network element is not locked to any reference signal (free-run or hold-over mode).

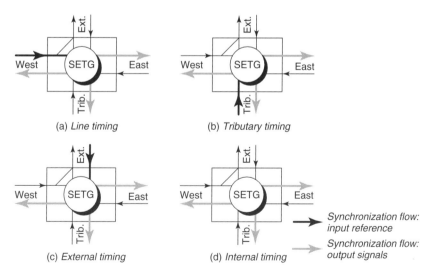

(a) *Line timing*   (b) *Tributary timing*

(c) *External timing*   (d) *Internal timing*

Figure 3.22. Timing modes of SETS [3.40]. (Adapted from Section 6.4.3/EG 201 793 [3.40], © ETSI 2000, by permission of ETSI)

In Figure 3.22, the external synchronization output T4 has not been highlighted as synchronization flow, because it may be set to supply timing derived from either the SETS or from a T1 point.

3.5.3 External Synchronization Physical Interface

The physical characteristics of the synchronization interface port (supplied by the SETPI function) are specified in Section 13 of ITU-T Rec. G.703 [3.41] (2048 kHz synchronization interface). The 2048 kHz signal is commonly referred to as a 'sinusoidal' analogue reference. Actually, the G.703 mask allows more irregular pulse shapes. Peak levels are between 0.75 V and 1.5 V for the coaxial cable interface (75 Ω).

3.6 SUMMARY

SDH multiplexers feature processes that generate jitter and wander on the tributary signals mapped: inclusion of additional bits in the mapping of tributaries into the STM frames, bit justification and pointer justification.

Three synchronization processes take place along an SDH transmission chain:

- the mapping of a PDH tributary into SDH frames in the first multiplexer, involving bit justification in the case of asynchronous mapping, carried out by the synchronizer;

- the re-synchronization of the VCs in intermediate nodes, by means of pointer justification, carried out by pointer processors;

- the demapping of the tributary from SDH frames in the terminating demultiplexer, carried out by the desynchronizer.

In designing SDH synchronizers, one of the main concerns is to effectively reduce the waiting time jitter due to bit justification. Methods based on stuff threshold modulation allow to force the waiting time jitter to fall outside the bandwidth of the PLL filter in the desynchronizer on the receiver side. Therefore, the amount of residual jitter at the output of the desynchronizer is reduced.

In designing SDH desynchronizers, bit-leaking techniques are often adopted because they can effectively reduce the jitter produced by pointer action. According to this approach, the desynchronizer is designed as the cascade of two elements. For example in the case of E4 demapping, the first has the task of spreading each of the incoming 24-bit phase steps into 24 1-bit phase steps in some way. This task is often referred to as *bit leaking* or even *phase spreading*. The second element may be a conventional second-order PLL that low-pass filters the 1-bit phase steps at the output of the first stage.

Re-synchronization of VCs in the pointer processor of intermediate nodes is accomplished by means of an elastic store, in which the VC bytes are written according to the timing signal extracted by the incoming STM-N signal and are read according to the local equipment clock. Pointer justification allows to compensate for the difference between the input and output frequencies. Also in designing advanced pointer processors, as for synchronizers, methods based on threshold modulation have been proposed, which allow to shift low-frequency components of the jitter to higher frequencies, where they can be more effectively filtered out.

A considerable effort has been devoted in standard bodies to the specification of the SDH Equipment Clock. SEC characteristics are addressed in several ITU-T Recommendations. In particular, ITU-T Rec. G.783 provides its detailed functional description and ITU-T Rec. G.813 specify thoroughly its accuracy and stability characteristics.

3.7 REFERENCES

[3.1] P. R. Trischitta, E. L. Varma. Jitter in digital transmission systems, Ch. 8. *Wander*. K. E. Plonty. Norwood, MA: Artech House, 1989.

[3.2] J. Gowar. *Optical Communications Systems*. Englewood Cliffs, NJ: Prentice Hall Inc., 1984.

[3.3] I. H. Malitson. Interspecimen comparison of the refractive index of fused silica. *Journal of the Optical Society of America* 1965; **55**(10): 1205–1209.

[3.4] L. G. Cohen, J. W. Fleming. Effect of temperature on transmission in lightguides. *Bell System Technical Journal* 1979; **58**(4).

[3.5] ITU-T Rec. G.783 *Characteristics of Synchronous Digital Hierarchy (SDH) Equipment Functional Blocks*. Geneva, April 1997.

[3.6] S. Bregni, M. D'Agrosa, L. Valtriani. Jitter testing technique and results at the desynchronizer output of SDH equipment. *IEEE Transactions on Instrumentation and Measurement* 1995; **IM-44**(3).

[3.7] A. Blanchard. *Phase-Locked Loops*. New York: John Wiley & Sons, 1976.

[3.8] F. M. Gardner. *Phaselock Techniques*. New York: John Wiley & Sons, 1979.

[3.9] R. E. Best. *Phase Locked Loops*. New York: McGraw–Hill Book Company, 1984.

[3.10] P. Horowitz, W. Hill. The art of electronics. Section 9.28 to 9.33. *Phase-Locked Loops*. Cambridge: Cambridge University Press, 1980.

[3.11] W. C. Lindsey, C. M. Chie. A survey of digital phase-locked loops. *Proceedings of the IEEE* 1981; **69**(4).

[3.12] V. F. Kroupa. Noise properties of PLL systems. *IEEE Transactions on Communications* 1982; **COM-30**(10): 2244–2252.

[3.13] H. Meyr, G. Ascheid. *Synchronization in Digital Communications. Vol. 1: Phase-, Frequency-Locked Loops, and Amplitude Control*. New York: John Wiley & Sons, 1990.

[3.14] Y. Takasaki. *Digital Transmission Design and Jitter Analysis*. Norwood, MA: Artech House, 1991.

[3.15] S. S. Abeysekera, A. Cantoni. A detailed analysis of phase locked loops used in clock jitter reduction. *Proc. of 4th UK/Australian International Symposium on DSP for Communication Systems (DSPCS '96)*, Perth, Australia, Sept. 1996, republished in *DSP for Communication Systems*, Eds. T. Wysocki, H. Ravazi, B. Honary. Boston: Kluwer Publishers, 1997: 243–253.

[3.16] R. C. Halgren, J. T. Harvey, I. R. Peterson. Improved acquisition in phase-locked loops with sawtooth phase detectors. *IEEE Transactions on Communications* 1982; **COM-30**(10): 2364–2375.

[3.17] D. L. Duttwailer. Waiting time jitter. *Bell System Technical Journal* 1972; **51**: 165–207.

[3.18] P. E. K. Chow. Jitter due to pulse stuffing synchronization. *IEEE Transactions on Communications* 1973; **COM-21**(7): 854–859.

[3.19] P. R. Trischitta, E. L. Varma. Jitter in digital transmission systems. Ch. 5. *Jitter Introduced by Digital Multiplexes*. Norwood, MA: Artech House, 1989.

[3.20] W. D. Grover, T. E. Moore, J. A. Eachern. Waiting time jitter reduction by synchronizer stuff threshold modulation. *Proc. of IEEE GLOBECOM '87*, 1987, pp. 13.7.1–13.7.5.

[3.21] W. D. Grover, T. E. Moore, J. A. Eachern. Measured pulse stuffing jitter in asynchronous DS-1/SONET multiplexing with and without stuff-threshold modulation circuit. *Electronic Letters* 1987; **23**: 959–961.

[3.22] W. D. Grover, J. A. Eachern. *Synchronization of Asynchronous Data Signals*. U.S. Patent 4811340, March 1989.

[3.23] R. G. Kusyk, T. E. Moore, W. A. Krzymien. Spectral analysis of waiting time jitter. *Electronic Letters* 1990; **26**(8): 526–528.

[3.24] G. L. Pierobon, R. P. Valussi. Jitter analysis of a double modulated threshold pulse stuffing synchronizer. *IEEE Transactions on Communications* 1991; **COM-39**(4): 594–602.

[3.25] S. S. Abeysekera, A. Cantoni, V. Sreeram. A comprehensive analysis of stuff threshold modulation used in clock rate adaptation schemes. *IEEE Transactions on Communications* 1998; **COM-46**(8): 1088–1096.

[3.26] S. S. Abeysekera. Optimum threshold modulation schemes for digital data transmission. *Proc. of IEEE International Conference on Acoustics, Speech and Signal Processing 1994 (ICASSP '94)*, Adelaide, Australia, April 1994, pp. IV 169–172.

[3.27] S. S. Abeysekera, A. Cantoni. A novel technique for the analysis of jitter resulting in pulse stuffing schemes. *Proc. of IEEE GLOBECOM '97*, Phoenix, AZ, USA, Nov. 1997.

[3.28] S. R. Norsworthy. Effective dithering of sigma-delta modulators. *Proc. of International Symposium on Circuits and Systems (ISCAS '92)*, 1992, pp. 1304–1307.

[3.29] H. Sari, G. Karam. Cancellation of pointer adjustment jitter in SDH networks. *IEEE Transactions on Communications* 1994; **COM-42**(12): 3200–3207.

[3.30] A. Reed. *British Telecom Contribution to ETSI/TM3,* Apr. 1989; *British Telecom Contribution to ANSI/T1*, June 1989.

[3.31] ITU-T Rec. G.823 *The Control of Jitter and Wander within Digital Networks which are Based on the 2048 kbit/s Hierarchy*. Geneva, March 1993.

[3.32] M. J. Klein, R. Urbansky. Network synchronization — a challenge for SDH/SONET?. *IEEE Communications Magazine* 1993; **31**(9): 42–50.

[3.33] R. Urbansky. Telecommunications network synchronisation using improved pointer processing techniques. *Proc. of 8th European Frequency and Time Forum*, Weihenstephan, Germany, March 1994, pp. 149–165.

[3.34] R. Urbansky. Telecommunications network synchronisation and wander generated by SDH equipment. *Proc. of 9th European Frequency and Time Forum*, Besançon, France, March 1995, pp. 210–216.

[3.35] M. J. Klein, R. Urbansky. SDH/SONET pointer processor implementations. *Proc. of IEEE GLOBECOM '94*, S. Francisco, CA, U.S.A., Nov. 1994.

[3.36] R. Urbansky. *Pointer Processing for Synchronization Signals in SDH Equipment*. ETSI TM3, The Hague, Netherlands, April 8–12, 1991.

[3.37] P. Sholander, H. Owen. SONET/SDH adaptive threshold pointer processor clocking strategies. *Proc. of IEEE GLOBECOM '94*, S. Francisco, U.S.A, Nov. 1994.

[3.38] P. Sholander, C. Autry, H. Owen. Mapping wander in SONET/SDH adaptive threshold modulation pointer processors. *Proc. of IEEE ICC '97*, Montreal, Canada, June 1997.

[3.39] R. Urbansky, W. Sturm. Design aspects and analysis of SDH equipment clocks. *European Transactions on Telecommunications* 1996; **7**(1): 39–48.

[3.40] ETSI EG 201 793 (ref. DEG/TM-01 080) *Synchronisation Network Engineering*. V.1.1.1, Oct. 2000.

[3.41] ITU-T Rec. G.703 *Physical/Electrical Characteristics of Hierarchical Digital Interfaces*. Geneva, Oct. 1998.

4

NETWORK SYNCHRONIZATION ARCHITECTURES

Network synchronization is a comprehensive expression that addresses in a wide sense any distribution of time and frequency over a network of clocks [4.1]–[4.3]. To give some examples, in telecommunications its goal may be either:

(1) to align the absolute time scales of network nodes, thus aiming for instance at aligning local clocks to the Universal Time Coordinated (UTC) (examples: synchronization of equipment real-time clocks, Internet NTP);

(2) to align the timing signals (or, more precisely, their significant instants) generated by local clocks, independently from a constant phase offset among them, thus aiming at minimizing phase fluctuations around such average phase offset (examples: the synchronization of SDH NEs in order to avoid pointer action, the synchronization of synchronous digital multiplexers or digital switching exchanges in order to avoid slips at input elastic stores);

(3) to equalize the frequencies of local clocks, without controlling their phase relationship (example: the distribution of a frequency-standard signal to FLL-based slave clocks).

In a network synchronized as in case (1), local timing signals are *synchronous* (cf. the definition provided in Section 2.1.2) and their total phases are aligned. Therefore, this network synchronization requires estimation and compensation of transmission delays on synchronization signals directed to each node.

In a network synchronized as in case (2), local timing signals are *synchronous* but there is no need to estimate transmission average delays of synchronization signals.

In a network synchronized as in case (3), finally, timing signals are just *mesochronous* (cf. the definition provided in Section 2.1.2).

In most cases dealt with in this book, network synchronization is intended as in case (2) and is achieved by transferring chronosignals (i.e., some pseudo-periodic signals such as sine or square waves), which carry a timing information with the uncertainty of the integer number of periods elapsed since the signal was generated (total transmission delay).

A *synchronization network* is the facility implementing network synchronization. It is able to provide all telecommunications networks with reference timing signals of required quality. Most modern telecommunications operators have set up one synchronization network to synchronize their switching and transmission networks.

Basic elements of synchronization network are *nodes* (autonomous and slave clocks) and *links* interconnecting them. An autonomous clock is a stand-alone device able to generate a timing signal, starting from some periodic physical phenomenon. A slave clock, on the other hand, generates a timing signal having phase (or much less frequently frequency) locked to a reference timing signal at its input. Slave clocks are usually implemented as PLLs (or FLLs).

Time and frequency are distributed by using the communications capacity of the links interconnecting the clocks (e.g. copper cables, optical fibres, radio links). However, network nodes may be many and spread over a wide geographical area. Therefore, two distinct issues must be faced:

- how to transfer timing from one node to the other (the *tactics* of point-to-point timing transfer);
- how to organize timing distribution among all nodes of the network (the *strategy* of network synchronization).

This chapter deals with both the above aspects. First, network synchronization is introduced from a historical point of view, pointing out how network synchronization issues evolved with telephone networks. The tactics of point-to-point timing transfer is addressed in particular, describing how it can be achieved on PDH and SDH links and the relevant architectures. Then, different strategies for network synchronization are presented and the synchronization-network standard architectures defined in the relevant ITU-T, ETSI and ANSI documents are outlined. Moreover, important aspects of synchronization-network planning, management, performance monitoring and protection are addressed. Finally, some real examples of synchronization networks realized worldwide by major telecommunications operators are described.

4.1   AN HISTORICAL PERSPECTIVE ON NETWORK SYNCHRONIZATION

The modern telecommunications networks result from a long evolution process, started since the end of the 19th century:

Network synchronization, at first an unknown issue as not relevant to network operation and performance, has played a role of increasing importance in telecommunications throughout this evolution process, especially since transmission and switching turned digital.

Transmission and switching are the two basic functions of any telecommunications network, and in particular of telephone networks.

Transmission is the action of conveying information point-to-point, for example from one node in a network to another one directly linked to it by a physical channel. Moreover, transmission can also be from one point to multiple points (*multicast*) or even from one point to all listeners on the medium (*broadcast*).

Switching, on the other hand, is the function of connecting a given input–output pair in nodes where multiple transmission links are terminated. It deals thus with the dynamic assignment of the transmission channels available in a network, on the basis of user connection requests.

To make an analogy with railways, transmission systems are tracks and switching nodes are shunts. Transmission and switching are the complementary foundations on which all telecommunications services are based. Both transmission and switching have been analogue at first, then one after the other turned to digital technology.

The evolution of digital transmission and switching technology for the public telephone networks began with isolated digital transmission links between analogue switching machines or analogue radio transmission systems. The fact that digital technology was being used was transparent to the interfaces. Thus, there was no need to relate the internal clock rate in one system with the internal clock rate of another system.

Even as higher level multiplexing systems were developed, there was no need (nor viable means) of relating the clock rates of the higher rate multiplexed signals with the clock rates of the lower rate tributaries. Indeed, transmission equipment based on PDH technology does not need to be synchronized, since the bit justification technique allows multiplexing of asynchronous tributaries with substantial frequency offsets.

Problems began to arise with such asynchronous architecture when digital technology moved to switching machines too. Digital switching equipment requires to be synchronized in order to avoid slips at input elastic stores. And while slips do not affect significantly normal phone conversations, they may be troublesome indeed on some data services! The introduction of the circuit-switched data networks and of the ISDN, therefore, yielded first the need for more stringent synchronization requirements.

As a matter of fact, however, the ongoing spreading of SDH/SONET technology in transmission networks has really made synchronization a hot topic in standard bodies since the 1990s. The need for adequate network synchronization facilities has become more and more stringent in order to fully exploit SDH/SONET capabilities: it is widely recognized that SDH/SONET transmission may rely on a suitable and dependable timing distribution to fully meet all its benefits, in particular because pointer action may yield excess jitter on transported tributaries.

Beyond SDH/SONET needs, anyway, nowadays network synchronization facilities are unanimously considered as a profitable network resource, allowing slip-free digital switching, enhancing the performance of ATM-based transport services and serviceable for improving the quality of a variety of services (e.g. ISDN, mobile cellular telephony, etc.).

For this reason, most major network operators have set up national synchronization networks, in order to distribute a common timing reference to each node of the telecommunications network. On the standardization side, ITU-T and ETSI bodies released new synchronization standards, suitable for the operation of modern (including SDH/SONET-based) digital telecommunications networks, specifying more stringent — and complex — requirements for jitter and wander at synchronization interfaces, for clock accuracy and stability and for the synchronization network architecture.

This section deals with network synchronization from a historical point of view, pointing out its different aspects and how network synchronization issues evolved with the telephone networks, beginning from old analogue FDM networks up to digital PDH, SDH/SONET, ATM and cellular mobile telephone networks. The relevant architectures for point-to-point timing transfer on PDH and SDH links are described.

4.1.1 Synchronization in Analogue FDM Networks

Until the introduction of digital techniques, any technological change impacted separately on transmission or switching, two well distinct functions implemented in different equipment and experiencing different evolution processes. The introduction of FDM multiplexing techniques enhanced enormously the capacity of transmission links, without bringing significant changes in the operation principles, the implementation techniques or the management and control systems of switching exchanges.

FDM is an *analogue* standard technique allowing multiple channels to share a common physical medium (see Section 2.1.5). Frequency shifting of channels is done through SSB modulation of a sine wave (*carrier frequency*). In the case of telephone signals, carriers are spaced 4 kHz while the single-signal net bandwidth is about 3 kHz (300 Hz ÷ 3400 Hz), in order to make easier channel separation by band-pass filtering.

As pointed out in Section 1.1.1, SSB demodulation must be coherent, i.e. it is based on multiplying the modulated signal by a sine wave with *same* frequency and phase of the carrier. Carrier synchronization was accomplished in the first FDM systems through a simple point-to-point strategy, i.e. limited to every single transmission system (multiplexer/line system/demultiplexer). Later on, when FDM systems spread to constitute larger networks, comprising links at different levels of the multiplexing hierarchy, the issue of an FDM network synchronization strategy had to be faced.

AT&T realized the first synchronization network in the 1970s [4.4]. The strategy adopted was based on deploying *carrier supplies*, i.e. equipment generating all the necessary carriers to be used by multiplexers and demultiplexers for all the hierarchical levels, and on synchronizing them by distributing a pilot frequency (usually multiples of 4 kHz) derived from a network master clock. Carrier supplies used PLLs to generate reference frequencies synchronous with the pilot received.

The main task of those PLLs was, on the one hand, to ensure adequate short-term stability by filtering phase fluctuations accumulated by pilots along the transmission links, and on the other to provide in any case an output reference frequency, even under loss of the input pilot, by *free-running* operation of the local oscillator. Free-run frequency accuracy requested to limit distortion in the demodulated signals was in the order of 10^{-7} [4.5][4.6]. Such a frequency accuracy was enough to ensure an adequate transmission quality of telephone channels, even under pilot frequency losses lasting for the mean time for restoring.

4.1.2 Synchronization and PDH Digital Transmission

As stated before, when digital transmission in the very beginning was limited to isolated links between analogue switching machines or analogue radio transmission systems, there was no need to relate the internal clock rate in one system with the internal clock rate of another system.

The need of exploiting at best the physical media led then to the development of digital TDM techniques, enabling to multiplex together thousands of telephone channels. Since the beginning of this evolution, a fundamental choice was made between the two options of digital synchronous multiplexing and digital asynchronous multiplexing: the latter was chosen to establish the PDH standard, in order to overcome the complexity of the issues related to synchronizing all the network nodes.

As explained in Chapter 2, PDH systems are based on bit justification, which allows multiplexing of asynchronous tributaries with substantial frequency offsets. Therefore, PDH transmission networks do not need to be synchronized. Every equipment clock is independent from the others, but their frequencies are just kept close to the nominal values within specified standard tolerance intervals.

It is now important to point out once more that PDH systems are transparent to the timing content of transported digital signals. An E1 signal[1], multiplexed with the other three asynchronous tributaries in an E2 and then so on in the upper PDH hierarchical level signals E3 and E4, once recovered at the end of the transmission chain has the *same* original average frequency as before the multiplexing/demultiplexing chain (of course with some jitter due to the transmission lines and to the justification process [4.7]–[4.9]), although the multiplexer clocks of the transmission chain are independent (cf. the *Fundamental Law of Digital Transmission Chains* in Section 3.2.4).

This property is remarkable indeed. The bit justification technique allows to *transfer the timing* content of a digital signal across a transmission chain where clocks are *asynchronous* instead, as shown in Figure 4.1. The thick grey links in the figure denote the signals which are synchronous with the master clock (i.e. transferring timing), while all the others are asynchronous. A 2.048 Mbit/s is generated by a digital switching exchange with the local clock driven by a master clock. The multiplex signals (thick black links) are not synchronous with it, but they embed, owing to bit justification, the 2.048 Mbit/s signal carrying timing. When the 2.048 Mbit/s is recovered, it is still synchronous with the master clock. Such a nice feature is exploited to transfer timing across PDH networks to synchronize clocks located in far locations.

4.1.3 Synchronization and Digital Switching

The advent of digital TDM techniques yielded a progressive integration of transmission and switching, as the PCM Primary Multiplex frame structure, made of 24 or 32 time

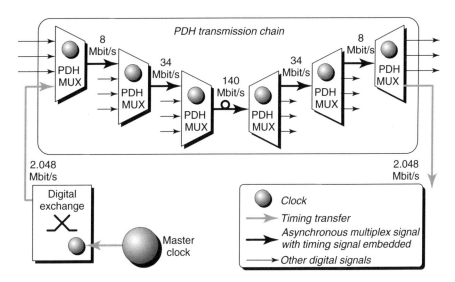

Figure 4.1. Timing transfer through a PDH transmission chain [4.3]

---

[1] In this example and in the next ones, we refer for the sake of simplicity to the European PDH, but obviously the same considerations apply to the North American PDH as well.

slots interleaved, allows to exploit the TDM principle for digital switching of circuit connections as well.

The European 2.048 Mbit/s PCM frame is made of 32 octets (time slots), 30 of which carry single 64 kbit/s telephone channels, while the North American 1.544 Mbit/s PCM frame is made of 24 slots. Digital switching and cross-connecting is based on moving octets (speech samples) from one time slot to another, from one input signal to another output signal. Time slot exchanging is basically done by delaying by a suitable time interval the incoming octets before retransmitting them in the output frames at the right place (time).

It is clear that digital switching can take place *only* if incoming frames (asynchronous as they can be generated by different pieces of equipment with different clocks) are made synchronous, with frame starts aligned, so that correspondent time slots at different inputs are perfectly time-aligned. Therefore, one of the tasks of the input line units of a digital switching exchange or cross-connect is to synchronize bits and frames of the incoming PCM signals, before feeding them into the switching fabric, as outlined in Figure 4.2. In this figure, for the sake of simplicity, only one frame per line is depicted (with alignment words shaded) and the time slot interchanging in the PCM frames is not pointed out.

Every incoming PCM signal is bit-synchronized according to the equipment local clock, so that all incoming frames can be then time-aligned. Bit and frame synchronizations[2] are accomplished according to the slip buffering technique explained in Section 2.3.2, for each input line. As noted in that section, any frequency offset between the write and read clocks yields slips: a controlled-slip size equal to one frame (256 bits) and a frequency tolerance of 50 ppm, with respect to the nominal value 2.048 Mbit/s, would yield a slip rate as high as 24 slips per minute (as it can be evaluated from Equation (2.9))!

Such a slip rate is not acceptable even for Plain Old Telephone Service (POTS). Since the introduction of the first telephone digital exchanges, therefore, the issue of controlling the slip rate was faced [4.10][4.11], by improving the accuracy of their clocks or through a network synchronization plan. In the beginning, equipping digital exchanges with high-precision independent clocks was generally preferred and considered the most promising

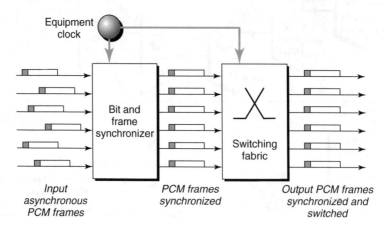

Figure 4.2. Bit and frame synchronization of PCM signals at the input of a digital switching exchange or cross-connect [4.3]

---

[2] Note that *frame synchronization* is used in this context with a different meaning than in Section 1.1.3.

solution, as few slips were tolerated and the cost of very accurate oscillators (quartz and atomic oscillators) was progressively decreasing. However, while single slips do not affect significantly normal phone conversations, they can be troublesome indeed on some data services.

4.1.4   Impact of Slips on Digital Services

The impact of slips on services carried on digital networks depends on the application. Studies reported throughout the 1970s and the 1980s [4.12]–[4.18] describe the effects of slips on services of various kinds as follows.

- For *uncompressed voice* (POTS), only a little percentage of slips leads to occasional audible clicks in the reproduced audio, which are not a serious impairment for speech [4.12]. Therefore, POTS is very tolerant of slips.

- *Compressed voice*, on the other hand, is less tolerant. A slip leads an audible click.

- A study on the effects of controlled slips on *Group 3 facsimile* transmission [4.14] found that a single slip caused distortion or missing lines. A slip can wipe out several scan lines (up to eight horizontal lines, equivalent to missing about 2 mm of vertical space). If slips occurred continuously, the affected pages would need to be retransmitted.

- For data transmitted on POTS channel (*voiceband modem*), a slip may cause a drop-out lasting from 10 ms to 1.5 s, depending on the data rate and coding [4.15].

- For *digital video* transmission (video teleconferencing, for example), tests [4.16] shown that a slip may cause segments of the picture to be distorted or frames to freeze for periods of up to 6 s. The seriousness and length of the impairment depends on the video coding and compression algorithm. The impairment is more serious for low bit rate encoding systems.

- In *data transport protocols*, slips reduce transmission throughput, owing to the need of retransmitting corrupted data units.

- *Encrypted services* are greatly impacted by slips [4.16]. A slip may result in the loss of the encryption key. The loss of the key causes the transmission to be unintelligible until the key can be resent and communication re-established. Therefore, all communications are halted. What's more, key retransmission adversely affects security.

4.1.5   Synchronization of Digital-Switching Equipment across PDH Links

As made clear in the previous section, in particular the introduction of circuit-switched data networks yielded the need for more stringent synchronization requirements and made network synchronization the most suitable strategy.

In networks using PDH for the transport infrastructure, it is necessary to synchronize primary-rate multiplexers and digital-switching equipment, such as 64-kbit/s switching exchanges and cross-connects, while PDH-transport networks do not need synchronization, as shown in Figure 4.3. Timing is commonly transferred across PDH links, exploiting their timing-transparency property, as said in the previous subsection (cf. Figure 4.1).

The scheme of synchronization of two digital switching nodes (e.g., exchanges) through a PDH transmission chain is outlined in Figure 4.4, where the same graphical notation of Figure 4.1 holds (i.e. spheres are clocks, thick grey links denote signals transferring timing, thick black links are asynchronous multiplex signals with timing signal embedded and thin links are other digital signals). The clock of the first exchange is enslaved to a master clock (e.g. the master clock of the whole digital exchange network) so that all the 2.048 Mbit/s signals output by this exchange are synchronous. The equipment clock of the second exchange is synchronized by means of one of these 2.048 Mbit/s (which may carry normal payload as well) assigned to transfer timing, which is transported across a PDH transmission chain from the first exchange to the second, multiplexed together with other signals.

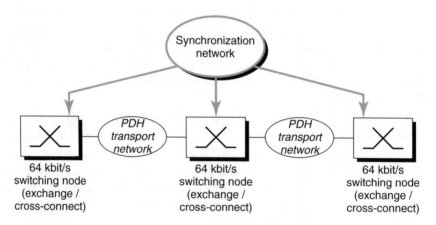

Figure 4.3.   Synchronization of digital-switching equipment in PDH networks

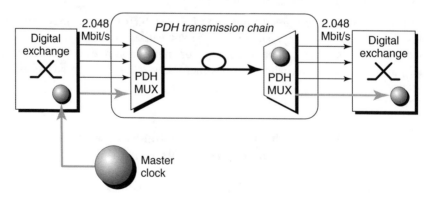

Figure 4.4.   Synchronization of two digital switching exchanges through a PDH transmission chain [4.3]

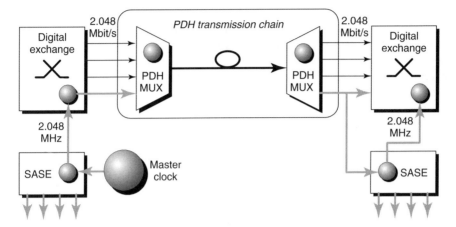

Figure 4.5. Synchronization of two digital switching exchanges served by SASE clocks through a PDH transmission chain [4.3]

A second scheme is based on the availability of a synchronization network based on the concept of *building clock*[3], i.e. a slave clock which serves an entire office building by supplying timing to all the equipment deployed there, including digital switching exchanges, digital cross-connects and in case terminal equipment and multiplexers (see Section 4.3).

This latter scheme is outlined in Figure 4.5. The clock of the first digital switching exchange is synchronized by the local building clock (SASE), synchronized by the master clock and distributing timing to the equipment of the first office building (usually by means of ITU-T Rec. G.703 [4.19] 2.048 MHz signals). The clock of the second exchange is not directly enslaved to a 2.048 Mbit/s transported through the PDH transmission chain. Conversely, the 2.048 Mbit/s carrying timing is used to synchronize the SASE clock that supplies timing to the equipment of the second office building, including the switching exchange.

The above schemes are currently the most applied worldwide to synchronize digital exchanges connected through PDH networks. Since the 1970s, most telecommunications operators have set up national network synchronization plans, to control slips in digital switching exchanges and Digital Cross-Connects (DXC), which are mostly based on the two schemes above of point-to-point timing transfer from one digital exchange to another.

4.1.6 Synchronization and SDH/SONET Digital Transmission

Contrary to PDH, SDH[4] transmission takes advantage from network synchronization and may rely on it, in order to limit jitter and wander generation on output tributaries. Although

---

[3] As it will be better treated later, such clocks are referred to as *Synchronization Supply Units* (SSU) or *Stand-Alone Synchronization Equipment* (SASE) in the ITU-T and ETSI standards, while they are known as *Building Integrated Timing Supplies* (BITS) in North America (ANSI standards).

[4] In this section, for the sake of brevity, we will refer to SDH, which is the ITU-T standard designed as superset of the ANSI standard SONET, understanding that all principle considerations apply to SONET too.

some modern equipment overcomes such dependence with an enhanced design [4.20] (see Section 3.4), yet the goal of guaranteeing jitter requirements at PDH/SDH boundaries in complex networks [4.21], where several PDH-to-SDH and SDH-to-PDH mapping/demapping processes take place and where equipment of different vendors is deployed, can be achieved only by an accurate synchronization of all the NEs aiming at avoiding any pointer action.

In networks using SDH for the transport infrastructure, therefore, it is necessary to synchronize not only primary-rate multiplexers and digital-switching equipment, but also the nodes of the SDH-transport networks, as shown in Figure 4.6.

Nevertheless, timing transfer in SDH networks cannot follow the same schemes as with PDH. Contrary to what was shown for PDH, in SDH networks it is definitely not advisable to carry timing on signals mapped in STM-N frames (e.g., 2.048 Mbit/s). The reason is that payload tributaries do not transport synchronization effectively, due to excess jitter exhibited in case of pointer justifications (see Sections 3.3 and 3.4).

The best and most straightforward way to transfer timing in SDH networks is to carry it directly on the multiplex STM-N signals. The quality of the timing recovered from STM-N signals is the best today achievable, as affected only by the line jitter (e.g. the jitter due to thermal noise and environmental conditions on the optical line) and not by bit justification or any other mapping issue.

The scheme of synchronization of two digital switching exchanges through an SDH transmission chain is outlined by Figure 4.7. Also here, the same graphical notation of Figure 4.1 holds: spheres are clocks, thick grey links denote signals transferring timing and thin links are other digital signals. Unlike the previous section, only the scheme based on the availability of a synchronization network with SASE clocks in every office building has been considered here (cf. Figure 4.5), as it is deemed the target solution for synchronization networks [4.22][4.23].

The SASE clock in the first office building synchronizes not only the digital switching exchange clock, but also the SDH Equipment Clock (SEC), so that the output multiplex signal is now synchronous with the network master clock, contrary to what was seen in the PDH case where the multiplex signal was asynchronous but embedding the signal carrying timing. At the receiver side, the SEC is not directly locked to the incoming STM-N signal, as might seem natural. A special function of the SDH equipment clock

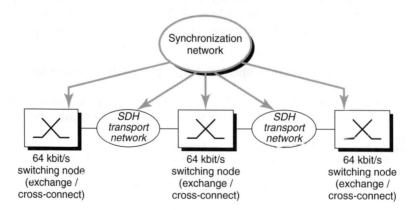

Figure 4.6.   Synchronization of digital-switching and transport equipment in SDH networks

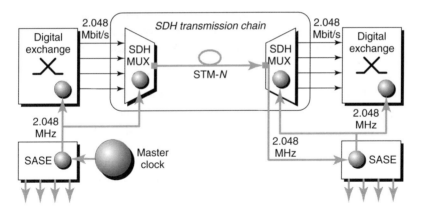

Figure 4.7.   Synchronization of two digital switching exchanges served by SASE clocks through an SDH transmission chain [4.3]

(see Section 3.5.1, Selector C) allows instead to extract timing from the incoming STM-N signal and to directly output it, not filtered, from the synchronization port as an ITU-T Rec. G.703 [4.19] 2.048 MHz signal, to synchronize the SASE of the second office building. This SASE distributes its timing to equipment of the office building, including the digital switching exchange and the SDH demultiplexer.

It is worthwhile noticing that this way of synchronizing the clocks of the second exchange may seem winding and unnecessarily complex, but it is definitely the best solution. Indeed, SASE clocks have much higher stability and filtering capabilities than simple SECs. Following this scheme, the clocks of the digital switching exchange and of the SDH (de)multiplexer in the second office building are synchronized by a timing signal which is much more stable. Moreover, if the STM-N signal should fail, the SASE guarantees a long-term output frequency in free-running operation which is much more accurate than the one of the SEC.

4.1.7   Synchronization in ATM Transport Networks

There is a rather common misunderstanding about the role of synchronization in Asynchronous Transfer Mode (ATM) networks. As the first word in the ATM acronym is *Asynchronous*, one is led to think that the natural operation of ATM equipment is in a non-synchronous environment, in networks where clocks are independent and not synchronized.

Actually, the word 'asynchronous' does not refer to the equipment clock operation, or to the *physical level* of information transfer, but to the information *transfer mode* instead, at an upper level of abstraction (logic level of information transfer, see Section 2.1.4). In other words, it is pointed out that the information sources are asynchronous, i.e. they start sending information in independent instants and not in preassigned time-slots, as it happens with a Synchronous Transfer Mode (STM[5]) like e.g. the PCM primary multiplex.

---

[5] Not to be confused with the same acronym STM already defined as *Synchronous Transport Module* (SDH frame).

Contrary to this popular misunderstanding, synchronization plays an essential role in ATM networks, and in particular in the integration of ATM equipment into existing telecommunications networks. To summarize, ATM equipment requires synchronization for mainly two reasons:

- to support *Constant Bit Rate (CBR) services*, based on AAL type 1 [4.24];
- to support *synchronous physical interfaces*, such as PCM primary multiplexes E1 (2.048 Mbit/s) and DS1 (1.544 Mbit/s) or SDH/SONET signals.

As far as the support of CBR services is concerned, Section 1.1.5 already outlined in some detail the relevant issues about packet jitter reduction in packet-switched networks to emulate circuit-like connections. The circuit-emulation CBR services require indeed that the timing of the carried service be maintained across the ATM network connection. To this purpose, the ITU-T accepted the SRTS timing recovery technique as the standard technique for AAL-1 circuit emulation on ATM [4.25]. It must be pointed out that this technique relies on the availability, in the ATM equipment where AAL connections are originated and terminated, of a synchronization signal traceable to a common network master clock. The performance of the SRTS technique and especially how it is affected by network synchronization impairments are still topics for research [4.26][4.27].

On the other hand, supporting physical interfaces such as PCM primary multiplexes or SDH signals entails that the ATM equipment must allow to be synchronized by means of an external network timing signal, in order to generate synchronous signals for ease of interworking with other equipment in the existing network, as already pointed out in the previous sections.

For the above reasons, international standards require that ATM equipment accepts external timing and is integrated into synchronization networks.

Synchronization of modern ATM equipment is accomplished through dedicated ports. Typically, synchronization physical interfaces are E1 (2.048 Mbit/s) or DS1 (1.544 Mbit/s) data signals or analogue 2.048 MHz signals, compliant with ITU-T Rec. G.703 [4.19]. Early ATM equipment did not feature such synchronization ports, as ATM synchronization requirements were not fully understood yet. As a matter of fact, the lack of a synchronization port made supporting CBR services more difficult.

The most straightforward way of synchronizing an ATM NE is to integrate it into a synchronization network based on the concept of building clock (SASE or BITS). In such an environment, ATM equipment should take the timing reference from the SASE/BITS, which supplies by definition the most accurate timing signal in the office building. This allows, moreover, to forget the burden of avoiding timing loops, which is the duty of the synchronization network manager.

If SASE clocks are not available, e.g. when there is no synchronization network or the ATM equipment is in the access network or in customer premises, then line-timing is the only feasible way of synchronizing. Line-timing means supplying synchronization to the ATM NEs by means of an incoming SDH, E1 or DS1 signal. In this case, care must be taken to avoid possible timing loops, and network planning to this purpose may become complex when many NEs are deployed.

4.1.8 Synchronization of Cellular Mobile Wireless Telephone Networks

Wireless networks pose similar yet different synchronization requirements as other wireline telephone networks. The generic architecture of a cellular mobile telephone network is shown in Figure 4.8: the cell site equipment (Base Transceiver Station, BTS) allows routing of calls from the 'air interface' to the network infrastructure, which includes Base Station Controllers (BSCs), Mobile Switching Centers (MSCs) and then on to the Public Switched Telephone Network (PSTN). Air interfaces include the European GSM/DCS (900/1800 MHz) [4.28] and the North American cellular systems (900/1900 MHz). BTSs, BSCs and MSCs are interconnected via trunk lines, carried for example on E1/T1 systems, PDH or SDH systems.

As with any other network element, BTSs, BSCs and MSCs need to be synchronized to ensure slip-free interconnection at trunk lines. Moreover, BTS needs to be synchronized to ensure frequency stability on the on-air wireless channels. More specifically, GSM [4.29] and Code-Division-Multiple-Access (CDMA) systems require at least about $5 \cdot 10^{-8}$ frequency accuracy, while the North American Time-Division-Multiple-Access (TDMA) system requires only $5 \cdot 10^{-7}$ frequency accuracy. In addition to this frequency synchronization, CDMA systems require also precise time synchronization, while GSM and other TDMA do not.

Owing to both these needs, CDMA systems mostly use Global Positioning System (GPS) receivers to time synchronize their elements as well as to produce a stable frequency reference for on-air signals. The North American CDMA standard IS-95 requires that all BTSs transmit their pilot sequences within 3 μs of time error. GPS receivers are equipped

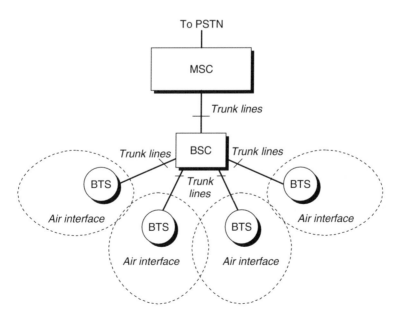

Figure 4.8. System architecture of a cellular mobile telephone network

with high-stability local oscillators, for provisioning precise time and frequency reference even under GPS signal loss (hold-over performance:. ±7 μs over 24 hours). MSCs and BSCs also need precise time synchronization, to assure uninterrupted handoffs when the user transits from one cell to another.

On the other hand, GSM/TDMA carriers require stable frequencies but do not require GPS for precise time synchronization. In order to fulfil such frequency-accuracy requirements, different technical solutions have been proposed.

A first solution is to deploy autonomous clocks in each Base Station. In this case, highly expensive clocks would be required, such as the network primary clocks specified by ITU-T Rec. G.811 [4.30] or the slave clocks specified by ITU-T Rec. G.812 [4.31] under the type-II clause. Alternatively, if using cheaper clocks, periodic recalibration would be needed.

A second and more common solution is to equip MSCs and BSCs with SASE clocks, which take their reference from a synchronization network, in order to maintain the frequency stability required for both slip-free interconnection and precise on-air carriers.

Whatever is the approach taken to synchronize network elements in a cellular telephone network, by means of SASE or a GPS receiver, the stringent requirements for cell site synchronization that are faced by modern operators can be summarized as follows.

- *Frequency stability* is required by CDMA, GSM and TDMA specifications to avoid that centre frequencies of on-air channels drift, resulting in co-channel interference and in problems at hand-off. In fact, poor control of the transmit frequency due to poor-quality frequency reference can lead to dropped calls, slow hand-off between cells and interference between channels. During call initialization and hand-off, the BTS broadcasts a series of instructions to the mobile handset for changes in frequency and time slot. Under timing transients, the BTS may send several corrective instructions, adjusting frequency and time slot parameters. Precision frequency synchronization at each BTS prevents these problems.

- *Time stability* is required by CDMA specifications (such as the North American IS-95) to keep pilot sequences in all cells within strict bounds of time misalignment and to assure uninterrupted handoffs when the user transits from one cell to another.

- *New services* such as locating mobile handsets and third-generation wireless systems (UMTS) will require even more stringent frequency and time stability.

- Trunk lines among BTSs, BSCs and MSCs are usually E1/T1 systems, but they may migrate in the future to *some other transport technology* that does not feature a good timing transparency, such as ATM, backhaul radio, etc. In this case, the availability of good synchronization external facilities will be even more needed.

4.1.9   Synchronization Today and Beyond

As summarized in the last sections, network synchronization was considered for the first time as an important issue to cope with when switching exchanges turned digital, in order to avoid slips in the input elastic stores. This has led the network operators, since the mid-1970s, to face the issue of synchronizing their digital switching networks.

Since the 1990s, as a matter of fact, all the major network providers have been considering the implementation of a modern national synchronization network a priority in their plans. ITU-T and ETSI standard bodies started to develop completely new synchronization standards, specifying more stringent requirements according to more complex criteria. The underlying reason is the progressive introduction of new digital techniques and services into the telecommunications networks, together with the awareness that network synchronization facilities are indeed a profitable network resource, which may be exploited to serve a wide range of equipment and services.

The examples provided in the previous sections show that a synchronization network is indeed a profitable shared resource, a 'background support technology' [4.16] serving a variety of equipment and services. Deploying a modern synchronization network, based on the BITS/SSU concept, makes available in every office building of the telecommunications network as many timing signals as requested, of greatest accuracy and traceable to a common network master clock, ready to serve equipment, services, or even customer premises. Today it is customary for national network providers to provide customer's private networks, which utilize some digital transport service with the rest of the world, with a timing signal traceable to the national reference, in order to achieve slip-free connection at switching interfaces between provider's and customer's networks. In a forthcoming scenario, indeed, even the *timing* itself might be sold as an advanced service to small network providers.

4.2 NETWORK SYNCHRONIZATION STRATEGIES

Assuming that the tactical issue of transferring timing from one node to the other has been solved successfully, then a suitable *strategy* of network synchronization should be established. Different approaches to organize timing distribution to all the nodes of the network can be chosen, depending on the specific requirements of the application.

Since the last decades, many network synchronization strategies have been envisaged, mainly to the purpose of synchronizing the nodes of telecommunications networks. The main features of such strategies will now be summarized in short, pointing out for each of them also its socio-political analogy. Such an analogy may be helpful to understand immediately the pros and cons of every strategy, as it may be perceived more straightforward to assess a form of government of human beings rather than of clocks.

4.2.1 Full Plesiochrony (Anarchy)

The full plesiochronous strategy is actually a *no-synchronization strategy*, that is it does not involve any synchronization distribution to the network of clocks. Each clock is independent from the others (*autonomous clocks*), as shown in Figure 4.9, hence the expression synchronization *anarchy*.

Anarchy is the easiest form of government, but it relies on the good behaviour of the single elements. Owing to the lack of any timing distribution, the synchronization of processes in different nodes is entrusted to the accuracy of the network autonomous clocks, which therefore must feature excellent performance. With pure synchronization anarchy,

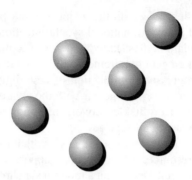

Figure 4.9. Full plesiochrony (anarchy)

clocks not only must be carefully trimmered to the nominal frequency at deployment, but they also must be often retuned to compensate frequency drifts due to ageing. Hence, the intrinsic robustness and simplicity of this strategy are counterbalanced by high deployment and maintenance costs, at least if the synchronization requirements are not trivial.

In the late 1960s, this strategy was generally considered the most promising for the synchronization of telecommunications networks, owing to the decreasing cost of atomic frequency standards and to the limited synchronization requirements envisaged then. Nevertheless, as the cost of such oscillators became stable and the new digital transmission and switching techniques began to demand increasing timing performance, this strategy was eventually abandoned for most applications.

4.2.2   Master–Slave Synchronization (Despotism)

The principle of Master–Slave (MS) strategies (despotism) is based on the distribution of the timing reference from one clock (*master clock*) to all the other clocks of the network (*slave clocks*), directly or indirectly, according to a star (two-level despotism) or tree (multi-level despotism) topology as shown in Figure 4.10. In star synchronization networks, a common master synchronizes all slave clocks through direct links. In tree synchronization networks, timing is distributed along routes that may encompass several slave clocks, organized in hierarchical levels.

While the master clock should be an expensive high-precision oscillator (usually based on an atomic frequency standard), slave clocks can be much cheaper. A slave clock is usually implemented as a PLL, much less frequently as a FLL, mostly based on a quartz-crystal oscillator and locked to the reference timing signal coming from the master. In PLLs, the output timing signal is kept synchronous with the input reference, by means of a feedback control on the phase error between them. In FLLs, the output timing signal and the input reference are just kept mesochronous, by means of a feedback control on the frequency error. Therefore, MS synchronization is usually implemented with PLL-based slave clocks.

The loop time constant of the PLL controls its filtering properties against the phase fluctuations on the input reference and on the internal oscillator. An important step in designing an MS synchronization network is therefore to design the loop time constants of the slave clocks, according to specific requirements. Further details on PLL properties will be given in Section 5.5. Moreover, chains of slave clocks are dealt with in Section 5.12.

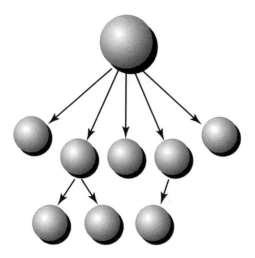

Figure 4.10. Master–slave synchronization (despotism)

Despotism is generally considered unethical, but it is certainly effective in ensuring a very tight control on the slaves: an MS network is synchronous with the master clock (at least if PLL slave clocks are used) and stable by definition. Therefore, MS-based strategies are currently the most widely adopted in many applications. Questions may arise, nevertheless, on what happens should the master fail. The Hierarchical MS strategy has been conceived to this purpose.

4.2.3 Mutual Synchronization (Democracy)

Mutual synchronization is based on the direct, *mutual control* among clocks, so that the output frequency of each one is result of the 'suggestions' of the others, as shown in Figure 4.11. Such a pure democracy looks appealing: there are no masters and no slaves, but a mutual cooperation. Though, the 'discipline' of the mutually controlled elements is hard to guarantee.

Modelling the dynamic behaviour of such networks, or even ensuring the stability of the control algorithms (the control algorithm must damp the transient impairments, avoiding

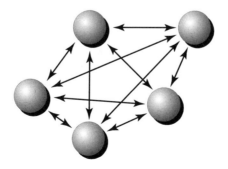

Figure 4.11. Mutual synchronization (democracy)

that they propagate indefinitely around the network synchronization loops), can be a very complex task [4.1][4.32][4.33]. Transient behaviour is very hard to predict, especially after protection reconfiguration.

Networks so designed tend thus to be quite expensive, due to the complexity of their management and control, but their results are extremely reliable. Hence, till now, the field of application of mutual synchronization has been mostly limited to special cases, e.g. military networks.

4.2.4   Mixed Mutual/Master–Slave Synchronization (Oligarchy)

In this mixed solution, the mutual synchronization strategy is adopted for a few network core clocks, while the MS strategy is adopted for the peripheral clocks, as shown in Figure 4.12.

Oligarchy is a compromise, which aims at mildening the absolutism of despotism. A greater reliability is achieved, while the peripheral MS synchronization simplifies substantially the system control issue compared to pure mutual synchronization.

4.2.5   Hierarchical Mutual Synchronization (Hierarchical Democracy)

A generalization of democratic strategy is the hierarchical mutual synchronization: in hierarchical democracy, some count more than others do.

Each of N network nodes is given a relative weight w_i $(0 \le w_i \le 1, \sum_i^N w_i = 1)$, as shown in Figure 4.13. When all weights are equal, this strategy turns a pure mutual synchronization. When the weight of one clock is equal to one and all the others to zero, this strategy turns an MS synchronization.

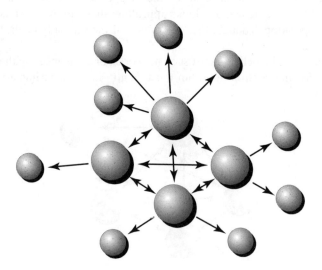

Figure 4.12.   Mixed mutual/master–slave synchronization (oligarchy)

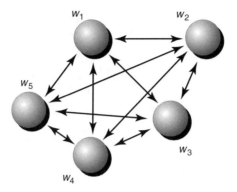

Figure 4.13. Hierarchical mutual synchronization (hierarchical democracy)

4.2.6 Hierarchical Master–Slave Synchronization (Hierarchical Despotism)

The Hierarchical Master–Slave (HMS) synchronization strategy is a variant of the pure MS strategy: a master clock synchronizes the slave clocks, directly or indirectly according to a tree topology, and these are organized in two or more hierarchical levels as shown in Figure 4.14. Protection mechanisms against link and clock failures are planned through alternate synchronization routes, not only between 'parent' and 'son' clocks, but also between 'brother' clocks and even between 'uncle' and 'nephew' clocks. If the master fails, another clock takes its place according to a hierarchical plan.

Under synchronization path failure, protection routes may be activated according to a static pre-selected protection routing table, or picked dynamically using the communication capacity among clocks to exchange messages about their current status of synchronization. If the current reference fails, a clock is able to select the best-supposed reference among those available at its inputs (at least according to what declared by candidate reference clocks).

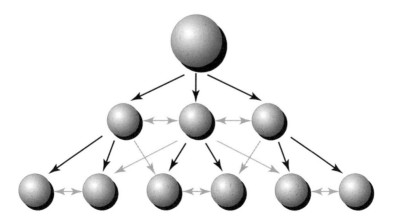

Figure 4.14. Hierarchical master–slave synchronization (hierarchical despotism)

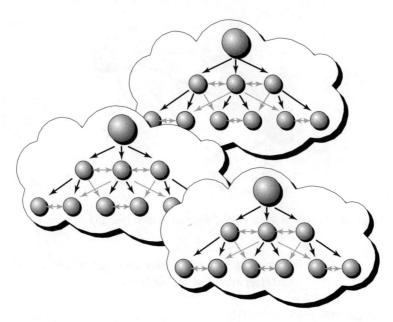

Figure 4.15.   Mixed plesiochronous/synchronous networks (independent despotic states)

HMS strategy is currently the most widely adopted to synchronize modern digital telecommunications networks, owing to its excellent timing performance and reliability, which can be achieved with limited cost. In particular, the latter protection approach (dynamic selection of the synchronization reference), based on a set of standard *Synchronization Status Messages* (SSMs) exchanged among clocks to establish priorities among inputs, is the standard solution recommended to synchronize SDH networks.

4.2.7   Mixed Plesiochronous/Synchronous Networks (Independent Despotic States)

Although most administrations adopted the HMS strategy to synchronize their national networks, it is evident how difficult is to lock all of them to a supranational timing reference.

Actually, the Global Positioning System (GPS) can take this role, from a *technical* point of view, but it is not accepted as first-choice primary reference by many national Administrations for *political* reasons, being it under the control of a foreign Administration (the US Dept. of Defense). Therefore, if we consider the networks of different countries, the most common situation is to have several national synchronous HMS networks each plesiochronous in relation to the others. The political analogy is a set of independent despotic states.

4.3   STANDARD ARCHITECTURES OF SYNCHRONIZATION NETWORK

As mentioned in the previous section, among the several network synchronization strategies conceived, the HMS strategy is the most widely adopted to synchronize modern digital telecommunications networks. Moreover, the ITU-T, ANSI and ETSI bodies have defined standard architectures for synchronization networks according to this strategy.

These HMS standard architectures do not differ in principle. They are organized hierarchically in levels:

- at level 0, one master clock (or more master clocks, for reliability) generates the network reference signal, by running in autonomous mode;
- at lower levels 1, 2, etc., slave clocks
  - are synchronized by signals coming from the upper level,
  - synchronize clocks at lower levels.

Therefore, in normal operation all the network clocks are traceable to the master clock. Synchronization is transferred from one clock to the other according to techniques as outlined in Section 4.1.

In the following sections, the standard architectures defined in the relevant ITU-T, ETSI and ANSI documents are outlined.

4.3.1 The ETSI and ITU-T Synchronization Network Architecture

The ETSI Technical Committee Transmission and Multiplexing (TM3, Working Group 6) began to work on new standards on digital network synchronization in the early 1990s. The main result is the family of documents, covering various aspects of network synchronization, EN 300 462 *Transmission and Multiplexing (TM); Generic Requirements for Synchronization Networks*, whose Part 2-1 defines the *Synchronization Network Architecture* [4.34]. Moreover, the EG 201 793 *Synchronisation Network Engineering* [4.23], drafted later by ETSI beginning from 1999, gives guidance on the manner in which specifications and principles, as laid out in various ETSI and ITU-T documents, can be applied to practical networks, both new and existing. It is a broad survey and provides guidelines for planning, implementing and maintaining the synchronization network.

In parallel with ETSI, in the early 1990s, ITU-T started to rewrite old Recommendations G.810, G.811, G.812 [4.35] (previously published in the Blue Book, 1988) dealing with PDH networks. These Recommendations are not in force anymore and have been substituted by the new coherent set of Recommendations G.810 [4.36], G.811 [4.30], G.812 [4.31] and G813 [4.37], specifying the characteristics of clocks suitable for operation in digital telecommunications networks (including SDH-based ones). Moreover, ITU-T worked on the synchronization network architecture, which is specified in Section 8 of Rec. G.803 *Architectures of Transport Networks Based on the Synchronous Digital Hierarchy (SDH)* [4.22].

Since the same people were working in the ETSI group as well as in the ITU-T committee (and the author of this book was with them for a few years), the two sets of standards are completely aligned and use the same notation, except for minor details. This is true not only for the synchronization network architecture, but also for all performance specifications on clocks. This is the reason why the title of this section refers to both bodies.

4.3.1.1   Scope of the ETSI and ITU-T Standards on Synchronization Network Architecture

The scope of the ETSI EN 300 462 and of ITU-T Rec. G.803 Section 8 is the design of synchronization networks suitable for the synchronization of SDH and PDH networks. Therefore, they apply to the design of new synchronization networks. It is understood that other existing networks (e.g., digital switching networks) may be served by such facility and take advantage of that.

4.3.1.2   Synchronization Strategies

The ETSI EN 300 462 and ITU-T Rec. G.803 Section 8 standards mention that two fundamental methods of synchronizing nodal clocks may be used: master–slave synchronization and mutual synchronization. However, only MS synchronization is judged 'appropriate for synchronizing SDH networks' and recommended for the standard architecture, while mutual synchronization is left for further study.

4.3.1.3   Clock Types and Functions in Synchronization Networks

The following *logical functions*, representing clocks at different levels in synchronization networks, are defined.

- *Primary Reference Clock* (PRC). A PRC is defined as a function that represents either an autonomous clock or a clock that accepts reference synchronization from some radio or satellite (e.g., GPS) signal and performs filtering. The PRC, thus, represents the network master clock. The same expression PRC denotes also the physical implementation of the logical function (i.e., stand-alone clock).

- *Synchronization Supply Unit* (SSU). An SSU is defined as a function that, in a network node:
  - accepts synchronization inputs from external sources, selecting one of them;
  - filters the timing signal derived from this selected source;
  - distributes the filtered timing signal to other elements within the node;
  - may use an internal timing source should all external synchronization references fail or degrade.

  The physical implementation of the logical function SSU may be integrated within an SDH network element or within a PSTN digital exchange, but most commonly is a piece of Stand-Alone Synchronization Equipment (SASE).

- *SDH Equipment Clock* (SEC). A SEC is defined as a function that, in an SDH network element:
  - accepts synchronization inputs from external sources, selecting one of them;
  - filters the timing signal derived from this selected source;
  - may use an internal timing source should all external synchronization references fail or degrade.

  The same expression SEC denotes also the physical implementation of the logical function (internal clock of a SDH network element).

4.3.1.4 Synchronization Network Architecture

Master–slave synchronization is recommended for the synchronization network architecture. Clocks are organized in a hierarchy, in which clocks at each level are synchronized by clocks at higher levels. Four hierarchical levels have been originally defined:

(1) *PRC*, whose characteristics are specified by ITU-T Rec. G.811 [4.30] and ETSI EN 300 462 Part 6-1 [4.34];

(2) *slave clock (transit node)*, most commonly a SASE, whose characteristics are specified by ITU-T Rec. G.812 [4.31] and ETSI EN 300 462 Part 4-1 [4.34];

(3) *slave clock (local node)*, most commonly a SASE, whose characteristics are specified by ITU-T Rec. G.812 [4.31] and ETSI EN 300 462 Part 4-1 [4.34];

(4) *SEC*, whose characteristics are specified in ITU-T Rec. G.813 [4.37] and ETSI EN 300 462 Part 5-1 [4.34].

The PRC is the highest-accuracy clock and the SEC is the lowest-accuracy clock in the hierarchy. In normal operation, all clocks of an MS network are traceable to the PRC.

The frequency accuracy recommended for a PRC is $1 \cdot 10^{-11}$ [4.30]. Therefore, two networks that derive timing from different PRCs will experience at most few slips per year. For this reason, some Administrations choose to establish a network architecture with multiple PRCs, instead of one single MS architecture.

The distinction between transit-node and local-node slave clocks deals with the position of the clock in the synchronization chain (cf. later Figure 4.17). Although still rather common in technical papers, this distinction has not been made anymore in the latest version of ITU-T Rec. G.812 [4.31]. In this Recommendation, slave clocks of three types are specified instead:

● the *Type I clock*, primarily intended for use at all levels in networks optimized for the 2048 kbit/s hierarchy (European networks);

● the *Type II and Type III clocks*, primarily intended for use at transit nodes and local nodes, respectively, in networks optimized for the particular 1544 kbit/s hierarchy that includes the rates 1544 kbit/s, 6312 kbit/s and 44 736 kbit/s (North American networks).

Additionally, this Recommendation includes specifications for other three types of clocks: the Type IV, Type V and Type VI clocks, deployed in existing networks adhering to earlier standards.

Two different categories of synchronization distribution are defined: intra-node and inter-node distribution.

● *Intra-node distribution* is within nodes containing an SSU, e.g. PSTN office buildings. A logical star topology is used. Within node boundaries, all network element clocks are synchronized by the highest hierarchical level clock in the node. Only that clock recovers timing from synchronization links incoming from other nodes. The synchronization network architecture for intra-node distribution is thus depicted in Figure 4.16(a).

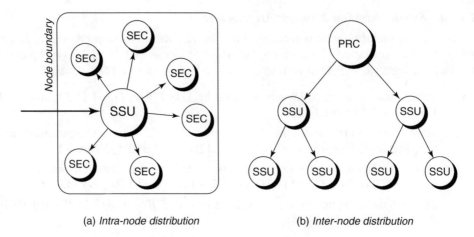

(a) *Intra-node distribution*   (b) *Inter-node distribution*

Figure 4.16. Synchronization network architecture for intra-node and inter-node distribution. (Adapted from Figures 8-1, 8-2/G.783 (4.22) and from Figures 3, 4/EN 462-2-1 (4.34), © ETSI 1999, by permission of ITU and ETSI)

- *Inter-node distribution* is among nodes and follows a tree topology, according to the synchronization network architecture depicted in Figure 4.16(b). For the correct operation of the synchronization network, clocks can be synchronized only by clocks of the same or higher hierarchical level, even under fault conditions. Moreover, synchronization loops must be avoided in any case.

4.3.1.5   Signals and Trails for Intra-Node and Inter-Node Timing Distribution

Intra-node distribution is carried out over short distances (typically less than 100 m). In countries where the European PDH is adopted, physical signals used are the analogue 2.048 MHz signal or the 2.048 Mbit/s E1 signal. In countries where the North American PDH is adopted, on the other hand, the 1.544 Mbit/s DS1 signal is mostly used. The ITU-T Rec. G.703 [4.19] specifies physical characteristics of all the above signals, that are transmitted most commonly over normal copper coaxial cables.

As far as inter-node distribution is concerned, timing is transferred from one node to another along *synchronization trails*, via a distribution network that may use the facilities of a transport network. Synchronization trails may contain SECs, but the timing transfer method should avoid intermediate pointer processing acting on the phase of the signal carrying synchronization. Therefore, the following two possible methods are indicated.

- To recover timing from an incoming STM-*N* signal (cf. Figure 4.7). As noted in Section 4.1.6, this method ensures the best quality of timing transfer, because the effect of unpredictable pointer action is on mapped tributaries and not on the multiplex signal.

- To recover timing from a synchronization trail not supplied by an SDH network, for example from a 2.048 Mbit/s signal carried along a PDH transmission chain (cf. Figures 4.1, 4.4 and 4.5).

4.3.1.6   Synchronization Network Reference Chain

In the master–slave synchronization-network standard architecture, timing is distributed from the PRC down to all the other nodes along chains of slave clocks in tandem connection.

For a better understanding of synchronization network chains, it is important to notice the following basic features that distinguish PRC, SSU/SASE and SEC:

- the PRC, being the network master clock, runs autonomously and must have the highest frequency accuracy $(1 \cdot 10^{-11})$ and stability; it complies with ITU-T Rec. G.811 [4.30] and ETSI EN 300 462 Part 6-1 [4.34];

- the SSU/SASE is a slave clock which distributes its timing to all the other clocks in the office building and to the other 'son' nodes in the HMS architecture; its main duty is then to filter effectively timing impairments on its reference and to supply a highly stable timing should all references fail; its lock time constant is therefore very long (typically, at least 1000 s), to filter out as much phase noise as possible; it complies with ITU-T Rec. G.812 [4.31] and ETSI EN 300 462 Part 4-1 [4.34];

- the SEC is a slave clock with poor long-term stability requirements and with short loop time constant (typically, not more than 1 s); it complies with ITU-T Rec. G.813 [4.37] and ETSI EN 300 462 Part 5-1 [4.34].

The ETSI and ITU-T standards defined thus the *reference chain* shown in Figure 4.17. The first clock in the chain is the PRC. From there, a chain of node clocks (SSUs) is built: intermediate SSUs are transit node SSUs, the last one may be a transit or a local node SSU. SSUs are interconnected via a variable number of SDH network element clocks (SECs).

In general, the quality of synchronization will deteriorate as the number of synchronized clocks in tandem increases. To meet synchronization requirements, the longest chain should not exceed K SSUs with up to N SECs interconnecting any two SSUs, where the values for the worst-case synchronization reference chain are $K = 10$ and $N = 20$, with the total number of SECs in the chain limited to 60. Such values have been derived from theoretical calculations and simulations. However, in practical synchronization network design, the number of network elements in tandem should be minimized for reliability and performance reasons.

In the latest standard documents [4.23], an additional element has been introduced in the chain: the Regenerator Timing Generator (RTG), which consists of a separate timing source for each signal direction of a bidirectional signal that is being regenerated.

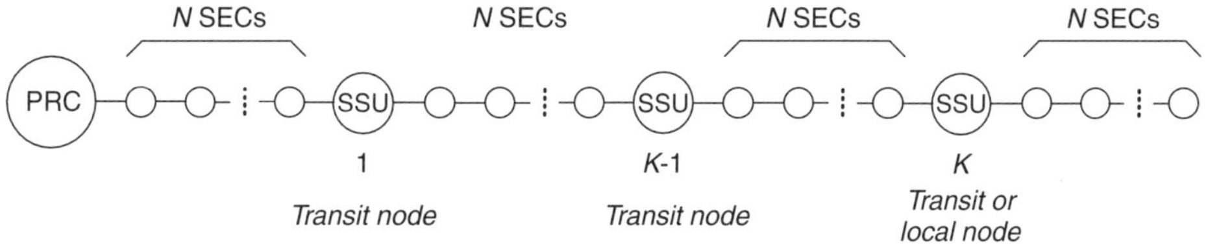

Figure 4.17.   ITU-T and ETSI synchronization network reference chain. (Adapted from Figure 8-5/G.783 (4.22) and from Figure 5/EN 462-2-1 (4.34), © ETSI 1999, by permission of ITU and ETSI)

Regenerators operate in *through timing mode*, in which the timing of the outgoing line signal is derived from the incoming line signal on the same direction. If a signal fails, the RTG supplies timing only for the generation of AIS downstream.

RTGs can be considered fully transparent to the timing of the regenerated signals, as their jitter contribution is usually negligible. Therefore, they are usually not mentioned in the reference chain and are omitted in counting the number of network elements along chains.

4.3.2   The ANSI Synchronization Network Architecture

North American standards on synchronization of digital networks date far before then ETSI and ITU-T standards. As a consequence, under several aspects, they are also less detailed, having paved the way to the development of newer ETSI and ITU-T standards. For this reason, in this section, we just summarize the key features of the ANSI standards on network synchronization.

ANSI released two documents relevant to network synchronization: T1.101 *Telecommunications — Synchronization Interface Standard* [4.38] and T1.105.09 *Telecommunications — Synchronous Optical Network (SONET) — Network Element Timing and Synchronization* [4.39]. Bell Communications Research (Bellcore, now Telcordia), on the other hand, released a few documents on the same subject which are more thorough and provide also planning guidelines: GR-1244 *Clocks For The Synchronized Network: Common Generic Criteria* [4.40], GR-436 *Digital Network Synchronization Plan* [4.41], SR-NWT-002224 *SONET Synchronization Planning Guidelines* [4.42].

The ANSI T1.101 [4.38] states that a commonality of all synchronization networks is the distribution of synchronization reference signals from one or more primary reference sources (master clocks) down to node clocks at lower levels. The master clock is named Primary Reference Source (PRS). From it, synchronization reference signals are distributed down to lower level nodes through Synchronization Distribution Facilities. At the lowest level, the last node distributes synchronization to customer equipment or other networks.

Bellcore, moreover, does not recommend any specific network architecture for synchronization distribution, since several architectures are deemed acceptable. Nevertheless, the following common criteria are indicated for designing a synchronization network architecture: avoidance of timing loops, minimization of length of clock chains and route diversity for alternate timing systems.

4.3.2.1   Strata

ANSI places clocks of synchronization networks into a hierarchy based upon performance levels (*stratum* levels): namely strata 1, 2, 3, 4 and 4E. Moreover, Bellcore has defined an additional level (stratum 3E), which is between ANSI stratum levels 2 and 3. In order to meet a certain level of performance, a clock must meet specific requirements for several aspects, namely free-run accuracy, hold-over, rearrangement timekeeping, hardware duplication and external timing capabilities (for details concerning the quantities used for characterization of clocks, the reader is referred to Chapter 5).

A *Primary Reference Source (PRS)* is a network master clock able to maintain a frequency accuracy of better than $1 \cdot 10^{-11}$ with verification to the UTC [4.38]. One class of PRS is a *stratum-1* clock, i.e. a completely autonomous clock that provides its timing without any external steering reference. Stratum-1 clocks usually consist of an ensemble of caesium atomic frequency standards (for details concerning the technologies of clocks, the reader is referred to Chapter 6). Two examples of PRSs that are not autonomous stratum-1 clocks are LORAN-C and GPS clocks. These systems use local rubidium or quartz oscillators that are steered by timing information obtained from LORAN-C or GPS navigational receivers. LORAN-C is a land-based system that was developed for maritime navigation and is operated by the US Coast Guard. NAVSTAR GPS is a satellite-based system developed for the US Dept. of Defense navigational needs and is operated by the US Air Force. GLONASS is the Russian system and is very similar to NAVSTAR. A GPS clock is able to maintain a frequency accuracy within few parts in 10^{-13} with verification to the UTC.

Stratum-2 clocks (analogous to ITU-T/ETSI transit node slave clocks) are typically found in toll switching exchanges and some digital cross-connect equipment. They are characterized by a free-run long-term frequency accuracy of better than $1.6 \cdot 10^{-8}$, mostly based on oven quartz or rubidium atomic oscillators. Although they may be not supposed to require a double input reference owing to the quality of their hold-over, Bellcore Technical Advisories recommend at least two external input references and automatic protection switching due to the greatest importance of a reliable network synchronization.

Stratum-3 clocks (analogous to ITU-T/ETSI local node slave clocks) are intended for use in local digital switching exchanges, most digital cross-connect systems, some PBXs and T1 multiplexers. They represent a large step down from stratum 2 in terms of hold-over performance and are characterized by a free-run long-term frequency accuracy of better than $4.6 \cdot 10^{-6}$. They are typically based on temperature compensated quartz oscillators. *Stratum-3E* clocks are used in timing signal generators for use in networks with SONET equipment and feature improved short-term stability.

Stratum-4 and *stratum-4E* (specific for customer-premises equipment) slave clocks are used in the inter-office synchronization distribution network and are found in most T1 multiplexers, PABXs, channel banks and echo cancellers. They are characterized by a free-run long-term frequency accuracy of better than $3.2 \cdot 10^{-5}$ and are not required to feature special hold-over performance.

The key requirements of strata slave clocks are summarized in Table 4.1, as specified in [4.38] and in Bellcore documents.

4.3.2.2 The Building Integrated Timing Supply (BITS)

According to ANSI [4.38] and Bellcore, the BITS is the recommended method for intra-office timing distribution: the master timing supply for an entire building. It is the most accurate and stable clock in the building (i.e., the lowest stratum number of the clocks in the building).

The BITS supplies DS1 and or Composite Clock (CC, a special 64-kHz signal format in use in North America for timing distribution) timing signals to all other clocks and digital equipment in the building that require synchronization. Therefore, the BITS clock is the only clock that takes reference from another office building.

The BITS concept is the ANSI equivalent of the ITU-T/ETSI SSU, which was defined later on its base. Two methods of implementing BITS have been in use in North-American networks: conceptual BITS and Timing Signal Generator (TSG) BITS.

Table 4.1   ANSI requirements of strata slave clocks

Characteristic	Stratum 2	Stratum 3	Stratum 4E	Stratum 4
Free-run accuracy	$1.6 \cdot 10^{-8}$	$4.6 \cdot 10^{-6}$	$3.2 \cdot 10^{-5}$	$3.2 \cdot 10^{-5}$
Hold-over stability	$1 \cdot 10^{-10}$/day	$3.7 \cdot 10^{-7}$/day[a]	not required	not required
Time interval error under reference rearrangement[b]	≤ 1 μs	≤ 1 μs	≤ 1 μs	not required
Phase change slope under reference rearrangement	≤ 20 ns in any 14 ms	≤ 20 ns in any 14 ms	≤ 20 ns in any 14 ms	not required
Hardware duplication	required	required	not required	not required
External timing inputs[c]	required	required	not required	not required

[a]Equivalent to less than 255 DS1 slips in the first 24 hours of holdover, as specified in [4.38].
[b]Timing reference rearrangement is defined as a clock switching its reference between two inputs or bridging a short reference interruption.
[c]An external timing input refers to a dedicated-for-timing clock input, used to feed timing directly into a clock when the timing reference does not terminate on a digital transmission system.

Conceptual BITS are made of two elements: a digital switch or a cross-connect system, acting as DS1 source, and a TSG, acting as CC source. The DS1 source uses one of its terminating DS1s from another office as its frequency reference. The TSG is timed by bridging off a DS1 output signal from the DS1 source and, in turn, supplies CC to all equipment that requires CC synchronization.

TSG BITS are made of one single element, TSG, timed by bridging off a DS1 coming from another office and supplying DS1 or CC timing signals to all equipment in the office building.

Bellcore recommends the TSG BITS implementation, analogously to the ITU-T/ETSI SASE implementation of the SSU. Bellcore calls for a highly reliable architecture for BITS clocks and recommends that any implementation with redundant BITS clocks needs to provide some means for assuring that the clocks remain phase aligned.

4.4   SYNCHRONIZATION NETWORK PLANNING

The goal of synchronization network planning is to determine the architecture for distribution of synchronization in a network, the level of clocks and the facilities to be used. Moreover, a performance and reliability assessment of the system designed is needed, to ensure that design requirements are met.

The HMS architecture has been adopted by most telecommunications administrations and is recommended by international standards, as stated before. The first choice is thus planning strategic sites for either one or more PRCs. In the latter case, the system designed will consist of several slightly plesiochronous networks.

Slave clock nodes are usually designed to accept two or more references. Only one reference is active at a time, while all the others are in standby. Suitable protection switching policies must be then selected, in order to ensure the highest synchronization reliability.

4.4.1 General Guidelines

The first rule for designing a good network synchronization plan should be to follow carefully all the applicable ITU-T Recommendations. Moreover, the ETSI EG 201 793 [4.23] provides excellent guidelines for planning, implementing and maintaining a synchronization network. These standards are the latest published and have been developed based on several years of experience on both sides of the ocean. They have been studied in particular for the needs of SDH and SONET networks, but they apply to generic digital networks as well. The standard synchronization network architecture recommended in ITU-T Rec. G.803 [4.22] should be adopted. Moreover, clocks compliant with (or better performing than) ITU-T Recs. G.811 [4.30], G.812 [4.31] and G813 [4.37] should be used. As noticed previously, the ETSI standard EN 300 462 [4.34] is completely aligned with ITU-T Recommendations and use the same notation.

In addition to this basic advice, some further general guidelines may be provided, aiming at achieving the best performance and robustness of the synchronization network designed.

- Design a *HMS plan* with enough levels to cover all strategic sites of the networks to synchronize. A second PRC for backup is recommended.

- *Avoid timing loops.* A timing loop occurs whenever a clock is synchronized by a timing reference that is traceable to that clock itself, directly or indirectly, as shown in Figure 4.18. Such a system of clocks is subject to unpredictable frequency instabilities: clocks in a timing loop tend to operate at the accuracy of their pull-in range (see Chapter 5), many times worse then in free-run or hold-over mode.

  Avoiding timing loops appears trivial when designing the network synchronization plan in normal operation mode, but it may be harder to guarantee under network automatic protection in consequence of failures. There is no simple way of detecting timing loops, as no alarms are generated on their creation. They can go undiscovered until service is affected by poor slip performance leading to investigation that eventually locate the loop. Therefore, automatic protection switching procedures must be carefully designed, so that timing loops cannot form under any circumstance. Timing loops can always be avoided in a properly planned network.

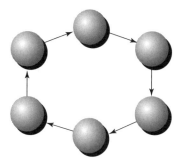

Figure 4.18. Timing loop

- *Maintain the clock hierarchy also after protection rearrangements.* Hierarchical master–slave architecture is recommended by international standards and should be adopted in network planning. Such hierarchy among clocks of different quality should be maintained in all situations, even after protection rearrangements in response to failures. Passing timing from better to worse clocks maximizes network performance especially in terms of hold-over stability, should any of the clocks in the chain fail.

- *Do not neglect the BITS/SASE concept.* The BITS/SASE is the clock designated to receive timing from outside the office building and to time all other clocks in the building. By using BITS/SASE clocks, the best clock controls the building equipment, cascading within the building is minimized and operation is easier. This concept is expounded further in Section 4.4.2.

- *Choose the best timing transfer facilities available.* A facility for timing transfer may be an E1/DS1 signal transported across a PDH network (cf. Section 4.1.5), a SDH STM-N signal, etc. The *best* facility may be defined as the reference with the fewest impairments (i.e., with the least time instabilities such as jitter and wander, phase hits, etc.), with the fewest micro-interruptions, and so on. Some guidelines to this aim are provided in Section 4.4.3.

- *Minimize the length of clock chains in transferring timing.* Cascading clocks in synchronization network chains should be minimized, as timing performance obviously degrades as timing is passed from clock to clock. Although the synchronization network reference chain has been studied to ensure that acceptable performance is met, it is evident that the more clocks are added to the chain and the more likely it becomes that failures make the downstream chain to enter hold-over.

- *Plan route diversity for alternate timing facilities.* If reference redundancy is planned for protection, whatever the facility chosen for timing transfer, alternate timing facilities must follow diverse routes.

4.4.2   Intra-Node Timing Distribution: the BITS/SASE concept

BITS/SASE clocks are recommended for intra-office timing distribution (cf. Section 4.3). They are designated to lock to an external timing reference, incoming from outside the building, and to synchronize all other clocks in the building. Therefore, the BITS/SASE should be the most robust clock, the clock with norrowest loop bandwidth (i.e., that one with the highest input phase noise filtering capability) and the best performing in hold-over among those in the building.

Some administrations rely on clocks equipping large switches or cross-connect systems for the BITS/SSU function. Nevertheless, *good practice suggests using dedicated equipment*, a timing signal generator whose sole purpose is for synchronization. BITS/SASE clocks currently on the market feature outstanding quality and performance monitoring functions that are not achieved by any equipment clocks.

The stratum level required in a BITS/SSU clock is determined by its hierarchical level in the synchronization network. ANSI Stratum-2 or ITU-T/ETSI SASE transit-node clocks are typically used in large toll offices, whereas ANSI stratum-3 (stratum-3E if there is SONET equipment in the location) or ITU-T/ETSI SASE local-node clocks are intended for use in smaller locations such as local offices.

For reliability, the BITS/SASE should receive two or more references from other locations. If dedicated facilities are used to time the BITS/SASE clock, the reference timing signal terminates directly on the clock interface. Whenever traffic-bearing facilities are used, the reference timing signal cannot terminate on the clock interface, but must be split or duplicated by means of a bridging repeater right before being terminated at the input interface of the receiving digital equipment.

Several methods are in use to supply synchronization to equipment within the office building, over distances typically not longer than 100 m: dedicated E1/DS1 synchronization signals, clock distribution units, synchronous clock insertion units, composite clock lines and plain square and sine wave signals.

Dedicated E1 or DS1 signals, at 2.048 Mbit/s or 1.544 Mbit/s respectively, are typically used to synchronize digital switching exchanges and non-SDH cross-connect systems. They are transmitted most commonly over normal copper coaxial cables. Their physical characteristics are specified by ITU-T Rec. G.703 [4.19]. If dedicated stand-alone equipment is used to implement the BITS/SSU function, as recommended, the BITS/SASE supplies the signals above directly from output synchronization cards. If a switch or cross-connect system includes the BITS/SSU clock, a bridging repeater must be used to supply timing signals duplicated from traffic-bearing signals output by the equipment.

A *Clock Distribution Unit* (CDU) is a device used to supply multiple reference signals, even of various physical characteristics, directly synthesized from one input reference. This device is not a BITS/SASE clock, as it does not include a slave oscillator and does not retime the signal. It is just something like a frequency synthesizer, which mainly acts as electronic buffer and pattern generator. It has been widely in use in the AT&T synchronization network, since the 1980s.

The signals output directly by the BITS/SASE clock, a bridging repeater or a CDU are connected to the external synchronization inputs of equipment in the office building. If some old equipment does not have external synchronization input, it can be synchronized by either dedicating traffic ports for synchronization use only or by using a *Synchronous Clock Insertion Unit* (SCIU), which is a device whose only function is to retime an E1/DS1 signal passed through it. By means of such device, BITS/SASE clocking can be injected on traffic-bearing lines and sent to equipment without external synchronization port. (It is needless to notice that slips will effect the retimed signal). Also this device has been in wide use in the AT&T synchronization network.

In modern equipment, the external synchronization port interface is the already mentioned standard sine-wave *2048 kHz signal* specified in Section 13 of ITU-T Rec. G.703 [4.19]. Some other equipment, in countries following the North American standard, accepts square or sine-wave *1544 kHz* or *8 kHz timing signals*. The 2048 kHz interface is the solution for intra-office timing distribution recommended by ITU-T and ETSI standards.

Another signal commonly adopted in North America for intra-office timing distribution is the so-called *Composite Clock* (CC), used for synchronizing equipment with DS0 interfaces (e.g., channel banks). The Composite Clock is a 64 kHz bipolar square wave, including 8 kHz timing in the form of a bipolar violation every 8 bits.

4.4.3 Guidelines for Inter-Node Timing Distribution

Timing is transferred from one node to another along *synchronization trails*, by means of facilities provided by a suitable transport network. In order to achieve effective timing transfer, it is essential to choose the best facility among those available, i.e. the reference with the fewest synchronization impairments or micro-interruptions. Some general guidelines to select the best facility for timing transfer are the following.

- Use optical STM-N/STS-N signals (cf. Section 4.1.6), in case along chains of SECs as in the synchronization reference chain defined in ETSI and ITU-T standards. This should be the solution of choice, because of the highest quality achieved by transferring timing on such optical carriers.

- If the above choice is not possible, the use of an E1/DS1 link transported across a PDH network is a good solution, because of the timing transparency of PDH networks on payloads (cf. Section 4.1.5). Some jitter is present on the tributaries recovered at the end of the PDH transmission chain, but the narrow bandwidth of BITS/SASE clocks in the receiving office building allows to filter out effectively most of it.

- Even though PDH transmission chains usually offer good timing transparency, for the E1/DS1 synchronization link a careful synchronization network planner should select routes that cross only fibre systems and with the minimum number of multiplexing/demultiplexing stages. Some administrations rely on dedicated timing-only links between office buildings, for improved ease of maintenance and supposed increased reliability and performance, but actually synchronization E1/DS1 links may carry traffic without particular drawbacks.

- As noticed in Section 4.1.6, payload signals carried on SDH/SONET should not be used to transfer timing, as they may exhibit excessive jitter especially under pointer action. Of course, the actual amount of jitter measured at signal interfaces depends on several factors, namely the type of payload (E1 vs DS1), the pointer adjustment rate along the chain of network elements, the particular design of synchronizers and desynchronizers actually deployed, etc. However, the basic fact is that using for network synchronization signals, which are severely affected by possible network synchronization impairments because of pointer action, is a logical vicious circle.

- Similarly, signals transported by ATM CBR services may exhibit excessive wander, even if SRTS is used for packet jitter equalization, and therefore should not be used for timing transfer.

- Other examples of transport facilities that should not be used are satellite links, owing to a diurnal Doppler-induced wander, and most radio relay systems, still for wander reasons.

4.4.4 Synchronization Network Planning in SDH and SONET Networks

Especially in the case of SDH and SONET networks, the first rule for designing a digital network synchronization plan, once again, should be to follow carefully all the applicable ITU-T recommendations. In particular, the specifications for the synchronization network reference chain (Figure 4.17) and for the PRC, SSU, SEC clocks should be kept in mind.

Synchronization network architecture in SONET-based networks may feature some peculiarity. In SONET networks, often there are no intermediate network element clocks in synchronization chains. Thus, the typical chain is made of BITS, receiving the reference from the input STS-N line through a SONET network element (analogously to the scheme of Figure 4.7) and forwarding the timing by synchronizing the SONET network element that outputs a STS-N line towards the next node. Most operators limit such chain length to not more than six nodes.

Bellcore recommends that BITS clocks in such a chain are stratum 2 or stratum 3E. These clocks have good phase noise filtering capabilities and good stability under rear-rangements and in hold-over.

Another peculiarity of SONET synchronization networks is the bandwidth of SONET network element clocks compared to that of SDH network element clocks (SECs). SECs are characterized by a broader bandwidth, in the order of $1 \div 10$ Hz, aiming at keeping phase transients short, a fact that is appreciated for example under synchronization rear-rangement along an SDH ring. SONET network element clocks, on the other hand, are not required to have such a speed constrain and have a filtering bandwidth in the order of 0.1 Hz. Hence, the synchronization rearrangement of a ring made of, say, 16 network elements can last as much as 5 min with SONET equipment, but only 10 s with SDH equipment.

4.5 SYNCHRONIZATION NETWORK MANAGEMENT

Analogously to traffic transmission and switching networks, modern synchronization networks do not consist merely of sole transport facilities and clock equipment, but are also provided with management and protection systems, which ensure the highest quality, reliability and dependability of the overall synchronization distribution. Simply stated, a management system provides the administrator with network view.

The most advanced synchronization networks are managed by systems designed according to the ITU-T *Telecommunications Management Network* (TMN) standard [4.43]. Some others are managed by systems that follow the *Simple Network Management Protocol* (SNMP) standard, originally designed for Internet equipment. All the latest synchronization network management systems, however, provide a rich set of management functions in order to allow full control and operation of the synchronization network.

This subsection just reviews some very basics of network management, referring in particular to synchronization network management. For a detailed treatise of OSI, TMN and SNMP standards, the reader may browse for example references [4.44]–[4.46].

4.5.1 Functional Areas and Abstraction Levels of Network Management

The OSI Management Framework first identified five functional areas of network management:

- *fault management*, dealing with the detection, isolation and correction of abnormal operation;

- *configuration management*, dealing with equipment configuration setting and identification;
- *accounting management*, dealing with the identification of costs for the use of
  hardware and software resources;
- *performance management*, dealing with Quality of Service (QoS) data logging;
- *security management*, dealing with the procedures enabling the manager to operate
  the functions that secure the managed resources from user misbehaviour and unauthorized access.

These five areas are now widely adopted to classify all the various management functions. To denote these areas, the term FCAPS is commonly used, from the five initial
letters of the functional areas.

The functional areas above are a *horizontal* classification of management functions. In
other words, various management functions are classified in that five sets according to
their different domains of application. On the other hand, management functions can be
also classified in a *vertical* fashion, according to what logical abstraction level they apply
to. Hence, management functions are commonly defined at the following hierarchical
abstraction levels:

- *network element management*, dealing for example with the configuration setting of
  tributary ports or the detection of hardware faults in single network elements;
- *network management*, dealing for example with establishing and releasing circuits
  along an SDH ring (configuration management), detecting circuit outages (fault
  management), logging Bit Error Rate (BER) data on circuits established (performance management) or recording information on the use of network resources
  (accounting management);
- *service management*, dealing with the configuration, the faults, the accounting, etc.
  of the service offered to a client, for example an internetworking service consisting
  of several connection and connectionless transport services among access points;
- *business management*, at the highest level of abstraction, dealing with the management of the overall set of services offered to customers.

At each level, as made clear by the examples above, the FCAPS classification still
applies. Moreover, management functions at one level rely on management functions at
lower hierarchical levels to provide their services. The model above can be thus illustrated
as in Figure 4.19. It is worthwhile noticing that, while technical complexity is higher at
lower levels, most revenues come from top levels.

The main management functions relevant to synchronization network management lie
in four areas and can be summarized as follows.

- *Fault management* is the set of facilities that enable the detection, isolation and
  correction of abnormal operation (fault). Possible causes of abnormal operation
  are design and implementation errors, external disturbances and lifetime expiration.
  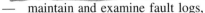Examples of faults are loss of reference signals at the inputs of clocks, hardware
  failures, etc. Therefore, fault management includes functions to:
  — maintain and examine fault logs,

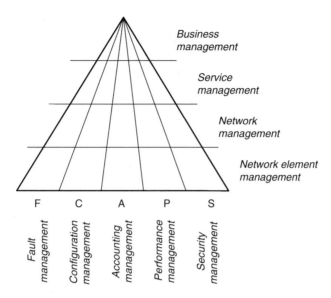

Figure 4.19. Functional areas and hierarchical abstraction levels of network management

   — accept and act upon fault notification (*alarms*),

   — trace and identify faults,

   — carry out diagnostic tests,

   — correct faults, whenever possible (e.g., start a synchronization trail rerouting for protection).

• *Configuration management* is the set of facilities that for example:

   — record current configuration data (e.g., loop time constant in slave clocks),

   — record changes in configuration data,

   — update configuration data,

   — identify network components upon turn-on and remove them from network image upon turn-off,

   — assign symbolic names at network equipment.

• *Performance Monitoring* is needed to optimize the QoS. To detect changes in the network's performance, statistical data are collected in real time and logged on a periodical basis. The use of such logs is not restricted to performance management, because also other management areas take advantage of these logs:

   — performance logs can be used by fault management to detect faults,

   — performance logs can be used by configuration management to decide when configuration changes are needed,

   — performance logs can be used by accounting management to document QoS with customers.

Moreover, to allow a meaningful comparison of performance logs, also the configuration existing at the time that performance logs were created must be recorded. Therefore, configuration information must be logged jointly.

- *Security management* facilities are the set of procedures ensuring that only authorized personnel are enabled to operate the various management functions. Under
this aspect, configuration functions are critical: only the synchronization coordinator should be enabled for example to change reference priorities for protection
switching, loop time constants, etc.

As far as the abstraction levels defined previously are concerned, synchronization
network management systems are focused on element management (i.e., for example
on the management of clocks, their configuration, their active references, etc.) and on
network management (i.e., for example on setting up synchronization trails). As self-
evident, a synchronization network management system should work in close interaction
with the management system of the transmission network, which is in charge of trans-
porting timing signals.

4.5.2   TMN Management

The term *Telecommunications Management Network* (TMN) was introduced by ITU-T and
is defined by Rec. M.3010 [4.43]. According to Rec. M.3010, '*a TMN is conceptually a
separate network that interfaces a telecommunications network at several different points*'.
   The general model of a TMN and its relationship with the telecommunications network
that is managed is illustrated in Figure 4.20. According to this figure, the *Operation
Systems* (OS) are connected with the network elements (switching nodes and transmission
equipment) of the telecommunications network managed via a *Data Communications
Network* (DCN).
   The OSs represent the 'intelligence' of the management system and perform most of the
management functions. They process the information collected from the managed systems,
take decisions and send commands to the managed systems. It is even possible that

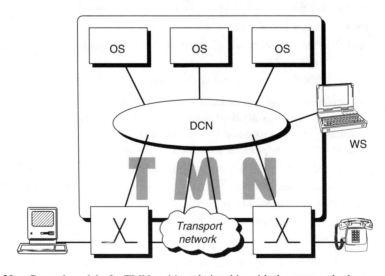

Figure 4.20.   General model of a TMN and its relationship with the managed telecommunications
network

management functions are performed by several OSs, which still exchange information through the DCN.

The DCN represents thus the communication medium between all the entities involved in the management system and the managed network. It is logically distinct from the telecommunications network managed, but physically it may even coincide with it. For instance, a practical implementation of the DCN may be a X.25 data packet-switched network, distinct from the telecommunications network managed. Conversely, in SDH networks the D1-D3 and D4-D12 channels of the SOH offer transmission capacity available to management message transmission; in this second example, therefore, the DCN physical implementation coincides with the telecommunications network managed.

The management system is not designed to operate completely autonomously, without human interaction and supervision. The *Work Station* (WS) is the communication medium with the human operator.

The TMN boundaries in Figure 4.20 are not accidental. Actually, the TMN includes the OSs, the DCN, part of the WS and part of the managed systems too. In fact, while the DCN-WS interface is specified by relevant TMN standards, the human-WS interface falls outside the scope of TMN. In an analogous fashion, only the interface between the DCN and the managed switching nodes and transmission equipment is part of the TMN.

4.5.3 TMN Functional Architecture, Reference Points and Interfaces

A set of macro-functions has been defined in TMN standards, to represent managing and managed entities, communication means, etc. The *Operation System Function* (OSF) represents the functions performed by the OS, as described above. The *Network Element Function* (NEF) represents the set of functions implemented in the Network Element managed, interacting with the TMN. The *Mediation Function* (MF) adapts, filters and synthezises the information flow between NEF (or its QAF, defined next) and OSF. The *Q-Adapter Function* (QAF) represents the interworking function between the TMN and non-TMN-standard equipment. The *Work Station Function* (WSF) provides the communication medium among OS, NEF, MF and the human operator. Finally, the *Message Communication Function* (MCF) and the *Data Communication Function* (DCF) represent the protocol stacks for communication between all the functions above: the MCF represents protocol stacks at OSI layers 4–7, the DCF the protocol stacks at OSI layers 1–3.

The functional architecture of the TMN, based on the blocks listed above, is thus depicted in Figure 4.21. Between blocks, *reference points* are defined which represent conceptual points where information is exchanged:

- *points f* are the points of information exchange between WFS and OSF or MF;

- *points x* are the points of information exchange between two OSFs of different TMNs;

- *points q* are the points of information exchange between, basically, the managed systems (i.e., NEFs or QAFs) and the management system (i.e., OSFs); two classes of points q are defined:

  — *points qx* between NEF and MF, MF and MF, QAF and MF,

  — *points q3* between NEF and OSF, MF and OSF, QAF and OSF, OSF and OSF (of the same TMN);

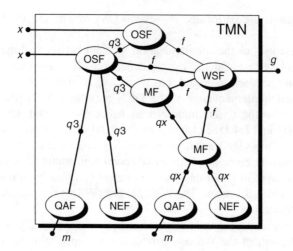

Figure 4.21.   TMN functional architecture and reference points

- *points m* (external to the TMN) are the points of information exchange between QAF and non-TMN entities;
- *points g* (external to the TMN) are the points of information exchange between WSF and human operator.

In correspondence with x, f and q reference points, X, F and Q *interfaces* have been defined respectively, for specification of standard communication protocols. On the other hand, points m and q, falling outside the scope of TMN, do not correspond to standard interfaces. As obvious, the communication interface between a TMN entity and a non-TMN network element (point m) cannot follow any standard specification. In an analogous way, the interface between the TMN WS and the human operator (point g) has not been specified in TMN standards, letting for instance software engineers free to design the graphical user interface they prefer.

At Q interfaces, three families of communication protocol stacks have been defined:

- *Qx interface* protocol stacks have been specified for management communication on local area (Network Elements, Mediation Devices and Q-Adapters are usually located in the same building);
- *Q3 interface* protocol stacks have been specified for management communication on geographical area (Network Elements, Mediation Devices and Q-Adapters are usually located in a different building than their managing Operation Systems);
- *Qecc interface* protocol stack has been specified [4.47] for management communication among SDH Network Elements (e.g., along a SDH ring) and is characterized by the use of SOH D1-D3 and D4-D12 channels as physical transmission medium.

4.5.4   Management Systems for Synchronization Networks

Modern synchronization networks can be managed by TMN-standard systems. Likewise, many PRC equipment and SASE currently on the market are provided with a Q-interface management port, to allow connection with TMN-standard systems. If clocks are not equipped with standard Q-interfaces, yet the same rich set of management functions is provided through proprietary interfaces. Also in this case, however, the synchronization-network management system can be provided with Q-interfaces to allow interworking with other management systems (for example, that one of the transport network).

The main management functions carried out by such synchronization network management systems are those already outlined in the first part of this section. Now, a few complementary notes will be given to better outline the main requirements of these systems.

4.5.4.1 Configuration Functions

The basic requirement of a synchronization-network management system, as far as the configuration-management area is concerned, is the capability of *displaying the current configuration* of network clocks (functional level of network element management). The network administrator, thus, should be able to retrieve all relevant configuration parameters (e.g., operational state, input priorities, switch states, etc.) from all pieces of equipment that are part of the synchronization network. These are the PRC and SASE clocks, but may also include the SECs of SDH equipment.

Second, the synchronization-network management system should be able to *display the current network synchronization topology* (functional level of network management), i.e. the tree-shaped graph consisting of clocks and active and non-active synchronization trails.

Configuration management is often referred to as *provisioning*. It is worthwhile to remark that in some telecommunications company, where we may say that a very traditional approach to network management is still followed, the feature of synchronization remote provisioning is judged not desirable. In this case, provisioning can be done for example by setting hardware dip-switches. The rationale of this choice is the fear that the possibility of remote provisioning may make more likely that incompetent personnel makes inappropriate configuration changes and the feeling that clocks should be configured at installation, or reconfigured solely under the synchronization coordinator supervision. Obviously, modern TMN systems offer enough security management procedures to make remote provisioning a secure and advisable feature.

A BITS/SASE clock may be equipped with hundreds of timing outputs, directed to equipment deployed on different floors in the building. Therefore, for synchronization troubleshooting, it is essential that the management system keeps track of clock resources, state and destination of clock outputs and so on. Also building maps and drawings should be stored in the system database and updated regularly, to be readily available in case of need.

4.5.4.2 Performance Monitoring Functions

Basic performance monitoring functions, in synchronization network management systems, simply keep track of the number of slips occurring at digital interfaces and sometimes record occurrence times of SDH/SONET pointer justification events. Advanced systems allow one to perform more sophisticated measures of synchronization performance and are able to detect timing degradations before they impact service. The topic of synchronization performance monitoring is treated in the next section.

4.5.4.3   Fault Management

Basically, fault management functions involve, on the one hand, the detection of failures and, on the other, the restoration of timing distribution (trouble-shooting).

Detection of synchronization faults can be carried out by monitoring directly synchronization signals and equipment for failures or by monitoring transmission facilities for impairments due to synchronization failures (e.g., slips). However, detecting slip and pointer adjustment alarms to trace synchronization failures can be complicated, as synchronization failures can be located remotely from the node experiencing slips or excessive pointer action.

Restoring inter-office timing distribution after a failure consists in setting up suitable algorithms for automatic reference switching in SASE/BITS clocks. In designing such algorithms, the aim is to avoid service degradations, by keeping all alternate references locked to the PRC, to make reference switching smooth, and carefully avoiding timing loops.

Moreover, the management system should allow any manual interaction in order to alleviate existing problems. For example, manual and forced selection of input references, as well as the capability of disabling input ports, are useful features.

4.5.4.4   Integration with Transmission Network Management System

As already noticed, synchronization network management systems should work in close interaction with the transmission network management system. Particularly remarkable is the case of integration between the synchronization-network management system and the SDH/SONET-transmission-network management system.

An example of implementation of such architecture is depicted in Figure 4.22. This example is an advanced architecture, because it assumes that SASE clocks are provided with Q-interfaces and managed by the same Element Manager that manages SDH Network Elements in an office building. More in detail, the role of the various blocks included in this example architecture can be summarized as follows.

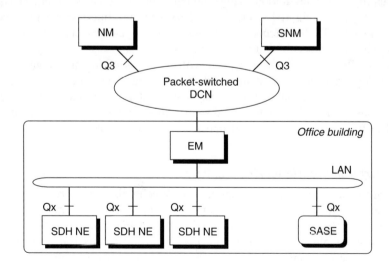

Figure 4.22.   Example of integration between a synchronization network management system and an SDH management system according to the TMN standard

The *Element Manager* (EM) implements functions belonging to the OSF and to the MF. It manages all the SDH network elements and the SASE clock deployed in the building where it is installed, working at the network element management level. Management communication, between the EM and the managed SDH NEs and the SASE clock, is carried out via a Local Area Network (LAN) through Qx-interface protocol stacks.

The *Network Manager* (NM) implements functions belonging to the OSF. It manages the overall national SDH transmission network, at a higher hierarchical level (network management level), through the lower level EMs deployed in remote buildings.

The *Synchronization Network Manager* (SNM) implements functions belonging to the OSF. Analogously to its peer-level NM, it manages the overall national synchronization network, at the network management level, through the lower level EMs.

All communication between the NM, the SNM and the EMs is carried out via a packet-switched public geographic network, implementing the DCN function. Q3-interface protocol stacks are used. As evident, severe security issues must be faced, when choosing to use a public communication network to transport management messages.

In another, more common example of integration between the synchronization-network management system and another network management system, the SASE clocks are managed by dedicated EMs or directly by the SNM, through proprietary interfaces. The SNM is then connected to the NM of the other network management system through a Q-interface, again via a public geographic network. In this case, integration is done only at the highest hierarchical level of management systems.

4.6 SYNCHRONIZATION NETWORK PERFORMANCE MONITORING

Advanced synchronization networks are provided with monitoring systems that allow to verify continuously, in real time, the performance achieved in timing distribution.

In principle, the synchronization performance at a certain office building can be verified by comparing the local timing (for example extracted from an E1 or DS1 line coming out from the local SASE) to the timing derived from a known, highly accurate reference, ideally to the UTC. Monitoring an E1/DS1 signal bearing traffic can be done in real-time and continuously by using a bridging repeater, to tap into the E1/DS1 line to monitor, and by performing some measure on it in comparison to some reference signal.

Synchronization performance data collected at each node are reported in real time to the synchronization network management system. It is then duty of the OS to process and synthetize raw performance data, to produce meaningful reports and to establish corrective actions if needed.

The rationale of synchronization performance monitoring is the need to be *proactive*, i.e. to detect timing degradations well before they impact service. There are several possible sources of severe timing impairments: maintenance and circuit provisioning activities, clock diagnostics and rearrangements, timing-transfer facility micro-interruptions, etc. Most of these phenomena yield abrupt phase hits on the timing signal, or even loss of it, and thus can be quite easily recognized. Nevertheless, being capable of detecting more subtle timing degradations promptly, such as for example slow frequency drifts or excessive phase wander with null average frequency error, may be the key factor enabling to guarantee that the quality of service will not be suddenly affected at a later time. Such subtle timing degradations are definitely not easy to diagnose with traditional alarm reporting systems.

It is needless to say that quality of service degradations due to some synchronization problem look always sudden, unexpected and of mysterious origin for almost everybody *but* the synchronization engineer. Therefore, a synchronization performance-monitoring system may be effective to automatically identify synchronization problems and to support correcting actions before the service is affected.

4.6.1 Synchronization Failures and Impairments

As already remarked in the introduction paragraphs, timing failures and impairments can be of different severity and origin. They can be classified in synchronization hard and soft failures.

Synchronization hard failures are those caused by equipment hardware or cable outages. Examples of equipment outages may be failures in micro-components (i.e., integrated circuits, capacitors, etc.) or in sub-systems (i.e., clock units, output cards, power supply, etc.). Cable outages are mostly caused by construction machines, which may damage the cables during road works, and are reported most often.

Most commonly, hard failures are detected by the downstream slave clock as *Loss of Signal* (LOS) or *Alarm Indication Signal* (AIS) alarms. The LOS is the state of alarm detected by an input card that does not sense a valid input signal anymore (that's what happens if the direct input cable is cut). The AIS is the alarm sent downstream by transmission equipment under severe state of alarm, for example if the upstream card detects a LOS.

Upon detecting a hard failure on the current reference input timing signal, the slave clock should select an alternative reference signal or enter hold-over mode. The slave clock should also report the hard failure to the network management system.

Possible causes of *synchronization soft failures*, on the other hand, are slow frequency drifts and excessive jitter and wander on the synchronization signals. Slow frequency drifts can be caused by timing loops, after inappropriate protection switching, or by intermediate transmission equipment entering hold-over mode. Most common sources of jitter and wander on digital signals used to carry synchronization, furthermore, have been discussed already in previous chapters.

Severe synchronization soft failures can yield occasional *Loss of Frame* (LOF) or slips at the input ports of transmission and switching equipment. Less severe soft failures may still be detected by plain slave clocks, when the synchronization input signal exceeds the frequency and wander error limits preset for the reference signal. Subtle synchronization soft failures may be deceitful and not manifest themselves until they do not worsen enough to suddenly affect the traffic signals. Hence the need of an accurate synchronization performance-monitoring system, which enables the network manager to proactively identify synchronization problems and to solve them before the service is affected.

4.6.2 Strategies for Synchronization Network Performance Monitoring

Performance monitoring of a synchronization network is based on *local* measurements of the timing performance, carried out in as many office buildings as possible.

In principle, such local measurements should be carried out by comparing the local timing to an UTC-traceable signal. In that way, a clock would know exactly its phase or frequency error compared to the ideal condition. Unfortunately, this approach is not practical: should the UTC be available in each office building, it may be better used directly as master reference for the building equipment, rather than to use it just to check the actual accuracy of the local timing derived from some other source.

Therefore, local synchronization quality measurements in an office building can be carried out by comparison among the sole available signals: the local timing signal, i.e. that generated by the local SASE clock, and one or more external timing signals incoming from the network. The next subsection will deal with algorithms that can be used to this purpose. Scope of this subsection, instead, is to discuss possible strategies for organizing synchronization performance monitoring in a wide area network.

According to the most general approach, SASE clocks are equipped with performance-monitoring cards that compare the local timing signal to one or more external references, for measuring the phase error between them. Two kinds of external reference can be used for synchronization performance monitoring at a certain SASE clock:

- *terminated synchronization signals*, i.e. input timing signals terminating a synchronization trail at that node;

- *return synchronization signals*, i.e. timing signals transported back to the node from the far end of synchronization trails originating from the node.

Using the former kind of reference corresponds to a *decentralized strategy* of synchronization network performance monitoring. Clocks located at terminal points of synchronization trails must be provided with performance-monitoring capability. The main advantages are that only one monitoring card is needed in each of those clocks, at least in principle if there are not several terminated synchronization signals for protection, and that a failure in a monitoring node would not affect the monitoring capability of the rest of the network.

Conversely, using the latter kind of reference corresponds to a *centralized strategy* of synchronization network performance monitoring. Clocks located in a central location carry out performance measurements on several return synchronization signals, as phase difference with the local timing. Compared with the decentralized strategy, in this case the main advantages are that fewer clocks need monitoring capability and that return signals are monitored at a central location, by direct comparison to the original reference they are supposed to trace. The main disadvantage is that central monitoring clocks must be equipped with several monitoring cards, because usually it is desirable to monitor as many return signals as possible.

It is evident that the centralized strategy allows to recognize rather easily synchronization network misbehaviours. If the central monitoring node detects synchronization impairments on a single return line, then it is reasonable to infer that the clock at the far end of that line is having problems. Conversely, should all the return lines exhibit synchronization impairments, then it is most likely that the central node reference is misbehaving.

Obviously, both strategies can be combined to set up a mixed strategy, optimized according to the specific case. Various examples of both strategies are shown in the scheme of Figure 4.23. Therein, wide grey arrows indicate how synchronization propagates from

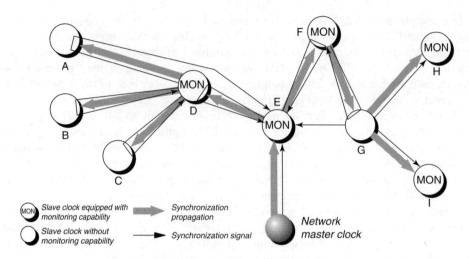

Figure 4.23. Centralized and decentralized strategies for synchronization network performance monitoring

the network master clock to slave clocks. Moreover, black arrows denote synchronization signals carried in some way along the transmission network: for example E1/DS1 signals carried on PDH systems (cf. Figure 4.5) or synchronization trails built as chains of SECs (cf. Figure 4.17).

Clocks marked with A, B, C and G are not equipped with monitoring capability. Clocks marked with D, E, F, H and I carry out performance monitoring: among them, clocks H and I perform measurements on terminated synchronization signals (decentralized strategy); clocks D, E and F perform measurements on return synchronization signals (centralized strategy). Monitoring return signals must not be done necessarily on signals returning from direct slave clocks in the timing distribution hierarchy: clock E monitors a return signal from clock D (direct slave) and a return signal from clock A (slave of clock D). Likewise, clock E monitors a return signal from clock F and a return signal from clock G (slave of clock F). Finally, note that clock E may perform measurements also on the terminated synchronization signal from the network master clock.

4.6.3   Methods for Synchronization Network Performance Monitoring

As explained in the previous paragraphs, local measurement of network synchronization performance is carried out in a network node by comparing the local timing signal, i.e. that generated by the local BITS/SASE clock, to one or more external timing signals incoming from the network. In particular, two timing signals are compared by measuring the phase or time error between them and then by evaluating suitable quantities for synchronization quality assessment. This function is usually integrated within the BITS/SASE clock and is implemented by special performance-monitoring input cards. The same performance-monitoring card is able to detect hard failures and alarms caused by soft failures.

4.6.3.1   Basics

According to the decentralized and centralized synchronization performance monitoring strategies outlined previously, comparing the local timing to the terminated synchronization signal means to measure the phase or time error[6] between the input reference and the output signal of the slave BITS/SASE clock (as shown in Figure 4.24(a)). On the other hand, comparing the local timing to a return synchronization signal means to measure the phase or time error between the clock output signal and another external reference, which is supposed to trace the monitoring clock (as shown in Figure 4.24(b)). Therefore, the performance-monitoring card in Figure 4.24 includes the following basic functions:

- measurement and acquisition of time error data between the clock output signal and each of the reference signals;

- evaluation of synchronization quality quantities based on the acquired data sets[7];

- communication with the management system for performance-monitoring reporting.

Slave clocks are treated in the next chapter, but since now it is worthwhile remarking some notes on what can be measured according to the schemes of Figure 4.24.

A PLL-based slave clock, locked to an external reference, can be modelled as a low-pass filter acting on the phase fluctuations present on the input timing signal. In most practical cases, such low-pass filters can be described with a simple two-poles transfer function $H(f)$, whose most important parameter (at least for this purpose) is the cut-off frequency f_c. Very coarsely speaking, this means that phase fluctuations on the input timing signal having Fourier frequency $f < f_c$ are transferred to the output timing signal, whereas phase fluctuations having Fourier frequency $f > f_c$ are smoothed or cancelled. In the time domain, the reciprocal $1/2\pi f_c$ is the *loop time constant*, i.e. some measure of the filter transient response duration, of how long the slave clock takes to track phase steps at the input[8].

4.6.3.2   Monitoring the Terminated Synchronization Signal

In the scheme depicted in Figure 4.24(a), performance monitoring is carried out by measuring the time error between the input and the output of the local slave clock. This configuration is standardized as *synchronized-clock measurement configuration* by ETSI EN 300 462 [4.34] and ITU-T G.810 [4.36]. Let

$$H(f) = \frac{\Phi_{\text{out}}(f)}{\Phi_{\text{in}}(f)} \tag{4.1}$$

be the low-pass transfer function of the PLL, where $\Phi_{\text{in}}(f)$ and $\Phi_{\text{out}}(f)$ are the Fourier transforms of the phase noise functions $\varphi_{\text{in}}(t)$ and $\varphi_{\text{out}}(t)$ on the input and output timing signals respectively. Then, the most remarkable facts are that:

---

[6] The *phase error* (difference) $\Delta\varphi$ between two timing signals is measured in [radians], often with the ambiguity of an integer number of full circles ($\Delta\varphi = \Delta\varphi + 2k\pi$). The *time error* is measured in [ns] and has no ambiguity: it means the total delay of one timing signal compared to the other. In clock stability characterization the time error is most commonly measured. Further details on phase and time error quantities will be provided in the next chapter.

[7] Most commonly, the *equivalent slip rate* and the standard stability quantities *MTIE* and *TDEV* are evaluated. The concept of equivalent slip rate will be expounded in the following of this subsection, while the MTIE and TDEV quantities will be defined in the next chapter.

[8] More precisely, the output signal goes to about 60% of the final value after one time constant and to about 99% of the final value after five time constants.

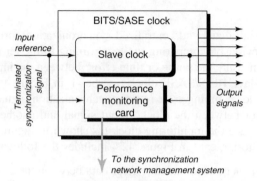

(a) *Comparing the output timing to the input reference*

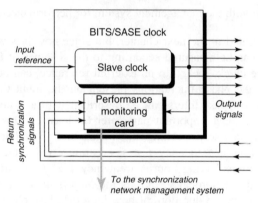

(b) *Comparing the output timing to return synchronization signals*

Figure 4.24. Synchronization performance monitoring in a BITS/SASE clock according to (a) decentralized and (b) centralized strategies

- the transfer function from the input phase to the input–output phase difference (as measured according to Figure 4.24(a))

$$\frac{\Phi_{in}(f) - \Phi_{out}(f)}{\Phi_{in}(f)} = 1 - H(f) \qquad (4.2)$$

is high-pass, i.e. only high-frequency $(f > f_c)$ phase noise from the input is revealed by measuring the input–output phase error;

- as it will be shown in the next chapter, internal oscillator phase noise is transferred high-pass filtered to the output phase $\varphi_{out}(t)$ and thus also to the phase error $\varphi_{in}(t) - \varphi_{out}(t)$; phase noise from other internal sources such as the phase detector, on the other hand, is transferred to the output low-pass filtered.

Therefore, to summarize and to give some empirical guidelines:

- by measuring the quantity $\varphi_{in}(t) - \varphi_{out}(t)$ according to the scheme of Figure 4.24(a), phase impairments present on the input reference and those generated by the slave clock itself are revealed mixed in the measurement results;

- for $f < f_c$, the slave clock is supposed to track input phase fluctuations; therefore, should slow phase changes be measured, they must be due to some clock internal misbehaviour (i.e., the slave clock is losing track and is generating the phase change);

- for $f > f_c$, the slave clock cuts off input phase fluctuations; therefore, should fast phase fluctuations such as steps or spikes be measured, they must be due most likely to the input reference line rather then to the local oscillator, which usually has a very good short-term stability (i.e., the slave clock is commonly assumed to work correctly, whereas the upstream clock does not).

4.6.3.3 Monitoring Return Synchronization Signals

In the scheme depicted in Figure 4.24(b), performance monitoring is carried out by measuring the time error between the output of the monitoring clock and a few return synchronization signals, coming from remote lower level nodes that are synchronized by the monitoring clock, directly or indirectly, and that are therefore supposed to trace it.

As in the previous case, generally it is not possible to distinguish unambiguously whether phase impairments revealed in the measurement results are due to the local oscillator or to some lower level clock in the downstream chains, from whose far ends return synchronization signals come back. Nevertheless, as already mentioned before, the availability of several return synchronization signals to monitor allows to apply a sort of majority voting in interpreting measurement results: reasonably, if a single return line exhibits impairments, then that slave clock chain will be assumed to misbehave; conversely, should all the return lines exhibit impairments, then the local clock will be probably assumed to misbehave.

Also in the scheme of Figure 4.24(b), measurements are carried out according to the synchronized-clock measurement configuration [4.34][4.36]. In the previous case, when monitoring the terminated synchronization signal (Figure 4.24(a)), the input and the output of the local slave clock were compared. In this case, instead, the input and the output of a lower level clock chain, slave of the local clock, are compared. Analogous considerations as in the previous case can be thus made, yielding the following empirical guidelines (for the sake of simplicity, all BITS/SASE clocks in the network are assumed to be characterized by the same cut-off frequency f_c):

- phase impairments generated by the local clock and by lower level clocks result mixed in the measurements; nevertheless, majority voting can help to interpret results;

- for $f < f_c$, the slave clock chain is supposed to track phase fluctuations generated by the local clock; therefore, should slow phase changes be measured, they must be due to some downstream misbehaviour (i.e., some element in the slave clock chain is losing phase-lock condition);

- for $f > f_c$, slave clocks cut off input phase fluctuations; therefore, should fast phase fluctuations such as steps or spikes be measured *on a single* return line, they must be due most likely to some clock on the trail from where that line comes back; should phase fluctuations be measured *on all* return lines, they must be due most likely to a local clock misbehaviour (e.g., an input reference rearrangement).

4.6.3.4   Role of the Synchronization Network Management System

From that stated above, it should be clear what the key role of the synchronization network management system is, which gathers performance data from all monitoring nodes.

The management system should allow not only the historical recording of raw measurement results, but also a synoptical graphical representation of data collected in every node (sequences of time error samples, stability quantities evaluated as functions of the observation interval, etc.). The aim is at helping the human operator to interpret raw figures and to ascertain what clocks are misbehaving, in case of impairments detected in some monitoring nodes. Possibly, a function of intelligent synthesis of data collected from several monitoring nodes may assist the human operator, automating in some way the majority-voting rule mentioned above.

4.6.3.5   Evaluation of the Equivalent Slip Rate

The performance-monitoring card acquires time-error data, measured between its input timing signals, and then evaluates some quantities for synchronization quality assessment. In advanced synchronization networks, the equivalent slip rate and the standard stability quantities MTIE and TDEV are most commonly evaluated, aiming at detecting timing degradations before they impact service.

Estimating the *equivalent slip rate* between two timing interfaces means to evaluate, by suitable algorithms, the controlled-slip rate that would be detected should two timing signals be used to drive the write and read processes in a bit-synchronizer buffer. The value of such a slip-monitoring feature is thus to emulate the possible effect of measured phase impairments on the operation of an actual network. Monitoring and counting slips occurred on digital interfaces of switching exchanges is *reactive*, as we record data loss when it has already happened. Measuring the time error among several, strategically chosen couples of nodes in the network and assessing the equivalent slip rate (or also more advanced stability quantities such as MTIE and TDEV) is *proactive* instead, because it provides a more general view of the network synchronization status: all nodes are under control if a good monitoring strategy is planned. Therefore, we are warned also about slips that *may* occur, should we feed a digital switch with signals coming from any two nodes of the network.

In the next chapter, the stability quantities MTIE and TDEV will be defined and algorithms for their estimation will be described. In this paragraph, instead, some hint for equivalent slip rate assessment is provided.

An algorithm for equivalent slip rate estimation evaluates, over certain time intervals, the expected number of controlled slips on a given timing interface between two primary rate digital signals. That is, given time-error data measured between two timing signals, the operation of an elastic store is emulated, in order to count the number of controlled slips that would occur, over certain time intervals, by using that timing signals for reading and writing.

Precisely speaking, the actual sequence of controlled slips depends not only on the sole time-error random process between read and write clocks, but generally also on:

- the *initial phase or time alignment* (i.e., the initial buffer fill level);
- the *hysteresis* value adopted in comparing the time error with thresholds upon taking the slip decision.

If time error is dominated by sole frequency offset, after an initial transient the slip rate does not depend on the initial buffer fill level neither on the hysteresis value (cf. Equation 2.9). Nevertheless, when zero-mean jitter and wander dominate, as between input and output of a system of slave clocks, the two factors above should be taken into account in emulating accurately the behaviour of actual buffers.

A possible way of dealing with the issue of the initial phase alignment in performance-monitoring cards is to emulate a set of multiple slip buffers, each with different initial fill level, uniformly chosen within the entire threshold spacing (equal to one PCM frame length, i.e. 125 μs in time units). For example, emulating an initially empty buffer and an initially full buffer yields the limit values of the slip rate. As far as the hysteresis value is concerned, ITU-T Recs. G.823 and G.824 [4.21] specify a minimum value of 18 μs (by the way, the designers of several widely deployed telephone digital switches, such as the AT&T 4ESS, adopted this value to set the hysteresis cycle in the input buffers).

Therefore, a simple algorithm, which emulates a slip buffer characterized by a given initial fill level *START*, could be the following, expressed with some pseudo-C programming language just for the purpose of explanation:

```
THR=125 μs;              /* 1 controlled-slip size on a PCM signal */
START=62.5 μs;           /* initial phase alignment, chosen in this example
                            equal to half controlled-slip size */

ΔH=18 μs;                /* hysteresis value */
get_measurement (x0);    /* acquisition of the first time-error sample */
Δx=START-x0;             /* time offset to add to each sample so to have
                            an initial phase alignment equal to START */

repeat {                 /* data acquisition and buffer emulation loop */
    get_measurement (x);        /* acquisition of a time-error sample */
    if (x+Δx>THR+ΔH) {          /* positive slip? */
        signal (SLIP+);             /* positive slip detected! */
        Δx=Δx-THR; }               /* time offset slipped back */
    else if (x+Δx<-ΔH) {        /* negative slip? */
        signal (SLIP-);             /* negative slip detected! */
        Δx=Δx+THR; }               /* time offset slipped forth */
} until (end_of_monitoring);
```

Obviously, the program above has been written just for the sake of providing an example. Practical implementations of equivalent slip monitoring algorithms may differ substantially, for instance in the treatment of the hysteresis process and of the issue of the initial phase alignment.

4.7   SYNCHRONIZATION NETWORK PROTECTION: SYNCHRONIZATION STATUS MESSAGING

The value of network synchronization is so high that the highest reliability and dependability must be ensured. Therefore, clocks are usually duplicated. Moreover, automatic synchronization-protection algorithms allow clocks to recover timing from at least two alternate synchronization trails, provided over diversely routed paths. Such synchronization-protection algorithms are designed so that the network hierarchy is always maintained: clocks recover timing only from clocks of upper or equal level.

An example of HMS architecture with protection trails is depicted in Figure 4.25. Primary synchronization trails are drawn as black arrows, whereas secondary protection trails are drawn in grey. Each clock is fed with at least two synchronization trails, coming from 'father', 'brother' or 'uncle' nodes in the hierarchy. Obviously, a priority scheme must be established, to allow clocks to switch from one reference to another following the rules of the protection algorithm designed.

Hard- and soft-failure detection at clock inputs is the most common switching trigger. For example, at the input of a SDH network element, detection of the following states of alarm on the synchronization interface usually initiates synchronization protection switching: *Loss of Signal* (LOS), *Alarm Indication Signal* (AIS), *Loss of Frame* (LOF), *Excessive Bit Error Rate* (E-BER), *Loss of Pointer* (LOP). Moreover, even slight synchronization quality degradations, detected by performance-monitoring cards, may be configured by the network manager to be a proactive protection-switching trigger.

4.7.1   Criteria for Designing a Synchronization Protection Algorithm

The availability of multiple references in network clocks implies that some selection algorithm has to be designed, in order to allow the network to restore proper synchronization distribution over alternate paths in case of failure. The ETSI EG 201 793 [4.23] recommends the following basic requirements that should be kept in consideration, when designing a synchronization protection algorithm:

- no timing loops may be created after protection rearrangements of synchronization distribution;

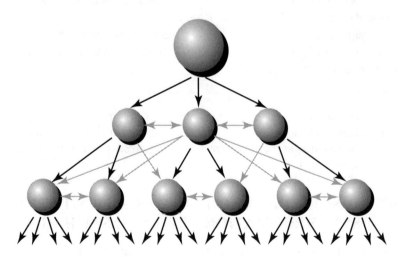

Figure 4.25.   Hierarchical master–slave synchronization with alternate protection trails

- when a clock enters hold-over more, it should not be a reference for clocks of better stability (higher hierarchical level);

- each network element should be synchronized by the highest-quality source among those available;

- the number of reference switches should be as small as possible.

4.7.2 Synchronization Status Messages

The ITU-T [4.22][4.48], ETSI [4.23][4.34] and ANSI [4.41][4.50] standard bodies have defined automatic mechanisms for a selection of the timing reference source and self-healing from synchronization failures, based on a quality marking scheme that provides an indication of the quality of the timing, by using *Synchronization Status Messages* (SSM) embedded in synchronization signals.

SSMs are identifiers embedded within digital signals that are used as synchronization sources for BITS/SASE clocks or other network elements. With SSMs, the originating node declares the quality of the synchronization source to which its signal is ultimately traceable (i.e., the type of clock to which it is synchronized directly or indirectly via a chain of network element clocks) or whether that signal is usable as synchronization source. This quality information is based on the type of source clock (PRC, SASE, SEC, etc.) and does not reflect any actual quality measurement. Therefore, it does not reflect degradations due to transport jitter and wander, incidental clock impairments, or so.

SSMs have been defined to allow clocks to select the best synchronization source (or better, the one *declaring* the best quality) among the available references according to a priority scheme. Their main purpose is to avoid timing loops, while allowing clocks to autonomously reconfigure after synchronization failures.

SSMs are designed to allow synchronization reference selection strategies compatible with most network topologies, such as point-to-point, linear, mesh or ring topologies. In particular, SDH/SONET rings have been deployed very widely and in increasing numbers, as they are inherently self-healing by automatic traffic rerouting if lines are interrupted. However, traffic restoration does not necessarily imply synchronization restoration. Support of SSM fills this gap, by allowing automatic recover also from synchronization failures.

4.7.3 Synchronization Status Messages in SDH Networks

In SDH networks (according to ITU-T and ETSI standards), SSMs are carried on STM-N signals in bits 5 through 8 of the MSOH byte S1 (cf. Figure 2.23 and Table 2.9), as defined by ITU-T Rec. G.707 [4.48]. Table 4.2 reports the bit values assigned to standard SSMs. Four patterns are assigned to the four G.803 hierarchical levels (cf. Section 4.3.1): PRC, SASE transit node, SASE local node, SEC. Moreover, two additional bit patterns are assigned: one to indicate that synchronization quality is unknown and the other to indicate that the signal should not be used for synchronization. The remaining codes are reserved for further quality levels that may be defined by individual administrations or vendors.

The *Quality unknown* message is defined to allow integration of non-messaging nodes into networks supporting SSM.

Table 4.2   Standard SSM for use in SDH networks according to ITU-T Rec. G.707 [4.48]

SDH S1 Bits b5-b8	SDH Synchronization Quality Level Description
0000	*Quality unknown*
0001	Reserved
0010	*Primary Reference Clock (PRC) (ITU-T Rec. G.811)*
0011	Reserved
0100	*SASE transit node clock (ITU-T Rec. G.812)*
0101	Reserved
0110	Reserved
0111	Reserved
1000	*SASE local node clock (ITU-T Rec. G.812)*
1001	Reserved
1010	Reserved
1011	*Synchronous Equipment Timing Source (SETS) (ITU-T Rec. G.813)*
1100	Reserved
1101	Reserved
1110	Reserved
1111	*Do not use for synchronization*

The reason of the choice of the '1111' pattern for the *Do not use for synchronization* message lies in the fact that such pattern may be emulated by equipment failures and in particular by a *Multiplex Section AIS* (MS-AIS) signal. This assignment choice is mandatory because the receipt of an MS-AIS is not necessarily interpreted as an indication of a failed synchronization-source physical interface. This assignment, therefore, allows this state to be recognized without interaction with the MS-AIS detection process.

4.7.4   Synchronization Status Messages in SONET Networks

Synchronization status messaging in North American SONET networks (according to ANSI standards) is quite similar to the mechanism defined by ITU-T and ETSI for SDH networks. However, the message interpretation and messaging requirements are somehow different, due to the SONET synchronization network peculiarities [4.51].

Also in SONET, SSMs are carried in bits 5 through 8 of the S1 byte, located in the STS-N overhead in the same position as the S1 byte in STM-N SOH. The main difference is that clock hierarchical levels associated with SSMs are the strata defined in Bellcore and ANSI standards [4.38][4.40][4.41]. Table 4.3 reports the set of bit values assigned to SONET SSMs, with their quality level as specified by ANSI. Also here, the remaining codes are reserved for further quality levels that may be provisioned by individual operators.

4.7.5   Synchronization Status Messages in PDH Networks

In PDH networks, it is possible to convey the SSM in the overhead bits of the primary multiplex signals E1 and DS1 used as timing distribution facilities, according to frame formats specified by ITU-T Rec. G.704 [4.52].

Table 4.3 Standard SSM for use in SONET networks

ANSI Quality Level	SONET S1 Bits b5-b8	ANSI SONET Synchronization Quality Level Description
1	0001	Stratum 1 (PRS)
2	0000	Traceability unknown
3	0111	Stratum 2
4	1010	Stratum 3
5	1100	SONET equipment internal clock
6	N/A	Stratum 4
7	1111	Do not use for synchronization

Table 4.4 Standard SSM for use in ANSI DS1 Extended Super Frame data link

ANSI Quality Level	DS1 ESF Data-Link Code Byte	ANSI SONET Synchronization Quality Level Description
1	00000100	Stratum 1 (PRS)
2	00001000	Traceability unknown
3	00001100	Stratum 2
4	00010000	Stratum 3
5	00100010	SONET equipment internal clock
6	00101000	Stratum 4
7	00110000	Do not use for synchronization

In the European E1 signal, one of the TS0 bits 4 through 8 of frames not containing the frame alignment signal (S_{a4}, S_{a5}, S_{a6}, S_{a7} and S_{a8} bits) can be used to convey SSM (cf. Section 2.2). The bit chosen to convey SSM constitutes thus a 4 kbit/s data-link channel. SSMs are usually coded with the same 4-bits words standardized for SDH networks (see Table 4.2) and are transmitted in that bit position over a multiframe spanning four E1 basic frames.

As far as the North American standard is concerned, two DS1 multiframe structures are standardized: the Super Frame (SF) over 12 DS1 basic frames and the Extended Super Frame (ESF) over 24 DS1 basic frames, as mentioned in Section 2.2. In the ESF structure, overhead F-bits are shared among an ESF alignment signal, a cyclic redundancy check for error control and a data link to service purposes. The SSMs can thus be transmitted in the ESF data link, using the code bytes reported in Table 4.4 [4.51].

Finally, we mention that SSMs have been defined also for PDH distribution trails carrying ATM, as specified by ITU-T Rec. G.832 [4.53].

4.7.6 Rules for Selecting the Active Reference after Failure

The full algorithm of SSM operation is described in ETSI EN 300 417-6 [4.49]. ETSI EG 201 793 [4.23] also summarizes its description.

First, the quality level of the incoming signals is considered and the references with highest available quality level are singled out. The quality level of a signal is determined by reading the SSM in its overhead. If an incoming reference fails (e.g., LOS), its quality level is set to 'do not use for synchronization'. After the signals with highest available quality level have been singled out, a priority assigned by the network administrator may be used to further order them.

Then, the reference with highest priority, among the group of references with highest quality level, is selected. In case some inputs have equal priorities assigned, the reference switch is made to an arbitrary input, except when the current reference is among this group; in that case, no switch is made.

4.7.7   Basic Rules for SSM Generation

The basic rules for SSM generation at the outputs of a network element are quite straight-forward.

- The quality of the current selected reference for the internal clock is retransmitted to all outputs, with the exception of the return direction of the input supplying the current reference. This output is set to 'do not use for synchronization', in order to avoid timing loops.
- If no references with adequate quality are available, the network element enters hold-over mode and sends SSM stating the quality level associated with its internal oscillator to all outputs.

4.7.8   Example of Automatic Synchronization Protection in an SDH Ring by Use of Synchronization Status Messages

In this section, we present a simple example of automatic procedure based on SSM for synchronization protection in an SDH ring. Several other examples are described in the ETSI EG 201 793 [4.23].

SDH and SONET rings have been very widely deployed worldwide, because of their inherent robustness to link and node failures. Several automatic procedures for traffic protection have been specified in relevant international standards [4.54][4.55]. Anyway, traffic restoration does not necessarily imply synchronization restoration. Therefore, SSMs have been defined purposely to fill this gap, by allowing automatic recover also from synchronization failures.

The example of synchronization protection procedure is outlined in three steps in Figure 4.26. Six nodes (SDH ADMs), numbered 1 through 6, are connected by two counter-rotating rings of fibre links, one for working traffic and one for protection traffic (unidirectional two-fibres ring). Node 1 is located in a central office building where a SASE clock (compliant with ITU-T Rec. G.812) is deployed. Clocks of SDH ADM equipment are compliant with ITU-T Rec. G.813.

In normal operation (Figure 4.26(a)), the ring is synchronized by node 1 (main node), synchronized by the local SASE clock, synchronized by a timing signal traceable to the network PRC (compliant with ITU-T Rec. G.811) through the synchronization network. All ring nodes, with the exception of the main node, can recover their timing on both sides from working- and protection-traffic links, circulating in the two directions along the ring. In this example, all nodes are normally locked to the working-traffic link, running clock-wise (see the thick grey arrows). Therefore, all nodes are transmitting to the clockwise

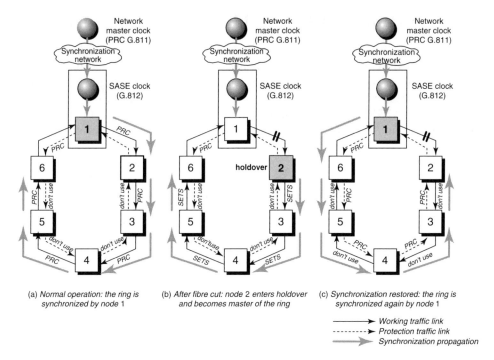

(a) *Normal operation: the ring is synchronized by node 1*

(b) *After fibre cut: node 2 enters holdover and becomes master of the ring*

(c) *Synchronization restored: the ring is synchronized again by node 1*

———————▶ *Working traffic link*

- - - - - - -▶ *Protection traffic link*

══════▶ *Synchronization propagation*

Figure 4.26. Synchronization protection in an SDH ring by use of SSM

next node the SSM 'PRC (G.811)', to indicate that their timing is traceable to the network PRC, and to the counter-clockwise next node the SSM *do not use for synchronization*, to prevent timing loops. Moreover, node 1 is sending the SSM 'PRC (G.811)' also to its counter-clockwise direction.

After a link failure on the working-traffic link between nodes 1 and 2 (Figure 4.26(b)), the ADM clock at node 2 loses its synchronization reference and, after a certain delay, enters hold-over mode. Then, it begins to send to the clockwise next node the SSM 'SETS (G.813)', to inform subsequent nodes that from now on they are recovering their timing from an SDH internal clock (SEC) running autonomously. Then, also node 3, after having received the SSM 'SETS (G.813)' from node 2, begins to send to the clockwise next node the SSM 'SETS (G.813)'. Nodes 4 and 5 do likewise.

Restoration of optimal synchronization (Figure 4.26(c)) begins when node 6, as soon as it receives the SSM 'SETS (G.813)' from node 5, switches reference and locks to node 1, from which it is receiving a better SSM 'PRC (G.811)' (synchronization priority). After some time to allow full reference switching, node 6 begins to transmit to node 5 (counter-clockwise direction) the SSM 'PRC (G.811)', to indicate that its timing is now traceable to the network PRC. Likewise, node 5, as soon as it receives the SSM 'PRC (G.811)' from node 6, switches reference from node 4, from which it is receiving the SSM 'SETS (G.813)', to node 6. In the same way, nodes 4, 3 and 2 switch reference at their turn. At the end of this reconfiguration process, optimal synchronization is restored: the ring is synchronized again by node 1, while timing is now distributed counter-clockwise carried on protection-traffic links.

4.8   EXAMPLES OF SYNCHRONIZATION NETWORKS

The aim of this final section is to provide a brief survey on some synchronization networks realized by major telecommunications operators worldwide and to show practical application of the concepts expounded throughout this chapter. Among the many possible examples, synchronization networks designed and realized in the United States of America by AT&T, in Switzerland by Swiss PTT, in Japan by NTT, in France by France Telecom, in Argentina by Telecom Argentina and in Italy by Telecom Italia have been chosen.

These networks have been selected for a short presentation in this book mainly considering their historical relevance in the technical development of synchronization networks, but also because of the availability of public information on their architecture and implementation. Actually, most operators are traditionally very discreet about releasing public technical reports that detail the characteristics of their networks and equipment. The lack of public information is the main reason why some undoubtedly important operators have not been mentioned in this section and why most examples expounded here are not the most recent.

4.8.1   The Synchronization Network of AT&T (USA)

The American Telephone and Telegraph Co. (AT&T) has been the first telecommunications operator worldwide to set up a synchronization network, in order to provide synchronization to its own national telecommunications network, to customer private digital networks, to most of the Bell Operating Companies (BOCs) and to many independent telephone companies. The brief information provided in this section is summarized from [4.16].

As mentioned in Section 4.1.1, the first synchronization network was deployed by AT&T in the 1970s, to solve the issue of distributing pilot frequencies to the FDM multiplexers and demultiplexers [4.4]. Pilot frequencies were distributed as 2.048-MHz analogue signals derived from a caesium-clock ensemble, known as the Basic Synchronization Reference Frequency (BSRF) and located in Hillsboro, Missouri. The BSRF reference signals were distributed on analogue radio and coaxial systems.

Even the most critical FDM systems could tolerate a 10^{-9} frequency offset, easily guaranteed by that BSRF analogue distribution systems even in the presence of transmission impairments. Nevertheless, the completion of the all-digital transmission and switching network led AT&T first to improve the existing analogue frequency distribution system and then to plan a new synchronization network.

This new synchronization network was planned and deployed by AT&T starting from the early 1980s. It is the first example of a synchronization network in the world designed according to modern principles, for synchronizing digital switching exchanges and SONET transmission equipment. A third, newer synchronization network has been planned for the beginning of the third millennium.

AT&T's synchronization network is designed according to a multiple-master HMS strategy, relying on GPS to provide long-term timing accuracy. This synchronization network has the capability of quality verification at several levels, by the deployment of a Timing Monitoring System (TMS) to verify the precision of timing distributed throughout the nationwide network.

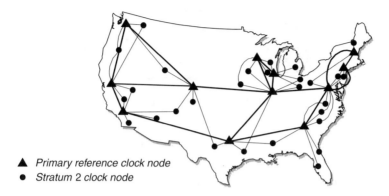

▲ *Primary reference clock node*
● *Stratum 2 clock node*

Figure 4.27. Pictorial representation of the topology of the AT&T synchronization network. (Adapted from [4.16], © 1989 IEEE, by permission of IEEE)

At the first level of the synchronization hierarchy, there are 16 PRCs deployed at strategical sites: 14 are located on the North American continent, one on the Hawaii islands and one in Puerto Rico, as shown in the pictorial representation of the network topology in Figure 4.27 (as reported in [4.16]).

PRCs consist of two secondary atomic frequency standards (rubidium oscillators) steered in the long run by a GPS receiver. Hence, the PRC's short-term stability is that provided by rubidium oscillators. Moreover, optimum hold-over performance is achieved.

The PRC reference signal is transmitted from each primary node to its neighbouring secondary nodes, which do not contain a PRC. Each PRC simultaneously monitors the reference it receives from at least two other PRC nodes as well as the secondary nodes to obtain long-term stability data. In this way, the long-term timing stability of each PRC location is tracked and verified to an accuracy of a few parts in 10^{-13}, or virtually that of UTC. The architecture is designed so that secondary nodes are independently monitored by different PRCs. The timing distribution facilities among PRCs are DS1 signals at 1.544 Mbit/s, carried over PDH optical transmission systems.

Within a primary node building, as shown in Figure 4.28, the PRC provides timing to a stratum-2 clock, incorporated in either a DACS II transmission system or a No. 4 ESS switching system. This clock, in turn, provides timing to all other clocks in that building, taking the role of BITS. The equipment incorporating the stratum-2 clock (viz. either a DACS II or a No. 4 ESS) outputs traffic-carrying DS1 signals for inter-office timing reference distribution. For intra-office timing, the stratum-2 clock supplies direct input to a Clock Distribution Unit (CDU, cf. Section 4.4.2), whose function is to provide multiple DS1 timing outputs to all clocks within the node requiring digital timing reference (e.g., No. 4 ESS, DACS I, No. 5 ESS, etc.) as well as to frequency-locked clocks for analogue FDM equipment.

At the second level of the synchronization hierarchy, secondary nodes contain only the stratum-2 clock, incorporated in either a DACS II transmission system or a No. 4 ESS switching system. The stratum-2 clock receives its DS1 timing references from diverse primary nodes via the so-called DSX-1 cross-connect panel, as shown in Figure 4.29. Thereafter, the intra-office reference distribution systems are the same for both primary and secondary nodes.

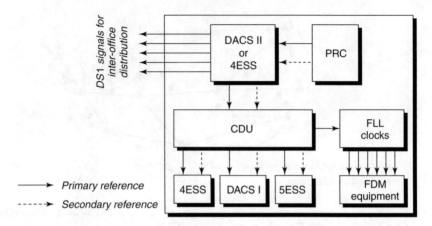

Figure 4.28. Timing distribution within a primary node of AT&T's synchronization network. (Adapted from [4.16], © 1989 IEEE, by permission of IEEE)

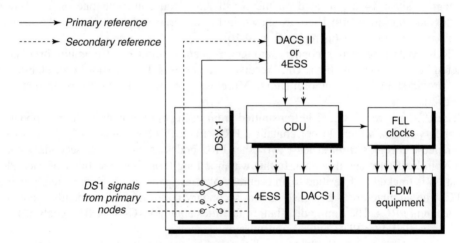

Figure 4.29. Timing distribution within a secondary node of AT&T's synchronization network. (Adapted from [4.16], © 1989 IEEE, by permission of IEEE)

4.8.2   The Synchronization Network of Swiss PTT

This section provides a brief overview of the Swiss synchronization network deployed in the decade 1975–85, as described in [4.56][4.57]. This network was planned according to the HMS strategy, as usual, but with some interesting peculiarity.

As shown in the scheme of principle of Figure 4.30, the synchronization network is divided in three timing regions of about equal importance. At the first level of the hierarchy, there are three caesium-controlled master centres, one per region. At the second level, there are building clocks (according to the same principle of modern SASE/BITS) realized with high-stability quartz PLLs with digital memory. At the third level, there are clocks of digital switching exchanges. At the fourth level, there are clocks of Private Automatic Branch Exchanges (PABX) and other terminal equipment.

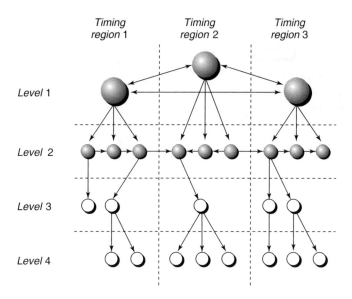

Figure 4.30.   Principle architecture of the Swiss PTT synchronization network

The three first-level master centres control more than 60 second-level clocks, located at all major nodes of the digital network. All international gateways are located at first- or second-level nodes.

The three regions operate in plesiochronous mode, although within very strict frequency tolerance ranges due to the etomic CS Master Clocks. In case of failure of a regional master, that region is automatically enslaved to one of the other masters. There is another fundamental reason to the choice of the number three of masters: three is the minimum number of clocks allowing an early detection of a fault by comparison and majority decision.

This principle has been applied also at the second hierarchical level. Second-level clocks are based on three high-stability quartz PLLs, each enslaved to a different reference line. A surveillance unit with majority decision allows to identify a clock showing excessive drift quickly and locally, i.e. without needing a further external reference.

The first- and second-level centres are connected through the long-distance trunk line network. The timing transfer facilities are 2.048-Mbit/s E1 signals, carried over PDH radio-relay and optical-fibre systems, operating at 34 or 140 Mbit/s and 140 or 565 Mbit/s respectively. The long-distance network has a meshed architecture, but the timing links are selected so to form tree structures with root on the master centres.

There are no more than two second-level centres in cascade. Each second-level centre receives its reference signals over three redundant links chosen to be diversely routed. The lower order centres are connected through the district and local networks.

The block scheme of a first-level node clock is depicted in Figure 4.31. The master oscillators comprise one caesium primary atomic frequency standard and two PLL units (PLL A and PLL C) equipped with a BVA quartz crystal oscillator. The two PLL units A and C are each enslaved to a reference signal received from the other two first-level network master clocks. The reference signals are tapped from 2.048-Mbit/s E1 signals by means of a Timing EXtraction unit (TEX), which generates a 2.048 MHz sine wave. The

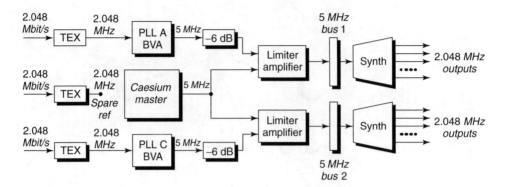

Figure 4.31.   First-level node clock of the Swiss PTT synchronization network [4.56][4.57]

TEX unit is also able to detect AIS on the reference line and to notify it to a supervision system.

The 5 MHz output signal from the caesium master is split and then separately combined with the outputs of each PLL on two 5 MHz bus units. The 5 MHz PLL output signals are both attenuated by a factor of two (−6 dB) before vector addition to the master signal, thus allowing any phase relationship between the two components (cancellation on π-phase opposition is prevented). Then, amplitude variations are suppressed in the bus units by means of a limiter. Slow phase variations on the vector sum, due to wander and systematic frequency differences, are limited to 33 ns peak-to-peak amplitude, according to the analysis and experiments carried out by Kartaschoff et al. [4.56][4.57].

In case of failure of the caesium unit, its output is squelched and cut off from the bus. The PLL units A and C then continue feeding the twin buses, each PLL remaining enslaved to one of the other two first-level network master clocks. Should all three caesium masters fail simultaneously, the three first-level node clocks fall into a mutual control mode. A third, spare reference line can be manually plugged into the PLL units in case of emergency.

The two 5 MHz buses feed a twin redundant pair of synthesizers, generating the 2.048 MHz output signals for distribution to multiplexing and switching equipment in the building.

The block scheme of a second-level node clock is depicted in Figure 4.32. Its structure is very similar to that of a first-level node clock. The only difference is the presence of three PLL units A, B and C instead of the caesium master with two PLL units characterizing first-level node clocks.

Each of the three PLL units receives a 2.048 MHz reference from a TEX unit connected to a separate 2.048 Mbit/s line, originating either from the regional master or from another second-order centre directly controlled by the regional master. The two 5 MHz buses are fed by sum signals of PLL units A + B and B + C. Therefore, if one of the PLL units fails, both buses remain active.

As in first-level node clocks, the two 5 MHz buses feed a twin redundant pair of synthesizers, generating the 2.048 MHz output signals for distribution within the building.

A control unit monitors phase differences between the input lines and those between the output lines. A normalized frequency departure of any output with respect to the others exceeding 4×10^{-11} causes the corresponding PLL unit to be cut off from the buses and

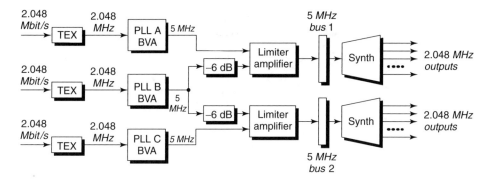

Figure 4.32.   Second-level node clock of the Swiss PTT synchronization network [4.56][4.57]

an alarm indication signal. Moreover, the status of the three PLL units is monitored and compared, allowing unambiguous decision by majority voting.

The third hierarchical level of the synchronization network is constituted by clocks of digital switching exchanges. These systems are provided by different suppliers and therefore exhibit different characteristics. Only general input and output signal parameters and the jitter transfer characteristics have been specified, according to relevant ITU-T Recommendations.

Clocks of digital switching exchanges are synchronized via an analogue 2.048 MHz reference signal from the second-order clock deployed in the same building, if any, or via at least two redundant 2.048 Mbit/s E1 lines from remote second-level nodes.

The lowest, fourth level of the synchronization hierarchy comprises customer-premises equipment such as digital PABX and all kinds of digital terminal equipment. All this equipment is synchronized by the digital signal incoming from the network.

4.8.3   The Synchronization Network of NTT (Japan)[9]

The synchronization network of Nippon Telephone and Telegraph Co. (NTT) has been operating since 1978. It has been designed to synchronize both POTS and data systems.

The NTT's synchronization network is designed according to the HMS strategy, with duplicated PRC for reliability, as shown in the scheme of principle of Figure 4.33. The two Master (M) nodes are located in Tokyo and in Osaka and are equipped with caesium primary atomic frequency standards.

At the second hierarchical level, there are several Sub-Master (SM) nodes, distributed over the geographical area of Japan. Below the second level, more than 2000 Slave (S) network clocks (BITS/SASE) are organized in another six levels. These slave clocks are called Clock Supply Module (CSM) by NTT. They were originally based on quartz oscillators and could accept sine-wave 1.544-MHz and 6.312-MHz reference signals on the synchronization interface.

The inter-office timing transfer facilities were synchronous 1.544-Mbit/s and 6.312-Mbit/s digital signals, carried via T1 and T2 transmission systems or multiplexed in higher order PDH systems. The timing transfer architecture was similar to that depicted in

---
[9] The brief information provided in this section has been collected with the help of Dr. Masami Kihara, with NTT Network Innovation Laboratories. The author warmly thanks him.

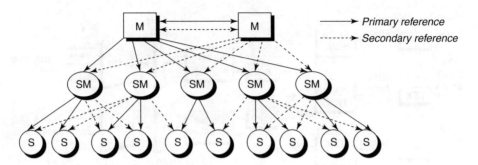

Figure 4.33. Principle architecture of the NTT synchronization network

Figure 4.5. In a master node, an analogue 1.544-MHz or 6.312-MHz reference signal from the BITS clock is supplied to the digital multiplexer, transmitting a synchronous 1.544-Mbit/s or 6.312-Mbit/s digital signal. In a slave node, the synchronized demultiplexer recovers the timing signal from the received digital signal and supplies it to the CSM clock, again as an analogue 1.544-MHz or 6.312-MHz signal.

On 1990, NTT upgraded the synchronization network architecture and most of the CSM clocks, improving the overall performance in order to meet the new requirements for synchronizing SONET transmission networks. Moreover, the management system of the synchronization network has been integrated with that of the SONET transmission network.

The new synchronization network architecture follows most of the modern guidelines outlined in the previous sections. Inter-office timing distribution is done via SONET transmission systems, using optical STS-N signals (cf. Section 4.1.6). Chains of SECs, as in the synchronization reference chain defined in ETSI and ITU-T standards, are not used yet. NTT planned to provide each SONET network element with direct reference from the local CSM clock. Therefore, as mentioned in Section 4.4.4, the typical clock chain is made of CSM clocks, receiving the reference from the input STS-N line through a SONET network element and forwarding the timing by synchronizing the SONET network element that outputs an STS-N line towards the next node. NTT is considering to transfer timing along chains of SECs solely in SONET rings.

New CSM clocks of NTT's network accept as synchronization reference an electrical sine-wave 6.312-MHz signal or an optical Returning-to-Zero (RZ) 155.520-MHz signal. Moreover, all new CSM clocks are equipped with rubidium secondary atomic frequency standards, featuring outstanding hold-over performance.

4.8.4   The Synchronization Network of France Telecom

The public telephone and data networks of France Telecom have been synchronized since the 1975 by a synchronization network made of more than 800 nodes. The brief information provided in this section is derived from [4.58] and may not include changes occurred in the last years.

The network architecture is the usual HMS, organized in four hierarchical levels. At level 0, there are two master clocks, located in Paris St. Amand and in Lyon Sevigne. Nodes of levels 1 and 2 are building clocks at transit offices. Nodes of lower levels are clocks of digital switching exchanges at local offices and primary multiplexers.

The two PRCs consist of duplicated caesium primary frequency standards. Therefore, in normal operation, the synchronization network consists of two slightly plesiochronous subnetworks. All other clocks at levels 1 and 2 of this network are made of duplicated quartz clocks.

Inter-office timing distribution is done via dedicated 2.028-Mbit/s E1 links between nodes of level 0 and level 1 and traffic-bearing E1 links between nodes of lower levels. All links are duplicated or further redundant to achieve high reliability.

Later, France Telecom began to upgrade the synchronization network in order to meet the new requirements for SDH synchronization. The new synchronization network consists of more than 1200 nodes and follows the same architecture as before. It has been planned to be provided also with a TMN management system and a performance monitoring system. The new SASE clocks are equipped with high-stability quartz oscillators. Input and output timing interfaces are both 2.048-Mbit/s and 2.048-MHz signals [4.19].

4.8.5 The Synchronization Network of Telecom Argentina

Beginning from the early 1990s, Telecom Argentina deployed a modern backbone SDH optical transmission network, connecting all the major towns of the country, and set up a state-of-the-art synchronization network, which was one of the most advanced in the world at the time of installation. The brief information provided in this section is summarized from [4.59].

The principle architecture of the Telecom Argentina synchronization network is shown in Figure 4.34. A classical HMS architecture, with three levels and lots of redundant links, has been planned.

At level 0, the PRC, located in one of most important office buildings of Buenos Aires, is equipped with two caesium primary atomic frequency standards and one GPS receiver. The PRC output timing signal is a weighted combination of these three references. To further increase the reliability of the system, the identical PRC of the other main national telephone operator Telefonica de Argentina can take its place as network master clock.

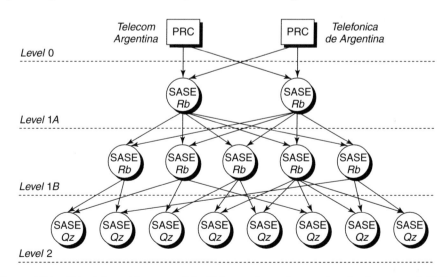

Figure 4.34. Principle architecture of the Telecom Argentina synchronization network [4.59]

At level 1, there are more than 11 SASE clocks, each equipped with two redundant rubidium secondary atomic frequency standards. Among them, two clocks (level 1A) are directly synchronized by both PRCs and act as main timing distributors for the whole network. The other level-1 clocks (level 1B) receive reference signals from both the level 1A clocks and synchronize different parts of the network, along multiple links for reliability.

At level 2, there are more than 35 SASE clocks, each equipped with a double redundant quartz high-precision oscillator.

Inter-office timing distribution among SASE clocks is mostly achieved along chains of SECs, according to the synchronization reference chain defined in ETSI and ITU-T standards, via STM-N signals (cf. Figure 4.7). Intra-office timing distribution, on the other hand, is performed by means of the ITU-T G.703 [4.19] standard analogue 2.048 MHz signal.

The whole synchronization network is managed by a modern management system. Each SASE clock is connected to the central OS through a packet-switched data network. Management functions comprise in particular fault, performance and configuration management. The scope of configuration management includes SASE timing distributors but not SASE oscillators, for security reasons.

The management system includes a performance-monitoring system as described in Section 4.6.3. SASE clocks can monitor the quality of incoming terminated reference signals as well as of return 2.048 Mbit/s signals from remote digital switching exchanges at lower levels of the synchronization hierarchy. Performance-monitoring cards acquire time-error data and transmit raw data measured to the central management system, located in Buenos Aires. The central system collects measurement data from remote monitoring nodes, stores and processes them, by evaluating standard stability quantities (such as MTIE, TIErms and TVAR defined in the next Chapter) and equivalent slip rates. All results are presented in graphical form to the network manager, for ease of proactive problem detection.

4.8.6   Synchronization of the Telephone Digital Switching Network of Telecom Italia

From the late 1970s to early 1990s, the Italian public network operators SIP, IRITEL and ITALCABLE, which later merged into the single public operator Telecom Italia, realized four different synchronization networks:

- the synchronization network of the telephone digital switching network (SIP);
- the synchronization network of the CDN (*Circuiti Diretti Numerici*, Direct Digital Circuits) network based on DXC 1/0 equipment, for provisioning data leased circuits (SIP);
- the synchronization network of the new SDH long-distance transmission network (IRITEL);
- the synchronization network of three switching exchanges for intercontinental traffic (ITALCABLE).

All these networks, different in size and in performance, are based on the HMS architecture and have the same PRC: the *Oscillatore Nazionale di Riferimento* (ONR, National

Reference Oscillator), administered by *Istituto Superiore delle Poste e Telecomunicazioni* (ISPT, Institute of the Italian Post and Telecommunications Office) and located in its premises in Rome. This PRC is made of a set of five caesium primary atomic frequency standards.

These networks, after the merger, went under the common administration of Telecom Italia. Among these networks, only the first one will be described in some detail, for the sake of brevity, based on the information provided in [4.58].

The synchronization of the telephone digital switching network of Telecom Italia follows the HMS architecture, as shown in Figure 4.35.

At level 1A, there are three SASE clocks located at three high-level transit offices (*Stadi di Gruppo di Transito*, SGT, in the Telecom Italia terminology) in Rome and acting as main timing distributors for the network. These nodes are synchronized by the PRC via direct links on optical fibre and coaxial copper cable, carrying an analogue 2.048 MHz timing signal. SASE clocks of level 1A are equipped with high-performance, redundant quartz oscillators (Oven-Controlled Crystal Oscillators, OCXO) and supply timing to all telephone digital switching exchanges and DXCs 1/0 deployed in the node.

At levels 1B and 2, the synchronization network is based on the equipment clocks of digital switching exchanges: transit offices (SGT) and local offices (*Stadi di Gruppo Urbano*, SGU, in the Telecom Italia terminology), respectively. Timing is distributed be means of traffic-bearing 2.048 Mbit/s digital signals, transported through the PDH transmission network. The recovery of synchronization distribution from level 1A to level 1B and to level 2 in case of link failures is entrusted to the normal traffic link protection procedures activated in the transmission network.

The level 3, finally, is constituted by clocks of terminal equipment in the access network, such as PCM multiplexers. They are synchronized by their respective head office of level 2 through the direct 2.048 Mbit/s line connecting them.

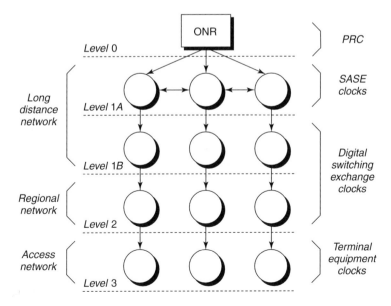

Figure 4.35. Principle architecture of synchronization of the telephone digital switching network of Telecom Italia

4.8.7 The New Synchronization Network of Telecom Italia

The new synchronization network of Telecom Italia was designed and deployed beginning from 1994, with the aim at substituting, at last, the four synchronization networks cited in the previous section with a newer one, modern and compliant with or exceeding the latest standard requirements. The brief information provided in this section is summarized from [4.58][4.60].

The new synchronization network has been designed according to the same HMS principle architecture sketched in Figure 4.35. Nevertheless, reliability has been a major concern in designing the topology: all nodes can be synchronized via multiple, diversely routed trails. The high-level architecture of the new synchronization network, in its first plan, is depicted in Figure 4.36 [4.58].

At level 0, the network PRC is the ONR located in Rome at ISPT premises. A second, spare PRC is the caesium primary frequency standard administered by *Istituto Elettrotecnico Nazionale Galileo Ferraris* (IENGF) and located in its premises in Torino.

At level 1A, there are three new SASE clocks, located in the same SGT buildings in Rome as the three SASE clocks of the former SIP synchronization network for the telephone digital switching network described in the previous section. These three SASE clocks are equipped with rubidium secondary frequency standards and with GPS receivers. Each level-1A SASE clock is synchronized by the main PRC in Rome, via two redundant direct 2.048 MHz links on optical fibres, but is connected also to the secondary PRC in Torino via two dedicated 2.048-Mbit/s links on PDH systems for protection. Moreover, the three SASE clocks are mutually interconnected by a full-mesh network of dedicated

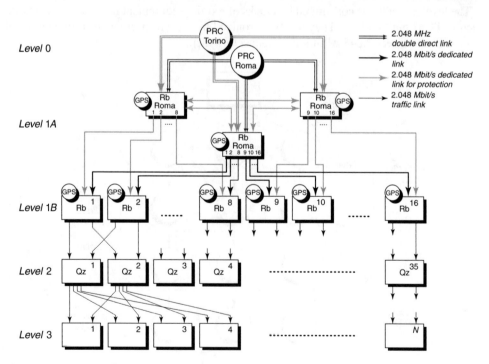

Figure 4.36. High-level architecture of the new synchronization network of Telecom Italia in its first plan

2.048-Mbit/s links on PDH systems for further protection, should both links to PRCs fail in a SASE node. To summarize, each level-1A SASE clock has extremely redundant reference inputs: two signals from the main PRC in Rome, two signals from the second PRC in Torino, two signals from the other two level-1A nodes and one signal from the local GPS receiver.SSMs are used to prevent synchronization loops after protection rearrangements. Rather surprisingly, the GPS reference has been assigned the least priority among the spare resources: it is used only should all other reference inputs fail (cf. Section 4.2.7).

At level 1B, there are another 16 SASE clocks, installed in SGT buildings spread over the Italian territory, equipped as the level-1A nodes with rubidium secondary frequency standards and GPS receivers. Each level-1B clock receives two references from two level-1A nodes via dedicated 2.048-Mbit/s links on PDH systems. Also in this case, SSMs are used to prevent synchronization loops after protection rearrangements and the GPS reference has been assigned the least priority among the spare resources. In normal operation, the same level-1A clock synchronizes all level-1B clocks.

At level 2, there are 35 (in the first synchronization plan) SASE clocks installed in other SGT buildings. These SASE clocks are equipped with high-precision quartz oscillators (OCXO) and receive two references from two distinct level-1B nodes via dedicated or, most commonly, traffic-bearing 2.048-Mbit/s links on PDH systems.

At level 3, there are a few hundred SASE clocks, installed in SGU buildings and equipped with the same quartz oscillators as level-2 clocks. Each level-3 clock receives two references from two distinct level-2 nodes via traffic-bearing 2.048-Mbit/s links on PDH systems.

The architecture described so far is that early planned on 1994. Since that year, nevertheless, it had been planned to carry timing on STM-N links for inter-office distribution, as described in Section 4.1.6, as soon as SDH would have replaced PDH in most systems of the long-distance transmission network.

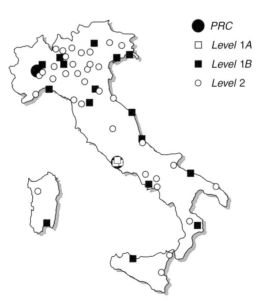

Figure 4.37. Pictorial representation of the geographical location of the main nodes of the new synchronization network of Telecom Italia [4.58]

4.9 SUMMARY

The goal of network synchronization in telecommunications may be either to align the absolute time scales of network nodes, or to align the significant instants of timing signals generated by local clocks independently from a constant phase offset among them, or to equalize the frequencies of local clocks without controlling their phase relationship. A synchronization network is the facility implementing network synchronization.

Network synchronization has played a role of increasing importance in telecommunications, especially since transmission and switching turned digital. Synchronization of analogue FDM networks was based on distributing a pilot frequency derived from a network master clock to local oscillators with limited free-run frequency accuracy. Digital transmission equipment based on PDH does not need to be synchronized, because bit justification allows multiplexing of asynchronous tributaries. Digital switching equipment, on the other hand, requires to be synchronized in order to avoid slips in the input elastic stores. Contrary to PDH, SDH transmission takes advantage from reliable network synchronization. Hence the need of more stringent synchronization requirements. Synchronization plays an essential role also in ATM networks and in cellular mobile wireless telephone networks.

Many network synchronization strategies have been envisaged. In the full plesiochronous strategy, each clock is independent from the others. In master–slave strategies, timing is distributed from a master clock to all slave clocks (usually realized as PLL) of the network, according to a star or tree topology. Mutual synchronization is based on the mutual control among clocks. In the mixed mutual/master–slave synchronization, the mutual synchronization strategy is adopted for a few network core clocks, while the master–slave strategy is adopted for the peripheral clocks. The hierarchical master–slave (HMS) strategy is a variant of the pure master–slave strategy: if the master fails, another clock takes its place according to a hierarchical plan.

The ITU-T, ANSI and ETSI bodies have defined standard architectures for synchronization networks according to the HMS strategy. ITU-T and ETSI standards are completely aligned and use the same notation, except for minor details. A central concept is the BITS (as denoted by ANSI) or SASE (as denoted by ETSI and ITU-T) clock, designated in an office building to receive timing from outside and to time all other clocks in the building.

The goal of synchronization network planning is to determine the architecture for distribution of synchronization in a network and the facilities to be used. The first rule should be to follow carefully all the applicable ITU-T and ETSI recommendations. Some further general guidelines are to avoid timing loops, to maintain the clock hierarchy also after protection rearrangements, to adopt the BITS/SASE concept, to choose the best timing transfer facilities available, to minimize the length of clock chains in transferring timing, to plan route diversity for alternate timing facilities.

Most modern synchronization networks are provided with management systems. The main management functions relevant to synchronization network management lie in the areas of fault, configuration, performance and security management.

Most advanced synchronization networks are provided with monitoring systems that allow to verify continuously, in real time, the performance achieved in timing distribution. The rationale of synchronization performance monitoring is the need to be proactive, i.e. to detect timing degradations well before they impact service.

The ITU-T, ETSI and ANSI standard bodies have defined automatic mechanisms for self-healing from synchronization failures, based on Synchronization Status Messages

(SSM) embedded in synchronization signals. With SSMs, the originating node declares the quality of the synchronization source to which its signal is traceable or whether that signal is usable as synchronization source. SSMs have been defined to allow clocks to select the proper synchronization source among the available references according to a priority scheme. Their main purpose is to avoid timing loops, while allowing clocks to autonomously reconfigure after synchronization failures.

Synchronization networks designed and realized in the United States of America by AT&T, in Switzerland by Swiss PTT, in Japan by NTT, in France by France Telecom, in Argentina by Telecom Argentina and in Italy by Telecom Italia were described in the last section.

4.10 REFERENCES

[4.1] W. C. Lindsey, F. Ghazvinian, W. C. Hagmann, K. Dessouky. Network synchronization. *Proceedings of the IEEE* 1985; **73**(10): 1445–1467.

[4.2] P. Kartaschoff. Synchronization in digital communications networks. *Proceedings of the IEEE* 1991; **79**(7): 1019–1028.

[4.3] S. Bregni. A historical perspective on network synchronization. *IEEE Communications Magazine* 1998; **36**(6).

[4.4] J. F. Oberst. *Keeping Bell System Frequencies on the Beam*. Bell System Record, Mar. 1984.

[4.5] ITU-T Rec. G.225 *Recommendations Relating to the Accuracy of Carrier Frequencies*. Blue Book, Geneva 1988.

[4.6] O. P. Clark, E. J. Drazy, D. C. Weller. A phase-locked primary frequency supply for the L multiplex. *Bell System Technical Journal* 1963; **42**: 319–340.

[4.7] P. R. Trischitta, E. L. Varma. *Jitter in Digital Transmission Systems*. Norwood, MA: Artech House, 1989.

[4.8] D. L. Duttwailer. Waiting time jitter. *Bell System Technical Journal* 1972; **51**: 165–207.

[4.9] P. E. K. Chow. Jitter due to pulse stuffing synchronization. *IEEE Transactions on Communications* 1973; **COM-21**(7): 854–859.

[4.10] H. L. Hartmann, E. Steiner. Synchronization techniques for digital networks. *IEEE Journal on Selected Areas in Communications* 1986; **SAC-4**(4): 506–513.

[4.11] ITU-T Rec. G.822 *Controlled Slip Rate Objective on an International Digital Connection*. Blue Book, Geneva 1988.

[4.12] AT&T. *Effects of Synchronization Slips*. ITU-T Contribution COM SpD-TD, No. 32, Geneva, November, 1969.

[4.13] R. Smith, L. J. Millot. Synchronisation and slip performance in a digital network. *British Telecommunications Engineering* 1984; **3**.

[4.14] J. E. Abate, H. Drucker. The effect of slips on facsimile transmission. *Proceedings of IEEE ICC'88*, 1988.

[4.15] H. Drucker, A. C. Morton. The effect of slips on data modems, *Proceedings of IEEE ICC'87*, 1987.

[4.16] J. E. Abate, E. W. Butterline, R. A. Carley, P. Greendyk, A. M. Montenegro, C. D. Near, S. H. Richman and G. P. Zampetti. AT&T's new approach to the synchronization of telecommunications networks. *IEEE Communications Magazine* 1989; **27**(4): 35–45.

[4.17] K. Inagaki *et al.* International connection of plesiochronous networks via TDMA satellite link. *Proceedings of IEEE ICC'82*, 1982.

[4.18] M. Decina, U. De Julio. Performance of integrated digital networks: international standards. *Proceedings of IEEE ICC'82*, 1982.

[4.19] ITU-T Rec. G.703 *Physical/Electrical Characteristics of Hierarchical Digital Interfaces*. Blue Book, Geneva, April 1991.

[4.20] M. J. Klein, R. Urbansky. Network synchronization — a challenge for SDH/SONET? *IEEE Communications Magazine* 1993; **9**: 42–50.

[4.21] ITU-T Recs. G.823 *The Control of Jitter and Wander within Digital Networks which are Based on the 2048 kbit/s Hierarchy*: G.824 *The Control of Jitter and Wander within Digital Networks which are Based on the 1544 kbit/s Hierarchy*; G.825 *The Control of Jitter and Wander within Digital Networks which are Based on the Synchronous Digital Hierarchy*. Geneva, March 1993.

[4.22] ITU-T Rec. G.803 *Architectures of Transport Networks Based on the Synchronous Digital Hierarchy (SDH)*. Sec. 8, Geneva, June 1997.

[4.23] ETSI EG 201 793 (ref. DEG/TM-01080) *Synchronisation Network Engineering*. v.1.1.1, Oct. 2000.

[4.24] ITU-T Rec. I.363 *B-ISDN ATM Adaptation Layer (AAL) Specification*. Geneva, 1996/2000.

[4.25] R. C. Lau, P. E. Fleischer. Synchronous techniques for timing recovery in BISDN. *IEEE Transactions on Communications* 1995; **COM-43**(2/3/4).

[4.26] K. Murakami. Jitter in synchronous residual time stamp. *IEEE Transactions on Communications* 1996; **COM-44**(6).

[4.27] K. Murakami. Waveform analysis of jitter in SRTS using continued fraction. *IEEE Transactions on Communications* 1998; **COM-46**(6).

[4.28] P. Mouley, M. B. Pautet. *The GSM System for Mobile Communications*. Lassay-les-Château: Europa Media Duplication S. A., 2000.

[4.29] ETSI EN 300 912 *Digital Cellular Telecommunication System (Phase 2+); Radio Subsystem Synchronization*.

[4.30] ITU-T Rec. G.811 *Timing Characteristics of Primary Reference Clocks*. Geneva, Sept. 1997.

[4.31] ITU-T Rec. G.812 *Timing Requirements of Slave Clocks Suitable for Use as Node Clocks in Synchronization Networks*. Geneva, June 1998.

[4.32] J. P. Moreland. Performance of a system of mutually synchronised clocks. *The Bell System Technical Journal* 1971; **50**(7).

[4.33] W. C. Lindsey, J. H. Chen. Mutual clock synchronization in global digital communication networks. *European Transactions on Telecommunications* 1996; **7**(1).

[4.34] ETSI EN 300 462 *Transmission and Multiplexing (TM); Generic Requirements for Synchronization Networks*; Part 1-1: *Definitions and Terminology for Synchronization Networks*; Part 2-1: *Synchronization Network Architecture*; Part 3-1: *The Control of Jitter and Wander within Synchronization Networks*; Part 4-1: *Timing Characteristics of Slave Clocks Suitable for Synchronization Supply to Synchronous Digital Hierarchy (SDH) and Plesiochronous Digital Hierarchy (PDH) Equipment*; Part 5-1: *Timing Characteristics of Slave Clocks Suitable for Operation in Synchronous Digital Hierarchy (SDH) Equipment*; Part 6-1: *Timing Characteristics of Primary Reference Clocks*.

[4.35] ITU-T Rec. G.810 *Considerations on Timing and Synchronization Issues*; G.811 *Timing Requirements at the Outputs of Primary Reference Clocks Suitable for Plesiochronous Operation of International Digital Links*; G.812 *Timing Requirements at the Outputs of Slave Clocks Suitable for Plesiochronous Operation of International Digital Links*; Blue Book, 1988.

[4.36] ITU-T Rec. G.810 *Definitions and Terminology for Synchronisation Networks*. Geneva, August 1996.

[4.37] ITU-T Rec. G.813 *Timing Characteristics of SDH Equipment Slave Clocks (SEC)*. Geneva, August 1996.

[4.38] ANSI T1.101 *Telecommunications — Synchronization Interface Standard*. January 1997.

[4.39] ANSI T1.105.09 *Telecommunications — Synchronous Optical Network (SONET) — Network Element Timing and Synchronization*. 1996.

[4.40] Bellcore GR-1244 *Clocks For The Synchronized Network: Common Generic Criteria*. June 1995.

[4.41] Bellcore GR-436 *Digital Network Synchronization Plan*. Revision 1, June 1996.

[4.42] Bellcore SR-NWT-002224 *SONET Synchronization Planning Guidelines*.

[4.43] ITU-T Rec. M.3010 *Principles for a Telecommunications Management Network*. Geneva, Feb. 2000 and related Recs. M.3000, M.60, M.3020, M.3100, M.3180, M.3200, M.3300 and M.3400.

[4.44] U. D. Black. *Network Management Standards — The OSI, SNMP and CMOL Protocols*. New York: McGraw–Hill Series on Computer Communications, 1992.

[4.45] U. D. Black. *Network Management Standards — SNMP, CMIP, TMN, MIBs and Object Libraries*. New York: McGraw–Hill Series on Computer Communications, 1995.

[4.46] W. Stallings. *Network Management*. IEEE Computer Society Press, 1993.

[4.47] ITU-T Rec. G.784 *Synchronous Digital Hierarchy (SDH) Management*. Geneva, June 1999.

[4.48] ITU-T Rec. G.707 *Network Node Interface for the Synchronous Digital Hierarchy (SDH)*, Geneva, March 1996.

[4.49] ETSI EN 300 417-6 *Transmission and Multiplexing (TM); Generic Requirements of Transport Functionality of Equipment; Synchronisation Layer Functions*.

[4.50] ANSI T1X1.3 *A Technical Report on Synchronization Network Management Using Synchronization Status Messaging*. TR No. 33, April 1994.

[4.51] S. Liu, R. Williams, F. Liu, T. -W. Leung, R. Subramanian. A practical overview of synchronization status messaging and its applicability to SONET networks. *Proc. of IEEE 14th Annual International Phoenix Conference on Computers and Communications*, Phoenix, AZ, USA, 1995.

[4.52] ITU-T Rec. G.704 *Synchronous Frame Structures Used at 1544, 6312, 2048, 8488 and 44736 kbit/s Hierarchical Levels*. Geneva, Oct. 1998.

[4.53] ITU-T Rec. G.832 *Transport of SDH Elements on PDH Networks: Frame and Multiplexing Structures*. Geneva, Nov. 1995.

[4.54] ITU-T Rec. G.841 *Types and Characteristics of SDH Network Protection Architectures*. Geneva, July 1995.

[4.55] ITU-T Rec. G.842 *Interworking of SDH Network Protection Architectures*. Geneva, April 1997.

[4.56] P. Kartaschoff, P. A. Probst, P. Vörös. A network timing concept for Switzerland. *Proceedings of SSC-ASMT '85*, Interlaken, Switzerland, 1985.

[4.57] P. Kartaschoff, P. A. Probst, P. Vörös. Network synchronization plan of Swiss PTT. *Proceedings of 17th PTTI*, Washington D.C., USA, 1985.

[4.58] A. Manzalini, A. Mariconda, L. Valtriani. Chapter 9: Le moderne reti di sincronizzazione. In *La sincronizzazione nelle reti numeriche di telecomunicazioni*. Torino, Italy: CSELT, 1996.

[4.59] S. Lorenzi, A. Barkasz, O. Rebollini. A new SDH synchronization network for telecom Argentina. *Proceedings of COMCON 5 — 5th International Conference on Advances in Communication and Control*, Rethymnon, Crete, Greece, June 1995.

[4.60] R. Bonello, A. Manzalini, A. Mariconda. Architecture and management strategies for the Italian synchronization network. *Proceedings of COMCON 5 – 5th International Conference on Advances in Communication and Control*, Rethimnon, Crete, Greece, June 1995.

5

CHARACTERIZATION AND MODELLING OF CLOCKS

Designing complex systems of clocks as synchronization networks is not an obvious task, especially if aiming at guaranteeing the quality of service demanded by modern applications. Several years ago, specifying the quality requirements of clocks for telecommunications was often limited to not more than specifying their fractional deviation from the nominal frequency. More recently, since the introduction of SDH and of advanced digital services, clocks have been requested to comply with much more stringent and complex requirements for accuracy and stability, as specified by international standard bodies. A knowledge effusion occurred from the world of oscillator specialists to the world of telecommunication engineers, that imported methods and models conceived for time and frequency metrology and for the characterization of state-of-the-art oscillators.

Therefore, in order to describe the behaviour of clocks in synchronization networks and to accurately specify their characteristics, first it is necessary to identify a proper mathematical model of the clock and of the timing signals generated and distributed. This chapter supplies this basic knowledge: models of autonomous and slave clocks are defined and the mathematical tools for time and frequency stability characterization are provided. Finally, chains of slave clocks are considered, in order to assess the performance of timing transfer along synchronization trails.

5.1 CLOCKS AND TIMING SIGNALS

A clock is a device able to supply a *timing signal*, or *chronosignal*, i.e. a pseudo-periodic (ideally periodic) signal usable to control the timing of actions. Under different forms, clocks are widely spread in the everyday life of everybody and penetrated a surprisingly wide range of applications. Among the many possible examples, clocks supply the timing to digital hardware systems (gates, chips or boards), where the operation of different modules must be synchronized to ensure the proper transfer of binary symbols, and to telecommunications systems, where digital signals are multiplexed, transmitted and switched. Chapter 1 outlined several different contexts in which clocks play a central role in telecommunications, driving synchronization processes at different levels.

In principle, clocks consist simply of a generator of oscillations based on any periodic physical phenomenon. Among the most common examples we may mention the swinging of a pendulum or a wheel in mechanical clocks, the vibration of atoms in a crystal around their minimum-energy position in quartz clocks or the radiation associated with specific quantic atomic transitions in atomic clocks. *Autonomous clocks* supply a timing signal generated from an internal oscillator, running independently of external influence. In *slave clocks*, conversely, the oscillator is controlled by an external reference.

Timing signals are usually described as sine waves for ease of mathematical analysis, without loss of generality. Thus, a general expression modelling the pseudo-periodic timing signal $s(t)$ at the output of a clock is given by [5.1][5.2]

$$s(t) = A(t) \sin \Phi(t) \tag{5.1}$$

where $A(t)$ is the instantaneous amplitude and $\Phi(t)$ is the *total phase*.

The instantaneous amplitude in the most common cases can be considered constant, or

$$A(t) = A_n + \varepsilon(t) \qquad \left| \frac{\varepsilon(t)}{A_n} \right| \ll 1 \tag{5.2}$$

where A_n is the nominal amplitude and $\varepsilon(t)$ is the deviation from the nominal value. However, $\varepsilon(t)$ can be safely neglected in our following analysis. In fact, output stages of clocks are able to control accurately the amplitude of output signals. On the other hand, the position of significant instants does not depend on signal amplitude modulation, at least if trigger events are properly chosen.

More interesting is to notice that the timing information of the signal $s(t)$ is carried by its total phase, or equivalently by its instantaneous frequency $v(t)$[1], which is given by

$$v(t) = \frac{1}{2\pi} \frac{d\Phi(t)}{dt} \tag{5.3}$$

5.2   TIMING SIGNAL MODEL AND BASIC QUANTITIES

Several models have been proposed in literature [5.1]–[5.4] to describe the behaviour of the instantaneous frequency $v(t)$ in actual clocks. A common and comprehensive model is the following

$$v(t) = v_0 + v_d(t) + v_a(t)$$

$$= v_n + \Delta v + \sum_{k=1}^{K-1} \frac{q_k t^k}{k!} + \frac{1}{2\pi} \frac{d\varphi(t)}{dt} \tag{5.4}$$

The *nominal frequency* v_n (design goal) and the *starting frequency offset* Δv (also called *syntonization error*) make up the starting frequency v_0 of the oscillator:

$$v_0 = v_n + \Delta v \tag{5.5}$$

The starting frequency offset Δv depends on the precision of the initial calibration.

---

[1] In most parts of the literature on time and frequency stability characterization, the symbol $v(t)$ is used to denote the instantaneous total frequency of a timing signal, while f is used to denote the Fourier frequency (see Section 5.6). In this chapter and in the following ones, we adhere to this notation.

The term $v_d(t)$ is the deterministic (for a given oscillator) time-dependent component of the clock total frequency, modelling as a power series the *frequency drift* mainly due to oscillator ageing, that is

$$v_d(t) = \sum_{k=1}^{K-1} \frac{q_k t^k}{k!} \qquad (5.6)$$

The coefficients q_k ($k = 1, 2, \ldots, K - 1$) are time-independent. They are random variables considering a set of clocks, but fixed parameters for a given oscillator. The frequency drift in real clocks is due to complex phenomena and strongly dependent on the particular clock under measurement [5.5][5.6], but for practical purposes and for the sake of simplicity the summation above is often truncated to the first term, so that

$$v_d(t) \cong q_1 t = D v_n t \qquad (5.7)$$

where D is the *linear fractional frequency drift rate*.

Finally, the term $v_a(t)$ is the random (aleatory) time-dependent component of the total instantaneous frequency, that is

$$v_a(t) = \frac{1}{2\pi} \frac{d\varphi(t)}{dt} \qquad (5.8)$$

where $\dot{\varphi}(\tau)/(2\pi)$ and $\varphi(t)$ are stochastic processes, respectively the random frequency deviation and the random phase deviation, modelling oscillator intrinsic phase noise sources.

From Equations (5.3), (5.4) and (5.7), approximating the frequency drift to its linear term, the model for the total phase results

$$\Phi(t) = 2\pi(v_n + \Delta v)t + \pi D v_n t^2 + \varphi(t) + \Phi_0 \qquad (5.9)$$

where $\Phi_0 = \Phi(0) - \varphi(0)$. In the ideal case, the total phase is linearly increasing with time, that is

$$\Phi(t) = 2\pi v_n t + \Phi_0 \qquad (5.10)$$

Two functions, strictly related to $\dot{\varphi}(t)$ and $\varphi(t)$, are used in treating random frequency and time fluctuations of clocks: the *random fractional frequency deviation* $y(t)$ and the *random time deviation* $x(t)$, defined as

$$y(t) = \frac{1}{2\pi v_n} \frac{d\varphi(t)}{dt}$$

$$x(t) = \frac{\varphi(t)}{2\pi v_n} \qquad (5.11)$$

Please note that the above two functions express the sole random phase and frequency fluctuations of the clock, apart from the frequency offset and drift. Together with the model and definitions provided previously, they have been conceived and widely adopted by specialists since the 1960s [5.1][5.7]. More recently, needs arisen in particular in the telecommunications field, for the design of synchronization equipment and networks, led to the introduction of the following other basic functions, more oriented to the timing aspects of clocks.

The generated *Time* function $T(t)$ of a clock is defined, in terms of its total instantaneous phase, as

$$T(t) = \frac{\Phi(t)}{2\pi \nu_n} \tag{5.12}$$

For an ideal clock, $T_{id}(t) = t$ holds, as expected.

For a given clock, the *Time Error* function $TE(t)$ between its time $T(t)$ and a reference time $T_{ref}(t)$ is defined as

$$TE(t) = T(t) - T_{ref}(t) \tag{5.13}$$

In practice, the reference clock is supposed to be much better performing than the clock under test and supplies the time reference for the measurement process itself. Thus, the Equation (5.13) should be actually rewritten as

$$TE[T_{ref}(t)] = T[T_{ref}(t)] - T_{ref}(t) \tag{5.14}$$

In the following, however, we will omit to point out that all quantities are actually measured at times dictated by the reference clock and we will write expressions as in Equation (5.13), to keep formulas simple. If we assume to have an ideal clock as reference, Equation (5.13) becomes $TE(t) = T(t) - t$.

It is worthwhile noticing, at this point, the relationship between the quantities $x(t)$ and $TE(t)$. Both represent a time deviation of the clock under test from a reference time, but the former expresses the sole random phase fluctuations, while the latter takes into account also the effects of any frequency offset and drift. Revealing the deterministic components in the TE measured data may be not straightforward and the result can highly depend on the parameter estimation technique adopted, as it will be shown in Chapter 7.

The *Time Interval* function $TI_t(\tau)$ is defined as

$$TI_t(\tau) = T(t + \tau) - T(t) \tag{5.15}$$

and represents the measure of a time interval τ, starting at time t, accomplished by the clock under test (again supposing to have an ideal reference clock). In other words, it expresses how long the clock under test perceives a time interval of ideal length τ starting at ideal time t.

The time error variation over an interval of duration τ starting at time t is called *Time Interval Error* function $TIE_t(\tau)$ and is defined as

$$TIE_t(\tau) = [T(t + \tau) - T(t)] - [T_{ref}(t + \tau) - T_{ref}(t)]$$
$$= TE(t + \tau) - TE(t) \tag{5.16}$$

It represents the error committed by the clock under test in measuring an interval τ, starting at time t, with respect to the reference clock. If the reference clock is ideal, the last equation can be also written as $TIE_t(\tau) = TI_t(\tau) - \tau$.

The example plots provided in Figures 5.1 and 5.2 clarify the meaning of functions $T(t)$, $TE(t)$, $TI_{t_0}(\tau)$ and $TIE_{t_0}(\tau)$ and the relationships among them.

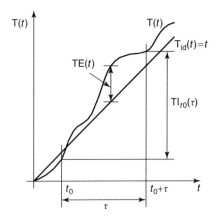

Figure 5.1.   Example plot of the functions $T(t)$, $T_{id}(t) = t$, $TE(t)$ and $TI_{t_0}(\tau)$

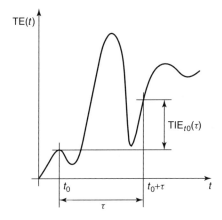

Figure 5.2.   Example plot of the functions $TE(t)$ and $TIE_{t_0}(\tau)$

5.3   BASIC CONCEPTS OF QUALITY OF CLOCKS: STABILITY AND ACCURACY

Characterizing the *quality* of a clock (or equivalently of its timing signal) is one of the most debated issues in practical applications of clocks. In the most common sense, one simple term is used to refer to clock quality: the *precision*, somehow denoting how much the clock under test is close to a reference clock (in time or in frequency). Nevertheless, the term 'precision' is even not included in the ISO International Vocabulary of Metrology [5.8] and its use is therefore not recommended.

The quality of a clock is usually defined by means of two other basic terms: the *stability* and the *accuracy*. Although it must be recognized that researchers and engineers working in different fields may understand these two terms with subtly different meanings, the definitions provided in the following paragraphs can be considered quite general and widely accepted.

5.3.1   Stability

The *stability* of a measuring instrument is its ability to maintain constant its metrological characteristics with time (ISO International Vocabulary of Metrology [5.8]). The *stability of a clock*, therefore, is its capacity of generating a time interval (or a frequency) with constant value (such a value may be different from the nominal value, but it is intended constant in time).

In other words, the stability of a clock deals with the measurement of random and deterministic variations of its instantaneous frequency (or of the time generated) compared to the nominal value (i.e., in practice, to that of a reference clock), *over a given observation interval*. For example, a stability measure of the instantaneous frequency based on the concept of variance σ^2, over an observation interval τ, informs about the instantaneous-frequency variance that is expected by collecting measurements along a time interval τ. With reference to the mathematical model of Equation (5.4), the clock time stability depends on the random phase noise $\varphi(t)$, the frequency offset $\Delta\nu$ and the frequency drift coefficients q_k. The relative weights of such parameters in affecting stability depend on the observation interval: if this is short, the frequency drift and even offset are negligible.

When the observation interval τ is small, the expression *short-term* stability is commonly used, otherwise the expression *long-term* stability applies. What should be the meaning of the word 'small' (i.e., where is the border between short and long term) depends on the specific application. For example, in the time metrology field, it is common to consider observation intervals longer than one day as long-term, while in telecommunications applications observation intervals above 100 s are definitely considered to fall in the long term as well. To further clarify the concept of short-term and long-term stability, Figure 5.3 shows two examples of instantaneous frequency plot. The timing signal in the upper graph is rather stable in the short term, but quite drifty in the long term; on the contrary, the one in the lower graph has a very poor short-term stability, but is very stable in the long term.

Since the 1960s, several quantities have been defined aiming at characterizing clock stability. With different properties, they highlight distinct phenomena in the phase noise or they are more oriented to specific applications. Details will be given in the next sections.

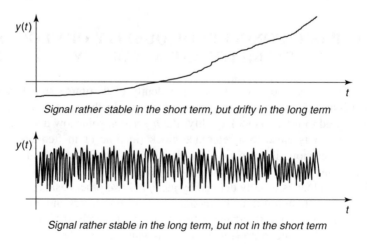

Signal rather stable in the short term, but drifty in the long term

Signal rather stable in the long term, but not in the short term

Figure 5.3.   Examples of long-term and short-term stability

5.3.2   Accuracy

The *accuracy of measurement*, on the other hand, denotes the closeness of the agreement between the result of a measurement and a true value of the measurand (ISO International Vocabulary of Metrology [5.8]). In the context of clock quality definition, thus, accuracy of a time scale means its agreement with UTC, while accuracy of the frequency delivered by a frequency standard means its agreement with the nominal value. Qualitatively speaking, accuracy tells how much a clock is 'right'.

According to common use in telecommunications, we define the *accuracy of a clock* as the maximum time or frequency error, which may be measured in general over the whole clock life (e.g., 20 years), unless differently specified. Therefore, the *time accuracy* means how well a clock agrees with UTC over the specified period, while the *frequency accuracy* is the maximum frequency error $\Delta\nu_{max}$ compared to the nominal value ν_n. It must be pointed out that also the accuracy depends in principle on the above mentioned parameters $\varphi(t)$, $\Delta\nu$ and the coefficients q_k, but in this case the observation interval is so long that in practice the only relevant quantities are the frequency offset and drift. The frequency accuracy is usually expressed by the a-dimensional ratio $\Delta\nu_{max}/\nu_n$ and is often measured in 10^{-6} units [μHz/Hz], in the engineering practice called also [parts per million] and abbreviated as [ppm][2].

Finally, to further clarify the difference between the concepts of stability and accuracy, let us point out that a clock could have a significant frequency error (and thus poor frequency accuracy), while still keeping constant this frequency error during its lifetime (and thus featuring perfect frequency stability).

For example, a hydrogen-MASER clock typically has better frequency stability than a caesium-beam clock from second to second or from hour to hour, but often not from month to month and longer. On the other hand, the typical caesium-beam clock is more accurate than the hydrogen-MASER clock. Quartz-oscillators can be very stable in the short term, but they drift in frequency and do not feature the frequency accuracy of atomic clocks [5.9].

5.4   AUTONOMOUS CLOCKS

An *autonomous clock* is a stand-alone device able to generate a timing signal, suitable for the measurement of time intervals, starting from some periodic physical phenomenon that runs independently of external influence.

Examples of autonomous clocks are the atomic frequency standards (such as the rubidium, or the caesium-beam or the hydrogen-MASER oscillators) and the crystal quartz oscillators. Some of them (the rubidium and the quartz oscillators) may also run locked to a reference timing signal and thus work as slave clocks.

A simplified model of autonomous clock is shown in Figure 5.4. Here, according to the model defined in Equation (5.9), the total output phase $\Phi(t)$ is made of three deterministic terms modelling the phase generated by an ideal oscillator, the frequency offset $\Delta\nu$ and the linear frequency drift D, together with the random phase noise $\varphi(t)$.

---

[2] The unit [ppm] is not an International System (SI) unit, and therefore its use is deprecated by time and frequency metrologists. In spite of this, we still use it in this book, following the common practice among telecommunications engineers.

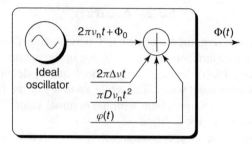

Figure 5.4. Simplified model of autonomous clock

The starting frequency offset Δv is as smaller as more accurate the clock calibration procedures have been during the production process. Such calibration procedures aim at making the clock generate an output frequency as closest as possible to the nominal frequency.

As far as the frequency drift coefficient D is concerned, it is worthwhile noticing that it is practically null in the case of caesium-beam oscillators, while it cannot be neglected in rubidium and quartz oscillators, at least in the most demanding applications. Moreover, accurate characterization of quartz oscillators needs to consider at least the first two drift coefficients (i.e., linear and quadratic).

Finally, details on the statistical characteristics of the random phase-noise process $\varphi(t)$, as it can be measured on actual clocks, will be provided in Section 5.9.

5.5 SLAVE CLOCKS

A *slave clock* is a device able to generate a timing signal, suitable for the measurement of time intervals, having phase (or much less frequently frequency) controlled by a reference timing signal at its input. The Phase-Locked Loop (PLL) and the Frequency-Locked Loop (FLL) are examples of slave clocks. In the former case, the slave clock outputs a timing signal that is synchronous with the input signal, owing to the feedback control on the phase error between them. In the latter case, the slave clock outputs a timing signal mesochronous with the input signal, because the feedback control acts on the frequency error between them.

Slave clocks are very widely employed in synchronization networks or in digital telecommunications equipment. Their major role is to recover timing from the input reference signal and to maintain output timing as close to the source-clock's timing as possible. This implies two basic functions. First, the slave clock must follow faithfully the source-clock's timing, recovering it from the input reference signal, even though this may be affected by transmission impairments of any kind. Second, the slave clock must be able to maintain adequate timekeeping even after a major timing-reference failure.

The widespread application of slave clocks in synchronization networks and in digital telecommunications equipment calls for a somehow detailed description of their models and properties. Therefore, the next subsections deal with the main characteristics of PLLs and the operation modes of slave clocks in synchronization networks.

5.5.1 Phase-Locked Loop Fundamentals

There is a huge literature dealing with PLL theory and application. Among the many books that have been written, the following ones can be definitely recommended. The pioneering work of Viterbi on the non-linear theory of PLLs is reported in [5.10]. The fundamental research of Lindsey, really a synchronization guru, is reported in [5.11], which, although not recent, contains a wealth of information on advanced topics of PLL theory that cannot be found anywhere else. The book by Blanchard [5.12] focuses on the application of PLLs to coherent receiver design. The books by Gardner [5.13] and Best [5.14] are aimed at engineers and stress practical aspects. The book by Horowitz and Hill [5.15][3], as the title itself says, deals with circuit implementation practice. The book by Meyr and Ascheid [5.16], finally, deals with phase and frequency control systems widely used in digital communications and provides a deep overview on the whole subject. In this section, the introductory sections of the book by Meyr and Ascheid have been followed, to provide an overview just on the very basics of PLL. We omitted most analytical details, in order to stress some practical results useful for PLL application in slave clocks for synchronization network. For further details the reader is referred to that book and other literature.

5.5.1.1 Automatic Phase Control and PLL Principle

A device implementing the phase locking to the reference timing signal is commonly based on a loop architecture, based on the negative feedback principle, where the output signal keeps tracking the phase fluctuations of the input reference. Such a device is called *Phase-Locked Loop* (PLL).

A phase-locked loop is a control system used to automatically adjust the phase of a locally generated signal $\hat{s}(t)$ to the phase of an incoming signal $s(t)$. As previously done, let us assume that the two signals are sine waves, respectively given by

$$s(t) = \sin[\omega_0 t + \theta(t)]$$

$$\hat{s}(t) = \sin[\omega_0 t + \hat{\theta}(t)] \tag{5.17}$$

where the phases $\theta(t)$ and $\hat{\theta}(t)$ are slowly varying with respect to the nominal angular frequency ω_0, i.e.

$$\left| \frac{d\theta(t)}{dt} \right| \ll \omega_0, \quad \left| \frac{d\hat{\theta}(t)}{dt} \right| \ll \omega_0 \tag{5.18}$$

Any difference in instantaneous frequency among the two signals is included in the time-varying function $\hat{\theta}(t)$, as

$$\hat{\omega}_0 = \omega_0 + \frac{d\hat{\theta}(t)}{dt} \tag{5.19}$$

The goal is to adjust the total phase of the output signal $\hat{s}(t)$

$$\hat{\Phi}(t) = \omega_0 t + \hat{\theta}(t) \tag{5.20}$$

---

[3] We cannot refrain from recommending the textbook by Horowitz and Hill as an outstanding source of information on an astonishing wide spectrum of electronic circuit design techniques. In that book, the so-called 'random-open and look' test always yields surprising results, enabling us to find unexpected and instructive topics.

to that of the input reference signal $s(t)$

$$\Phi(t) = \omega_0 t + \theta(t) \tag{5.21}$$

Such two signals can be phase-aligned by an automatic control system if we are able to generate a control signal as a function of the phase error between them. An easy way to achieve this is to simply multiply the two signals, as

$$\sin[\omega_0 t + \theta(t)] \cdot \sin[\omega_0 t + \hat{\theta}(t)]$$

$$= \frac{1}{2}\cos[\theta(t) - \hat{\theta}(t)] - \frac{1}{2}\cos[2\omega_0 t + \theta(t) + \hat{\theta}(t)] \tag{5.22}$$

Since the phase $\theta(t)$ is slowly varying compared to $2\omega_0$, the second cosine term can be easily removed by low-pass filtering. The first cosine term provides a measure of the phase difference $\theta(t) - \hat{\theta}(t)$, but, being an even function, does not provide the key information whether $\theta(t)$ is larger than $\hat{\theta}(t)$ or *vice versa*. A way to get an odd function of the phase difference $\theta(t) - \hat{\theta}(t)$ is to advance the locally generated signal $\hat{s}(t)$ by $\pi/2$ before multiplication by the input reference signal, to yield

$$\sin[\omega_0 t + \theta(t)] \cdot \sin\left[\omega_0 t + \frac{\pi}{2} + \hat{\theta}(t)\right]$$

$$= \frac{1}{2}\sin[\theta(t) - \hat{\theta}(t)] + \frac{1}{2}\sin[2\omega_0 t + \theta(t) + \hat{\theta}(t)] \tag{5.23}$$

Then, if the phase error $\theta(t) - \hat{\theta}(t)$ is non zero, an error signal with the same sign as the phase error is produced. This error signal is filtered and then applied to an oscillator whose output frequency can be varied by the voltage applied to it: a Voltage-Controlled Oscillator (VCO). When the control voltage equals zero, the VCO runs at its quiescent (free-run) frequency ω_0. A positive or negative control voltage causes the VCO to increase or decrease, respectively, its instantaneous angular frequency $d\hat{\theta}(t)/dt$, thereby forcing the phase error to decrease or increase accordingly.

Hence, note that, in such a control system, zero-phase error occurs when the locally generated VCO signal and the input signal $s(t)$ have a phase difference of $\pi/2$. In order to output a signal $\hat{s}(t)$ with the same phase as the input signal $s(t)$, a phase shift of $\pi/2$ must be applied to the signal directly output by the VCO.

5.5.1.2   Scheme of PLL and Baseband Model

The automatic phase control system presented above is generally referred to as Phase-Locked Loop (PLL). Its block diagram, in the most basic form, is shown in Figure 5.5. The building blocks are a multiplier (*phase detector*), a *low-pass filter* having transfer function $F(s)$ (in the Laplace domain) and a *voltage-controlled oscillator* (VCO).

Please note that a signal multiplier is not the only way to generate an output functioning as measure of phase error between two inputs. Two broad categories of phase detectors are *multiplier-type* circuits and *sequential-logic* circuits. The ideal multiplier is a useful analytical model for phase detector, but in its basic form is rarely used in actual equipment. Multiplier-type phase detectors fully utilize the signal waveform and are capable of operating on signals deeply buried in noise. An example of sequential-logic phase detector, on the other hand, is the reset−set flip-flop. Sequential-logic phase detectors operate on signal level crossings only and provide a linear phase detection characteristic. For the

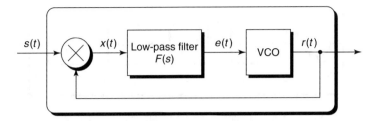

Figure 5.5. Block diagram of a basic phase-locked loop

sake of simplicity, in the following analysis PLL systems based on an ideal multiplier phase detector will be considered.

If the input reference signal and the signal output by the VCO are given respectively by[4]

$$s(t) = \sqrt{2}A \sin \Phi(t)$$

$$r(t) = \sqrt{2}K_1 \cos \hat{\Phi}(t) \tag{5.24}$$

where A [V] and K_1 [V] are their root-mean-square values, then the output of the multiplier, discarding the sum-frequency term, is given by

$$x(t) = AK_1K_m \sin[\theta(t) - \hat{\theta}(t)] \tag{5.25}$$

where K_m [V^{-1}] is the multiplier gain. The frequency of the VCO is function of the low-pass filtered error voltage $e(t)$. Therefore

$$\frac{d\hat{\Phi}(t)}{dt} = \omega_0 + K_0 e(t) \tag{5.26}$$

where K_0 [s^{-1} V^{-1}] is the VCO gain factor. From the definition of $\hat{\Phi}(t)$, we get also

$$\frac{d\hat{\theta}(t)}{dt} = K_0 e(t) \tag{5.27}$$

The analysis of the model yields thus to the following dynamic equation for the phase error $\phi(t) = \theta(t) - \hat{\theta}(t)$

$$\frac{d\phi(t)}{dt} = \frac{d\theta(t)}{dt} - KA \int_0^t f(t - u) \sin \phi(u) \, du \tag{5.28}$$

where $K = K_0 K_m K_1$ [s^{-1} V^{-1}]. The product KA is usually called *loop gain*.

Therefore, the equivalent mathematical model of the phase-locked loop represented in Figure 5.5 is the control system depicted in Figure 5.6, obeying the dynamic Equation (5.28). This model is a *baseband* model of the PLL, because centred on the

---

[4] The only difference between the signal $r(t)$, output directly by the VCO, and the output signal $\hat{s}(t)$ defined before is the $\pi/2$-phase shift and the peak value $\sqrt{2}K_1$.

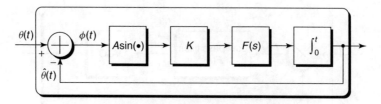

Figure 5.6. Baseband model of the phase-locked loop

phase error $\phi(t)$ independently of the quiescent frequency ω_0. Moreover, it is a *nonlinear* system model, which makes difficult its mathematical analysis.

5.5.1.3 Linear Model of PLL and Transfer Functions

If the phase error $\phi(t)$ is small (PLL locked to the reference) i.e. $\phi(t) \ll 1$ rad then the approximation

$$\sin \phi \approx \phi \qquad (5.29)$$

can be used and the non-linearity can be disregarded, as shown in the plot of Figure 5.7 where the two functions are compared for $-\pi \leq \phi \leq +\pi$.

Under this approximation, the system in Figure 5.5 becomes linear and thus can be analysed with standard techniques. Hence, the linearized baseband model of the phase-locked loop, in the Laplace domain, is depicted in Figure 5.8. The functions in the time domain $\phi(t)$, $\theta(t)$ and $\hat{\theta}(t)$ have been replaced with their Laplace transforms, denoted as $\phi(s)$, $\theta(s)$ and $\hat{\theta}(s)$, respectively.

From Equation (5.28), linearized by the approximation of Equation (5.29), we get the *closed-loop transfer function* of the PLL

$$H(s) = \frac{\hat{\theta}(s)}{\theta(s)} = \frac{KAF(s)}{s + KAF(s)} \qquad (5.30)$$

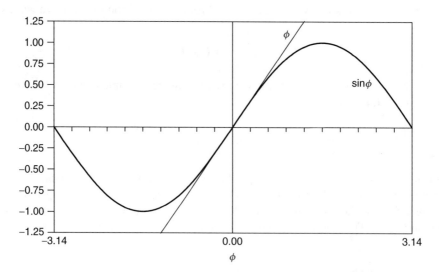

Figure 5.7. Compared plot of the functions $\sin(\phi)$ and ϕ for $-3.14 \leq \phi \leq 3.14$

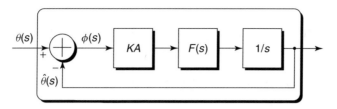

Figure 5.8.   Linearized baseband model in the Laplace domain of the phase-locked loop

By contrast, the *open-loop transfer function*

$$G_0(s) = \frac{KAF(s)}{s} \tag{5.31}$$

yields $\hat{\theta}(s)/\theta(s)$ when the feedback from the integrator to the sum node is open. Hence, the two important basic transfer functions of the phase-locked loop are the following

$$\frac{\hat{\theta}(s)}{\theta(s)} = H(s) = \frac{G_0(s)}{1 + G_0(s)} \tag{5.32}$$

$$\frac{\phi(s)}{\theta(s)} = \frac{\theta(s) - \hat{\theta}(s)}{\theta(s)} = 1 - H(s) = \frac{1}{1 + \dfrac{KAF(s)}{s}} = \frac{1}{1 + G_0(s)} \tag{5.33}$$

5.5.1.4   PLL Type and Order

The characteristics of the phase-locked loop are determined by the kind of filter $F(s)$.

A PLL without filter at all, i.e. with $F(s) = 1$, is called a first-order loop. One with a single-pole low-pass filter is a second-order loop. In general, a loop with a filter having $k - 1$ poles is called a kth-order loop. The order of the PLL is thus the degree of the polynomial at the denominator of the closed-loop transfer function $H(s)$ in Equation (5.30).

In control theory, feedback systems are also distinguished by their *type*. If the open-loop transfer function $G_0(s)$ has k poles at the origin ($s = 0$), i.e. k integrators, the phase-locked loop is a type-k system. Loop order can be higher then type, since all poles contribute to the order, but not to the type if they are not at the origin.

In the PLL model of Figure 5.8, the VCO is represented by an integrator. Therefore, the loop filter must have $k - 1$ poles at the origin to yield a type-k loop.

5.5.1.5   Steady-State Phase Error

The steady-state phase error can be evaluated from Equation (5.33) by application of the final-value theorem, which states that

$$\lim_{t \to \infty} \phi(t) = \lim_{s \to 0} s\phi(s) \tag{5.34}$$

Three notable cases to consider are the phase step, the frequency step and the frequency ramp as input signal $\theta(t)$.

In the case of *phase step* on the input, we have

$$\theta(t) = \Delta\theta \leftrightarrow \theta(s) = \frac{\Delta\theta}{s} \tag{5.35}$$

and

$$\lim_{t \to \infty} \phi(t) = \lim_{s \to 0} s \frac{\Delta\theta}{s} \frac{1}{1 + \dfrac{KAF(s)}{s}} = 0 \quad \text{for } F(0) \neq 0 \tag{5.36}$$

Therefore, the PLL damps any input phase step to zero, even if the loop filter $F(s)$ is a pure constant.

In the case of *frequency step* on the input, the input phase is a ramp as

$$\theta(t) = \Delta\omega t \leftrightarrow \theta(s) = \frac{\Delta\omega}{s^2} \tag{5.37}$$

and we have

$$\lim_{t \to \infty} \phi(t) = \lim_{s \to 0} s \frac{\Delta\omega}{s^2} \frac{1}{1 + \dfrac{KAF(s)}{s}} = \lim_{s \to 0} \frac{\Delta\omega}{s + KAF(s)} \tag{5.38}$$

Therefore, to keep the steady-state phase error small, the value $F(0)$ should be as large as possible. In particular, it is clear that the PLL is able to cancel any ramp phase error, resulting from a frequency step applied at $t = 0$, if $F(s)$ has at least one pole at $s = 0$, i.e. the loop is of type 2.

In the case of *frequency ramp* on the input, as it happens for example receiving a constant-frequency signal from a source moving with constant radial acceleration relative to the receiver, the input signal is

$$\theta(t) = \frac{\Delta\dot{\omega}t^2}{2} \leftrightarrow \theta(s) = \frac{\Delta\dot{\omega}}{s^3} \tag{5.39}$$

and we have

$$\lim_{t \to \infty} \phi(t) = \lim_{s \to 0} s \frac{\Delta\dot{\omega}}{s^3} \frac{1}{1 + \dfrac{KAF(s)}{s}} = \lim_{s \to 0} \frac{\Delta\dot{\omega}}{s^2 + sKAF(s)} \tag{5.40}$$

Therefore, the PLL can track an input frequency ramp with zero steady-state phase error only if $F(s)$ has at least two poles at $s = 0$, i.e. the loop is of type 3. If a single-pole filter $F_1(s)$ is used, a residual phase error equal to

$$\lim_{t \to \infty} \phi(t) = \frac{\Delta\dot{\omega}}{KAF_1(0)} \tag{5.41}$$

remains.

5.5.1.6   Stability of the Feedback System

A basic requirement for any feedback system is *stability*, which for our purposes can be defined in the sense that a system should respond to any bounded input signal with a bounded output signal. From basic control theory, the simplified Nyquist criterion states that a feedback system is stable if the *phase margin*[5]

---

[5] In this and in the following formulas, the *frequency response* $G(\omega)$ of the system is used, which is obtained by replacing the complex variable s in the transfer function $G(s)$ by the pure imaginary part $j\omega$. Therefore, $G(\omega)$ is a complex number for any given real angular frequency $\omega = 2\pi f$. The name *frequency response* is

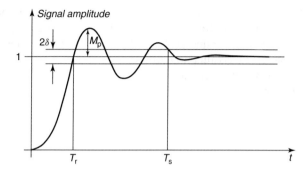

Figure 5.9. Example of system response to a unitary step input signal in the time domain and descriptive measures of the transient behaviour

$$\phi_R \hat{=} \arg[G_0(\omega_c)] - (-\pi) \qquad (5.42)$$

is positive, where ω_c is the *crossover frequency*, at which the log magnitude of the open-loop frequency response function $G_0(\omega)$ crosses the 0-dB gain axis, i.e.

$$|G_0(\omega_c)| = 1 \qquad (5.43)$$

5.5.1.7 Transient Response of the System

The exact transient response of the system for any input signal can be evaluated by solving the dynamic Equation (5.28). Under the linear approximation, the response can be also evaluated by convolution of the input signal with the inverse transform of $H(s)$ (5.30), as

$$\hat{\theta}(t) = \theta(t) * h(t) \qquad (5.44)$$

Typically, a step input signal is considered to evaluate the system response in the time domain. Descriptive measures of the transient response to a step input signal are the rise time T_r, the settling time T_s and the peak overshoot M_p, as illustrated in Figure 5.9. The rise time T_r is the time needed for the transient response to reach 100% of the step amplitude. The settling time T_s is the time needed for the transient response to settle within an error not higher than $\pm\delta\%$ around the step amplitude (a small value such as $\delta = 5\%$ is customary). The peak overshoot M_p is the maximum deviation of the transient response from the step amplitude.

---

due to the fact that if a sinusoidal signal

$$s_{in}(t) = \text{Re}\{A_{in}e^{j\omega t}\} = A_{in}\cos\omega t$$

is input into a linear system characterized by the transfer function $G(s)$, then the steady-state response is the signal

$$s_{out}(t) = \text{Re}\left\{|G(\omega)|A_{in}e^{j\omega t + j\arg[G(\omega)]}\right\} = A_{out}\cos(\omega t + \phi_{out}).$$

In other words, amplitude and phase of the output sine wave are related to the input amplitude and phase by the frequency response $G(\omega)$ as

$$A_{out} = |G(\omega)|A_{in}$$

$$\phi_{out} = \arg[G(\omega)]$$

What is interesting is that the transient response measures above can be approximately estimated, knowing the crossover frequency ω_c and the phase margin ϕ_R of the open-loop transfer functions. To this aim, we note that Equation (5.32) can be rather crudely approximated as a single-pole low-pass function around the crossover frequency ω_c (a slope of $G_0(\omega)$ greater than 40 dB/decade about $\omega \cong \omega_c$ would not yield sufficient phase margin to guarantee system stability) and thus

$$H(s) \approx \frac{1}{1 + s/\omega_c} \leftrightarrow h(t) \approx \omega_c e^{-\omega_c t} \quad \text{for } t \geq 0 \tag{5.45}$$

Hence, for $\delta = 5\%$, we obtain the approximated value of the settling time

$$T_s \cong \frac{3}{\omega_c} \tag{5.46}$$

Moreover, qualitatively, we notice that zero phase margin (i.e., $G_0(\omega_c) = -1$) corresponds to a pair of imaginary poles at $s \pm j\omega_c$ and therefore to an everlasting undamped sinusoidal oscillation as response to a step input signal (100% peak overshoot). A small phase margin ϕ_R corresponds to a large peak overshoot. A large phase margin ϕ_R means safe system stability and thus small peak overshoot.

5.5.2   Second-Order Phase-Locked Loop

Second-order PLLs are the most common in practical systems. As shown in the previous section, a single integrator in the loop filter allows to track a constant frequency offset between the input signal and the local oscillator without residual steady-state phase error. Also in this section, the book by Meyr and Ascheid [5.16] has been mostly followed to provide a simple overview on the topic.

Two types of single-pole filter are particularly worthwhile to consider in this analysis:

$$F_1(s) = \frac{1 + sT_2}{sT_1} \tag{5.47}$$

$$F_2(s) = \frac{1 + sT_2}{1 + sT_1} \tag{5.48}$$

The former filter $F_1(s)$ contains a perfect integrator and may be realized with active circuitry, i.e. a dc high-gain operational amplifier (*active filter*). On the other hand, a *passive filter* realization can approximate perfect integration by a low-pass filter with a pole at $s = -1/T_1$, as in the latter filter $F_2(s)$. A PLL incorporating passive filter is called an *imperfect second-order loop* and, in steady state, yields a fixed phase error in response to an input frequency offset.

With active filter implementation $F_1(s)$, from Equation (5.30) the overall closed-loop transfer function results

$$H_1(s) = \frac{KA(1 + sT_2)}{s^2 T_1 + sKAT_2 + KA} \tag{5.49}$$

while, in case of passive filter $F_2(s)$, the transfer function

$$H_2(s) = \frac{KA(1 + sT_2)}{s^2 T_1 + s(1 + KAT_2) + KA} \tag{5.50}$$

is obtained.

More generally, in second-order PLLs the single-pole loop filter is of the form

$$F(s) = \frac{A + sB}{C + sD} \tag{5.51}$$

The most common practical realizations of the loop-filter general form above are the following four:

- passive filter with a pole at $s = -1/T_1$
  ($A = 1$, $B = 0$, $C = 1$, $D = T_1$);

- passive filter with a pole at $s = -1/T_1$ and a zero at $s = -1/T_2$ (5.48)
  ($A = 1$, $B = T_2$, $C = 1$, $D = T_1$);

- active filter with a pole at $s = 0$
  ($A = 1$, $B = 0$, $C = 0$, $D = T_1$);

- active filter with a pole at $s = 0$ and a zero at $s = -1/T_2$ (5.47)
  ($A = 1$, $B = T_2$, $C = 0$, $D = T_1$)

A more practical way to write general second-order closed-loop transfer functions is the form

$$H(s) = \frac{Es + F}{s^2 + 2\zeta\omega_n s + \omega_n^2} \tag{5.52}$$

where ω_n is the *natural angular frequency*, ζ is the *damping ratio*, E and F are constants dependent on ω_n and ζ. The natural angular frequency ω_n is the frequency at which, in the PLL open-loop transfer function, the extension of the line with -40 dB/decade slope crosses the 0-dB line. As it will be shown in the next paragraphs, ω_n and ζ are fair descriptors of the PLL transient response, because they are related to the crossover frequency ω_c and the phase margin ϕ_R.

The formulas giving the values of the parameters E, F, ω_n, ζ, T_1 and T_2 are provided in Table 5.1, for the four cases of loop-filter transfer function above.

The magnitude of the closed-loop transfer function (5.52) is plotted in the graph of Figure 5.10, in the case of a perfect second-order PLL (active filter), as function of the normalized angular frequency ω/ω_n and for six values of the damping ratio ζ.

It is evident that $H(s)$ is low-pass type, with asymptotic slope -20 dB/decade for $\omega \gg \omega_n$: the PLL is able to cut off high-frequency phase fluctuations on the input signal. A practical measure of the PLL *bandwidth* is the cut-off frequency B [Hz] for which input phase fluctuations are transferred to the output attenuated by a factor $\sqrt{2}$, i.e.

$$20 \, \text{Log}_{10}|H(2\pi B)| = -3 \, \text{dB} \tag{5.53}$$

This measure is very common in practical measurements on slave clocks. In terms of natural angular frequency ω_n and damping ratio ζ, it results:

$$B = \frac{\omega_n}{2\pi}\sqrt{2\zeta^2 + 1 + \sqrt{(2\zeta^2 + 1)^2 + 1}}$$

$$\cong \frac{\omega_n}{\pi}\zeta \qquad \text{for } \zeta > 2 \tag{5.54}$$

Also of interest is the *overshoot* around $\omega \cong \omega_n$ (maximum gain): here, phase fluctuations on the input signal are amplified, instead of being reduced. It is evident that this

Table 5.1 Parameters of second-order PLL transfer function $H(s)$ (5.52) in the cases of perfect (active filter) and imperfect (passive filter) integrator

	Passive Filter		Active Filter	
Loop filter $F(s)$	$\dfrac{1}{1+sT_1}$	$\dfrac{1+sT_2}{1+sT_1}$	$\dfrac{1}{sT_1}$	$\dfrac{1+sT_2}{sT_1}$
E	0	$2\zeta\omega_n\left(1-\dfrac{\omega_n}{KA}\dfrac{1}{2\zeta}\right)$	$2\zeta\omega_n$	$2\zeta\omega_n$
F	ω_n^2	ω_n^2	ω_n^2	ω_n^2
ω_n	$\sqrt{\dfrac{KA}{T_1}}$	$\sqrt{\dfrac{KA}{T_1}}$	$\sqrt{\dfrac{KA}{T_1}}$	$\sqrt{\dfrac{KA}{T_1}}$
ζ	$\dfrac{1}{2}\sqrt{\dfrac{1}{KAT_1}}=\dfrac{\omega_n}{2KA}$	$\dfrac{T_2}{2}\sqrt{\dfrac{KA}{T_1}}\left(1+\dfrac{1}{KAT_2}\right)$ $=\dfrac{T_2\omega_n}{2}+\dfrac{\omega_n}{2KA}$	0	$\dfrac{T_2}{2}\sqrt{\dfrac{KA}{T_1}}=\dfrac{T_2\omega_n}{2}$
T_1	$\dfrac{KA}{\omega_n^2}$	$\dfrac{KA}{\omega_n^2}$	$\dfrac{KA}{\omega_n^2}$	$\dfrac{KA}{\omega_n^2}$
T_2	0	$\dfrac{2\zeta}{\omega_n}\left(1-\dfrac{\omega_n}{KA}\dfrac{1}{2\zeta}\right)$	0	$\dfrac{2\zeta}{\omega_n}$

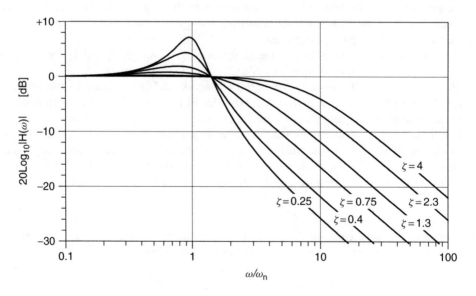

Figure 5.10. Magnitude of the closed-loop transfer function (5.52) of a perfect second-order PLL (active filter), as function of the normalized angular frequency ω/ω_n and for $\zeta = 0.25, 0.4, 0.75, 1.3, 2.3, 4$

is an unwanted property in slave clocks. Along a chain of slave clocks with the same ω_n, cumulated phase noise around this frequency gets considerably amplified and may become annoying even if the overshoot of the single clock is small. Hence, the overshoot size is usually less than 0.1 dB in slave clocks for synchronization networks.

As far as loop stability is concerned, we note that a second-order phase-locked loop is stable for all values of KA, because all poles of the closed-loop transfer function $H(s)$ lie in the left half-plane for all values of KA (necessary and sufficient condition for system stability).

As far as the transient response is concerned, we remind that the measures T_r, T_s and M_p introduced previously can be estimated as functions of the crossover frequency ω_c and the phase margin ϕ_R. These two quantities are related to the natural frequency ω_n and to the damping ratio ζ. Leaving the proof to [5.16], in second-order PLLs the following approximated relationships hold

$$\omega_c \cong 2\zeta\omega_n \qquad \text{(active filter)}$$

$$\omega_c \cong 2\zeta\omega_n\left(1 - \frac{\omega_n}{KA}\frac{1}{2\zeta}\right) \qquad \text{(passive filter)} \tag{5.55}$$

$$\zeta \cong \frac{\phi_R}{100} \quad \text{for } \zeta < 0.7 \tag{5.56}$$

with ϕ_R expressed in degrees (e.g., if $\phi_R = 60°$, then approximately $\zeta \cong 0.6$). For $\zeta > 0.8$, the damping ratio ζ significantly departs from $\phi_R/100$ and becomes much higher than it. As stated before for a generic PLL, we point out again that large phase margin (i.e., large damping ratio) means safe system stability and thus small peak overshoot.

5.5.3 Third-Order, Type-3 Phase-Locked Loop

Loops with type higher than 2 are sometimes used, because they feature improved tracking capability of high-dynamics phase fluctuations. We remind that a type-3 PLL, characterized by a loop filter $F(s)$ with two poles at $s = 0$, can track a frequency ramp on the input with zero steady-state phase error. Also in this section, the book by Meyr and Ascheid [5.16] has been followed to provide a simple overview on the topic.

Loop stability considerations yield that two zeros are needed in the open-loop transfer function if $F(s)$ has two poles at $s = 0$. Therefore, the open-loop transfer function is of the form

$$G_0(s) = \frac{KAF(s)}{s} = \frac{KA(1 + sT_2)(1 + sT_3)}{s(sT_1)^2} \tag{5.57}$$

and the closed-loop transfer functions result

$$H(s) = \frac{\hat{\theta}(s)}{\theta(s)} = \frac{\omega_c(s + \alpha\omega_c)(s + \alpha_1\omega_c)}{s^3 + \omega_c(s + \alpha\omega_c)(s + \alpha_1\omega_c)}$$

$$1 - H(s) = \frac{\phi(s)}{\theta(s)} = \frac{s^3}{s^3 + \omega_c(s + \alpha\omega_c)(s + \alpha_1\omega_c)} \tag{5.58}$$

where

$$\omega_c = KA \left(\frac{T_2}{T_1}\right)\left(\frac{T_3}{T_1}\right), \quad \alpha = \frac{1}{\omega_c T_2}, \quad \alpha_1 = \frac{1}{\omega_c T_3} \tag{5.59}$$

Also in this case, ω_c is the crossover frequency.

As far as stability is concerned, system analysis reported in [5.16] shows that, while first- and second-order PLLs are unconditionally stable, a third-order, type-3 PLL is stable if the gain KA is larger than a minimum value, evaluated as

$$KA > \frac{1}{2T_2}\left(\frac{T_1}{T_2}\right)^2 \tag{5.60}$$

in the common case in which the two zeros are coincident ($T_2 = T_3$). Therefore, to ensure stability in practical systems, the input signal amplitude A must be kept above a minimum value by means of some automatic gain control system.

As far as the transient response is concerned, analysis carried out in [5.16] yields an interesting result: the transient response of a third-order type-3 loop is well approximated by that of a second-order type-2 loop having equal crossover frequency ω_c and phase margin ϕ_R. Actually, this is true also for systems of any order, provided that the response is primarily due to a *dominant* pair of complex poles in the closed-loop transfer function[6].

This result is useful in first-draft designing third-order loops by second-order approximation. Given for example settling time T_s and peak overshoot M_p, the designer can derive the corresponding second-order loop damping ratio ζ from numerical tables (an example graphical plot is reported in [5.16]). Using this value of ζ, as shown in [5.16], the open-loop transfer functions of a third-order type-3 PLL and of a second-order PLL, both characterized by same crossover frequency ω_c and phase margin ϕ_R, are given respectively by (frequencies normalized to ω_c)

$$G_{0/3}(s) = \frac{\left(\dfrac{s}{\omega_c} + \alpha\right)^2}{\left(\dfrac{s}{\omega_c}\right)^3}, \quad G_{0/2}(s) = \frac{\dfrac{s}{\omega_c} + \left(\dfrac{1}{2\zeta}\right)^2}{\left(\dfrac{s}{\omega_c}\right)^2} \tag{5.61}$$

with

$$\omega_c = 2\zeta\omega_n, \quad \zeta = \frac{1}{2}\sqrt{\frac{1-\alpha^2}{2\alpha}} \tag{5.62}$$

Therefore, using the value of ζ obtained to solve the equation above for α, the third-order loop transfer function can be readily designed from the second-order approximation.

5.5.4 PLL Performance with Input Additive Noise

Characterizing the PLL behaviour in the case the input signal is affected by some additive noise $n(t)$ is not a simple task, at least in the general case. Meyr and Ascheid in their

---

[6] As a rule of thumb, a *complex conjugate pair of poles* in the closed-loop transfer function $H(s)$ can be considered *dominant* if at least one of the following conditions is fulfilled:

- the absolute values of the real parts of the additional poles are at least 10 times larger than the absolute value of the real part of the dominant pole pair;
- in the case a pole is located near the dominant pair, there is a zero close to it, forming then a zero-pole doublet.

book [5.16] analyse this problem with the fairly general assumption of narrow-band, band-pass noise.

The problem can be greatly simplified by the further assumption that the input noise has small amplitude, so that the non-linearity in the phase detector model can be approximated by a linear characteristic, as in Equation (5.29). The linearized baseband model of the PLL with input additive noise is then shown in Figure 5.11. The only difference with the scheme of Figure 5.8 is the input noise that the PLL has to track, modelled by the additive noise term $n'(t)$[7] inserted *after* the phase-detector node, and some constant factors grouped and moved around the loop. Here, the VCO gain factor K_0 [s^{-1} V^{-1}] and the phase detector gain $K_d = A K_m K_1$ [V] are used.

In this analysis, the noise $n(t)$ added on the input signal is assumed to be narrow-band, band-pass, Gaussian and zero-mean. A narrow-band band-pass noise process is defined in the frequency domain by the property that its spectrum is negligibly small everywhere except in the frequency range $|\omega - \omega_0| < \pi B_{IF} \ll \omega_0$, where ω_0 is the central frequency and B_{IF} is the equivalent noise bandwidth, which will be defined formally in the next paragraph. Then, the input additive noise $n(t)$ can be expressed as

$$n(t) = \sqrt{2} n_c(t) \cos \omega_0 t - \sqrt{2} n_s(t) \sin \omega_0 t$$
$$= \sqrt{2} \text{Re}\{\vec{n}_L(t) e^{j\omega_0 t}\} \quad (5.63)$$

where $n_c(t)$ and $n_s(t)$ are two zero-mean, stationary, Gaussian, base-band (i.e., slowly variant) random processes. The complex vector

$$\vec{n}_L(t) = n_c(t) + j n_s(t) \quad (5.64)$$

is called *complex envelope* of the noise process.

The equivalent noise bandwidth B_{IF} [Hz] is so defined that, if we approximate the noise power spectral density $S_n(\omega)$ by two rectangles for the positive and negative frequency axes of width $2\pi B_{IF}$ and amplitude $S_{n_c}(0)$, the area will be the same as the area under $S_n(\omega)$, i.e.

$$\int_{-\infty}^{\infty} S_n(\omega) d\omega = S_{n_c}(0) 2\pi B_{IF} + S_{n_c}(0) 2\pi B_{IF} \quad (5.65)$$

Hence,

$$B_{IF} = \frac{1}{2\pi} \int_0^{\infty} \frac{S_n(\omega)}{S_{n_c}(0)} d\omega \quad \text{[Hz]} \quad (5.66)$$

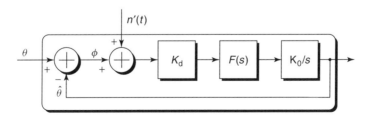

Figure 5.11. Linearized model of phase-locked loop with input additive noise

---

[7] The context should make clear that, in this case, the prime sign does not denote derivative.

Note that B_{IF} is measured in cycles per second and not in radians per second, as made evident by the factor $1/2\pi$ in the formula above. The value $S_{n_c}(0)$ is often denoted as $N_0/2$, where N_0 is the noise one-sided spectral density if this is constant.

Getting back to the scheme of Figure 5.11, the additive noise process $n'(t)$ injected onto the phase error can be viewed as angular phase disturbance that replaces the narrow-band band-pass noise $n(t)$ added on the input signal when we move to the baseband model. It results [5.16]

$$n'(t) = \frac{n_c(t)}{A}\cos\hat{\theta} + \frac{n_s(t)}{A}\sin\hat{\theta}$$

$$= \frac{1}{A}\text{Re}\{\vec{n}_L(t)e^{-j\hat{\theta}}\} \tag{5.67}$$

In [5.16] it is shown that the statistical description of $n'(t)$ is approximately independent on the VCO output phase $\hat{\theta}(s)$, at least if the loop is locked to the input reference. Hence the sole independent variable t in the notation $n'(t)$. In other terms, the phase $\hat{\theta}(s)$ can be considered nearly constant over the short time interval $1/B_{IF}$, where a significant correlation between two noise samples $n(t)$ and $n(t + 1/B_{IF})$ exists (the noise power spectrum is assumed non-zero only within the interval wide $2\pi B_{IF}$ and centred on ω_0).

Owing to the superposition principle of linear systems, to assess the effect of the input additive noise we may set $\theta = 0$ and then use n' instead of θ in Equation (5.30), to yield

$$S_{\hat{\theta}}(\omega) = |H(\omega)|^2 S_{n'}(\omega) \tag{5.68}$$

As far as the noise power is concerned, when $\theta = 0$ we have $\phi = -\hat{\theta}$ and the phase error variance (the additional phase error variance generated by the insertion of the noise $n'(t)$) is

$$\sigma_\phi^2 = \frac{1}{2\pi}\int_{-\infty}^{\infty}|H(\omega)|^2 S_{n'}(\omega)d\omega \tag{5.69}$$

In many cases of practical interest, the power spectral density of $n'(t)$ is nearly constant for all frequencies of interest (i.e., where $|H(\omega)|^2$ is significantly different from zero) and thus the following approximation can be used

$$S_{n'}(\omega) \approx \frac{S_{n_c}(0)}{A^2} = \frac{N_0}{2A^2} \tag{5.70}$$

We obtain then

$$\sigma_\phi^2 = \frac{N_0}{2A^2}\frac{1}{2\pi}\int_{-\infty}^{\infty}|H(\omega)|^2 d\omega = \frac{N_0 B_L}{A^2} \tag{5.71}$$

having defined the equivalent loop noise bandwidth

$$B_L = \frac{1}{2\pi}\int_0^{\infty}\frac{|H(\omega)|^2}{|H(0)|^2}d\omega \quad \text{[Hz]} \tag{5.72}$$

analogously to the equivalent noise bandwidth B_{IF}.

5.5.5 PLL Performance with Internal Noise Sources

The characterization of internal noise sources in a PLL is a very important topic, aiming at evaluating slave clock performance by analytical tools. Kroupa, in his paper [5.17], provided a general survey on noise properties in PLL systems.

Assuming that the noise internally generated has small amplitude and that the PLL is working locked to the reference (i.e., the input noise, if any, has small amplitude as well), the PLL can be modelled again as a linear system. Thus, a simplified version of the linear model approximating the PLL behaviour with internal noise sources is depicted in Figure 5.12. In this figure, the main noise sources are modelled as phase random processes injected as additive terms: ϕ_{VCO} [rad] is the phase noise generated by the VCO and V_{DF} [V] is the tension noise produced cumulatively by the phase detector and the loop filter. Moreover, ϕ_{in} [rad] is the phase noise on the input signal and ϕ_{out} [rad] is the overall phase noise resulting on the output signal. Please note the difference between this scheme and that of Figure 5.11: in the previous subsection, the effect of noise *added* on the input signal was analysed; in this subsection, instead, the effect of noise *modulating the phase* of signals circulating in the loop is considered.

The analysis of this linear model, by considering one input at a time among ϕ_{in}, ϕ_{VCO} and V_{DF} and owing to the superposition principle, yields the following three transfer functions (cf. Equations (5.32) and (5.33)):

$$H(s) = \frac{\phi_{\text{out}}(s)}{\phi_{\text{in}}(s)} = \frac{G_0(s)}{1 + G_0(s)} = \frac{K_0 K_d F(s)}{s + K_0 K_d F(s)} \tag{5.73}$$

$$H_A(s) = \frac{\phi_{\text{out}}(s)}{V_{\text{DF}}(s)} = \frac{H(s)}{K_d} = \frac{K_0 F(s)}{s + K_0 K_d F(s)} \tag{5.74}$$

$$H_B(s) = \frac{\phi_{\text{out}}(s)}{\phi_{\text{VCO}}(s)} = 1 - H(s) = \frac{s}{s + K_0 K_d F(s)} \tag{5.75}$$

As usual, the expression $K_0 K_d = K_0 A K_m K_1 = KA$ [s^{-1}] denotes the loop gain. Being the loop filter $F(s)$ low-pass, it is important to notice the following two basic facts.

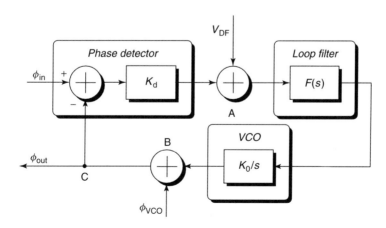

Figure 5.12. Linear model of a phase-locked loop with the main internal noise sources

- The transfer functions $H(s)$ and $H_A(s)$ are low-pass. Therefore, the phase noise on the input reference signal $s_{in}(t)$ and the internal tension noise produced by the phase detector and the loop filter are low-pass filtered to the output signal $s_{out}(t)$.

- The transfer function $H_B(s)$ is high-pass. Therefore, the internal phase noise produced by the VCO is high-pass filtered to the output signal $s_{out}(t)$.

5.5.6 Operational Ranges of Slave Clocks

Four key parameters describe the performance of a PLL in terms of operational limits of the input frequency: the *hold-in range* $\Delta\omega_{HI}$, the *pull-out range* $\Delta\omega_{PO}$, the *lock-in range* $\Delta\omega_{LI}$ and the *pull-in range* $\Delta\omega_{PI}$. Typically, we have $\Delta\omega_{LI} < \Delta\omega_{PO} < \Delta\omega_{PI} < \Delta\omega_{HI}$. Standard recommendations on slave clocks include specifications for some or all the parameters above. The definitions provided in this section are those in use in ITU-T standards on synchronization. For a more formal and mathematical treatise, the reader is refered to [5.16].

5.5.6.1 Hold-In Range

The hold-in range $\Delta\omega_{HI}$ is defined as the largest offset between the frequency ω_{in} of the input reference signal $s_{in}(t)$ and a specified nominal frequency (approximately the VCO free-run frequency ω_F), within which the slave clock maintains lock as the frequency varies slowly (rigorously speaking with $d\omega_{in}/dt \to 0$) over the frequency range.

In other terms, within the hold-in range, the PLL can track slow (*quasi-stationary*) variations of the input frequency. For this reason, the hold-in range may be called also the hold range in *static* conditions.

5.5.6.2 Pull-Out Range

The pull-out range $\Delta\omega_{PO}$ is defined as the largest offset between the input frequency ω_{in} and a specified nominal frequency (approximately the VCO free-run frequency ω_F), within which the slave clock stays in the locked mode and outside of which the slave clock cannot maintain locked mode, irrespective of the rate of the frequency change.

In other terms, within the pull-out range, the PLL can track *arbitrarily fast* variations (e.g. steps) of the input frequency. For this reason, the pull-out range may be called also the hold range in *dynamic* conditions.

5.5.6.3 Lock-In Range

The lock-in range $\Delta\omega_{LI}$ is defined as the largest offset between the input frequency ω_{in} and a specified nominal frequency (approximately the VCO free-run frequency ω_F), within which the PLL locks 'fast' (i.e., in a time in the order of $1/\omega_n$ seconds for a second-order PLL) to the new input frequency[8].

5.5.6.4 Pull-In Range

The pull-in range $\Delta\omega_{PI}$, is defined as the largest offset between the input frequency ω_{in} and a specified nominal frequency (approximately the VCO free-run frequency ω_F), within

---

[8] More precisely, the lock-in range is the interval within which the PLL locks to the new input frequency without *cycle slips* before settling at a stable equilibrium point. A PLL is said to slip a cycle when the phase error increases by $+2\pi$ (positive cycle slip) or -2π (negative cycle slip).

which the slave clock achieves locked mode, irrespective of how long it takes to lock[9]. The reader might be surprised that $\Delta\omega_{PI} < \Delta\omega_{HI}$. To understand this, we should consider the fundamental difference between *tracking* and *acquisition*. Tracking presumes that the PLL is already in lock and the input frequency is changed until the loop loses lock. During acquisition, the PLL is initially out of lock.

5.5.7 Operation Modes of Slave Clocks in a Synchronization Network

A slave clock may operate in the following three modes in a synchronization network, depending on the availability of a trusty reference signal at its input.

5.5.7.1 Locked Mode

The slave clock is tracking the input frequency, which keeps within the hold-in and pull-out ranges ($\omega_{in} < \Delta\omega_{PO} < \Delta\omega_{HI}$). Should the input frequency move outside the pull-out range ($\Delta\omega_{PO} < \omega_{in} < \Delta\omega_{HI}$), the slave clock stays in locked mode as well, provided that the input-frequency change rate $d\omega_{in}/dt$ is slow enough.

The locked mode may be *ideal*, if the reference signal is always available and stable, or *real* (stressed), if the reference is affected by impairments of various kind.

Even though the ideal locked mode is not typical of real synchronization network operation, understanding clock performance under ideal operation is important to get bounds for clock performance. Under ideal conditions, the slave clock operates in strict phase lock with the incoming reference. For observation intervals shorter than the PLL time constant, the clock stability is determined by the short-term stability of the local oscillator as well as by quantization effects, for example in the phase detector, if the PLL is realized with digital techniques. As it will be shown better in a later section, in the absence of input reference impairments and due to internal noise sources, the output timing signal exhibits a phase noise approximately behaving as white noise phase modulation.

Stressed operation, on the other hand, is the typical mode of operation of slave clocks receiving timing over a network facility that may introduce little impairments such as micro-interruptions. Micro-interruptions are short time intervals in which the signal becomes not usable as timing reference. In real networks, they may last tens of microseconds and can be expected even in a number of 1 to 100 per day.

All interruptions affect slave clock operation! When reference has been restored (or all the more reason if the interruption persisted and clock had to switch reference), we find some residual time error between the slave clock and the input reference, cumulating at each interruption. The actual amount of timing error produced by each reference interruption depends on the clock design, but is specified by international standards to be at worst not greater than 1 μs. The overall timing error, sum of independent random terms, behaves as a random-walk noise phase modulation (i.e., white noise frequency modulation, as it will be shown in a later section).

Moreover, slave clocks with poor performance in response to micro-interruptions, such as some cheap clocks typically deployed at customer premises, after a reference interruption simply switch reference or enter autonomous mode of operation, producing large

---

[9] Within the pull-in range, the PLL locks to the new input frequency even with one or more cycle slips. The larger the initial frequency difference, the more cycles the loop will slip before it settles. The pull-in process is also called *unaided acquisition of frequency* and is a slow and unreliable process.

phase hits on the output. This is due to the fact that most cheap clocks intended for customer premises application do not incorporate phase build-out features to limit time-keeping error on transients after micro-interruptions. The magnitude of such phase hits may be so large to make downstream clocks un-lock or switch reference and thus may cause slips in the customer network equipment.

5.5.7.2 Free-Run Mode

If the reference signal fails, or the input frequency moves outside the hold-in range (or fast outside the pull-out range) and keeps outside the pull-in range so that the clock cannot re-lock, the clock has to work *autonomously*, supplying the free-run frequency ω_F of the internal oscillator (VCO) working with null control tension. Of course, the nominal free-run frequency ω_F is designed equal to the nominal reference frequency.

Actually, real slave clocks for synchronization networks never enter plain free-run mode. Hold-over is the standard mode of autonomous operation in practical equipment.

5.5.7.3 Hold-Over Mode

As in the free-run mode, if the reference signal fails the clock works autonomously, supplying the frequency generated by the VCO. In this case, nevertheless, the last control tension value at the input of the VCO before the reference failure is maintained, so to hold the last output frequency value over. More sophisticated clocks even store several samples of the VCO control tension, so that after entering the hold-over mode the VCO is controlled with a variable tension extrapolated from the last data. An excellent hold-over stability can be achieved in this way, even under a substantial non-linear drift of the VCO frequency.

Hold-over mode is thus the slave-clock mode of operation in the rare cases that all input references failed or are judged not reliable (for example, because providing an input frequency outside the pull-in range). Hold-over performance is mainly characterized by the residual initial frequency offset and frequency drift. The initial frequency offset is caused by the settability resolution of the local oscillator frequency and the noise on the timing reference on clock un-locking. Frequency drift is due to ageing of the local oscillator.

5.6 FREQUENCY-DOMAIN AND TIME-DOMAIN STABILITY CHARACTERIZATION

The characterization of clock stability is usually carried out by characterizing, by means of suitable analytical tools, the random processes $\varphi(t)$, $x(t)$, TE(t), $y(t)$ or $\nu(t)$ (cf. Equations (5.4)(5.11)(5.13)). There is a huge literature on time and frequency stability characterization for precision oscillators. An essential list of selected publications on this subject is reported in the bibliography [5.1]–[5.4][5.18]–[5.33].

Historically, a dichotomy between the characterization of such processes in the *Fourier-frequency domain* and in the *time domain* were established: the inadequacy of measurement equipment strengthened the barriers between these two characterizations of the same noise process. Although these barriers are mainly artificial, nevertheless it is not always possible to translate unambiguously from any quantity of one domain to anyone of the other.

Examples of stability measures in the frequency domain are the one-sided *Power Spectral Densities* (PSDs, or more simply spectra) of the phase, time and frequency fluctuations, since they are functions of the Fourier frequency f.

Note that the word 'frequency' may be used with two different meanings that should not be confused. The symbol $v(t)$ denotes the time-dependent instantaneous frequency of the clock. The symbol f indicates the time-independent Fourier frequency that is argument of spectral densities and Fourier transforms: in base-band representation, positive values of f denote frequencies above the central frequency v_n, negative values denote frequencies below v_n.

On the other hand, variances of the same fluctuations, averaged over a given observation interval, are examples of stability measures in the time domain, since they are functions of the observation interval τ (time).

In general, both frequency-domain and time-domain stability measures can be evaluated starting from samples of time error or of instantaneous frequency.

5.7 CLOCK STABILITY CHARACTERIZATION IN THE FREQUENCY DOMAIN

Analysis in the Fourier frequency domain is of great importance both for theoretical purposes, in terms of richness of information included, and for practical application purposes, since it expresses the power spreading in the frequency domain around the ideal carrier frequency. For this reason, spectral densities of various kinds have been widely used for clock stability characterization.

In particular, the analogue measurement of PSD functions directly in the frequency domain (spectral analysis) has been for a long time the main technique for studying the behaviour of oscillators.

More recently, the introduction of high-resolution digital instrumentation for the measurement in the time domain (time counters) made more appealing the time-domain measures in most applications. In fact, the recent telecommunications standards recommend, as standard stability measures, time-domain quantities evaluated from samples of time error measured directly in the time domain.

However, spectral analysis [5.34] of clock noise is still considered one of the main tools for clock characterization [5.1][5.2][5.4][5.25][5.27]. Nowadays, digital spectral analysers evaluate the PSD of interest by Fast Fourier Transform (FFT) computation on data samples measured in the time domain by a time counter.

5.7.1 Power Spectral Density of the Timing Signal

The most straightforward and intuitive way to characterize the accuracy and stability of the timing signal $s(t)$ supplied by a clock is to evaluate directly its two-sided PSD, or more precisely its low-pass (base-band) representation, often denoted as $S_s^{RF}(f)$ (spectrum in radio frequency).

The RF spectrum $S_s^{RF}(f)$ is a continuous function, proportional to the timing-signal power delivered to a matched load per unit of bandwidth centred on f. It is measured in [W/Hz] or [V²/Hz] units.

In case of ideal timing signal (constant frequency $v(t) = v_n$), the radio-frequency (RF) spectrum $S_s^{RF}(f)$ is a Dirac pulse $P_s\delta(f)$, where P_s is the signal power, as shown in

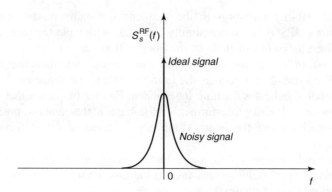

Figure 5.13.   Power spectral density of ideal and noisy timing signals

Figure 5.13. In the real case of timing signal affected by some phase and amplitude noise, the PSD ideal pulse spreads and exhibits some spectral content also at frequencies $f \neq 0$. A *pedestal* around the ideal frequency $f = 0$, therefore, is typically found in power spectral densities measured on real oscillators.

The spectrum $S_s^{\mathrm{RF}}(f)$ is definitely *not* a good tool to characterize clock frequency stability: unfortunately, given $S_s^{\mathrm{RF}}(f)$, it is not possible to determine unambiguously whether the power at various Fourier frequencies is the result of amplitude rather than phase fluctuations in the timing signal $s(t)$. Nevertheless, if the power due to amplitude modulation noise is negligible and the phase fluctuations are small (i.e., the mean square value $\langle \varphi^2(t) \rangle$[10] is much less than 1 rad^2) as it happens in practical clocks, then the spectrum $S_s^{\mathrm{RF}}(f)$ has approximately the same *shape* of the phase-noise spectrum $S_\varphi(f)$. However, it is still difficult to obtain quantitative results about $\varphi(t)$ from it.

5.7.2   Power Spectral Densities of Phase and Frequency Fluctuations

In the Fourier frequency domain, the most commonly used stability measures are the one-sided PSDs of $\varphi(t)$, $x(t)$ and $y(t)$ denoted as $S_\varphi(f)$ [rad^2/Hz], $S_x(f)$ [ns^2/Hz] and $S_y(f)$ [Hz^{-1}] respectively, because they describe directly the time and frequency stability characteristics.

The PSDs above are equivalent representations of the same noise process. In fact, the following relationships hold among them:

$$S_y(f) = \frac{f^2}{v_{\mathrm{n}}^2} S_\varphi(f)$$

$$S_x(f) = \frac{S_\varphi(f)}{(2\pi v_{\mathrm{n}})^2} \tag{5.76}$$

$$S_y(f) = (2\pi f)^2 S_x(f)$$

---

[10] The symbol $\langle \cdot \rangle$ denotes the infinite *time-average operator* on the argument function. For example, in the case of continuous-time argument function $s(t)$, it is defined as

$$\langle s(t) \rangle = \lim_{T \to \infty} \frac{1}{2T} \int_{-T}^{T} s(t)dt$$

The PSD $S_y(f)$ was first recommended in 1971 by IEEE as standard frequency stability measure in the Fourier frequency domain [5.1][5.7]. It is worthwhile noticing that, under the assumption of Gaussian stationary random processes, the power spectral density (or equivalently the autocorrelation function, which is its inverse Fourier transform) contains maximum information about the random process. The time-domain variances that will be defined in the next sections are related to the spectral density, by some integral relationships, but do not include full characterization of the process.

Spectral densities of phase and frequency fluctuations, in principle, can be measured by a spectrum analyser following some kind of phase and frequency demodulator.

5.7.3   Other Spectral Measures

The other two spectral measures of phase and frequency stability that are sometimes encountered are the PSD of non-normalized frequency fluctuations $\Delta v(t)$[11]

$$S_{\Delta v}(f) = v_n^2 S_y(f) \tag{5.77}$$

measured in [Hz^2/Hz], and the single-sideband measure of phase noise

$$\mathscr{L}(f) = \frac{1}{2} S_\varphi(f) \tag{5.78}$$

(read as *script 'ell'*) valid only for values of the Fourier frequency far enough from the carrier at $f = 0$, or more precisely in the interval $f_A < f < \infty$ where

$$\int_{f_A}^{\infty} S_\varphi(f)df \ll 1 \text{ rad}^2 \tag{5.79}$$

The quantity $\mathscr{L}(f)$ is usually measured in [dBc/Hz] units (decibel of carrier-over-noise power per bandwidth unit), as

$$\mathscr{L}(f)\Big|_{\frac{\text{dBc}}{\text{Hz}}} = 10 \text{ Log}_{10} \mathscr{L}(f) \tag{5.80}$$

This quantity is found rather commonly in technical specifications of commercial oscillators. The reason of its appeal lies in the physical interpretation of $\mathscr{L}(f)$: a signal-to-noise ratio that can be assessed readily on a spectrum analyser. Its use is not recommended by relevant standards.

5.8   CLOCK STABILITY CHARACTERIZATION IN THE TIME DOMAIN

While frequency-domain characterization proves to be very meaningful and complete in studying the behaviour of oscillators, it is important to point out that the main concern in digital telecommunications lies in controlling time deviations *over given observation intervals*. For example, the buffer-fill level in digital transmission and switching equipment

---

[11] Unfortunately, there is a lack of uniformity in the notation among authors. The symbol $\Delta v(t)$ used herein should not be confused with the symbol Δv used in Equation (5.4), denoting the time-constant starting frequency offset of the clock. In Equation (5.4), the symbol $\Delta v(t)$ was denoted as $v_a(t)$.

is proportional to the time error cumulated between the write and read clocks. Controlling the time error over a certain observation interval means to control buffer overflows and underflows over time intervals of that length.

Time-domain characterization of time and frequency stability answers the obvious question: what is the stability (i.e., time and frequency deviation) of the timing signal under test over a given time interval τ (determined by the particular field of application)? Time-domain stability quantities are basically a sort of prediction of the expected time and frequency deviations over an observation interval τ. Therefore, they are more oriented to telecommunications purposes than spectral densities. Moreover, time-domain analysis is often much more efficient in providing meaningful measures of long-term performance.

Instabilities in oscillators are time variations of the quantities of interest (i.e., phase and frequency). Time-domain characterization is thus based on the measure of variations that occur over a specified time interval τ. Since phase and frequency of real oscillators are random phenomena, measurement must involve some time-averaging to yield statistically meaningful quantities.

In the following, the most important time-domain stability quantities defined in literature are presented. For each of them, the theoretical definition is given. Finally, the standard estimators of the time-domain quantities defined by ITU-T and ETSI, for the specification of timing interface requirements, are provided in the last subsection.

Among the list of publications already cited, for a general but thorough scientific treatise on time-domain quantities we recommend especially the excellent survey papers [5.1][5.2][5.22]. The third one, in particular, provides an outstanding review, for both broadness and depth, of quantities defined for time and frequency stability characterization. Other fundamental papers for time-domain characterization are, amongst all, [5.3][5.4][5.20][5.21][5.26][5.28]. Finally, papers [5.31][5.32][5.33] focus on characterization and measurement of telecommunications clocks.

5.8.1 Basic Measurement of $y(t)$ in the Time Domain

The oscillator *instantaneous* frequency $v(t)$ and fractional frequency $y(t)$, defined in Equations (5.4) and (5.11) respectively, are not actually observable in practice, because any frequency measurement technique does involve a finite time interval over which the frequency is measured and averaged, for example by counting the number of cycles of the input signal during the time interval.

Therefore, the average value of the frequency, measured over an observation interval τ beginning at a generic instant t_k, is a quantity more directly related to the experimental result. In particular, the averaged sample of the normalized, fractional frequency $y(t)$ is very widely used. It is defined as [5.1][5.22]

$$\overline{y}_k(\tau) = \langle y(t_k) \rangle_\tau = \frac{1}{\tau} \int_{t_k}^{t_k+\tau} y(t)dt \qquad (5.81)$$

From Equation (5.11), we get moreover

$$\overline{y}_k(\tau) = \frac{\varphi(t_k + \tau) - \varphi(t_k)}{2\pi v_n \tau} = \frac{x(t_k + \tau) - x(t_k)}{\tau} \qquad (5.82)$$

It is worthwhile remarking, at this point, that the operator in the discrete-time domain that corresponds to the derivative operator defined in the continuous-time domain is the *difference* operator. The first difference y_k of a sequence of samples $\{x_k\}$, evenly spaced with sampling period T, is given by

$$1^{\text{st}} \text{ difference}: \quad y_k = \frac{x_{k+1} - x_k}{T} \tag{5.83}$$

equivalent in the discrete-time domain to the first derivative $y(t) = x'(t)$ in the continuous-time domain. Analogously, the second difference z_k is given by

$$2^{\text{nd}} \text{ difference}: \quad z_k = \frac{y_{k+1} - y_k}{T} = \frac{x_{k+2} - 2x_{k+1} + x_k}{T^2} \tag{5.84}$$

equivalent to the second derivative $z(t) = y'(t) = x''(t)$. Further-order differences are defined similarly.

The quantity $\overline{y}_k$ is easily related to experimental results, produced for example by time-counting techniques. Thus, it has been used to define most time-domain quantities that will be presented in the following. One sample $\overline{y}_k$ is given by one single measurement of duration τ accomplished starting at time t_k. Right now, it is worthwhile noticing that several samples $\overline{y}_k$, given by repeated measurements, are needed to allow statistical characterization of instability over the time interval τ.

Due to random fluctuations of $y(t)$ in real oscillators, repeated measurements of $\overline{y}_k$ yield random results (or better, different samples of a random variable). The fundamental issue of time and frequency characterization in the time domain is thus to identify suitable statistical measures of $\overline{y}_k$. In particular, a statistical measure of the dispersion of the $\overline{y}_k$ samples provides a time-domain measure of instability over τ.

5.8.2 Classical Variance of $y(t)$ (True Variance)

The plain, classical variance σ^2 and its square root σ (standard deviation) are widely used statistical tools to measure the dispersion of samples of a random variable. In our case, under the assumption that $y(t)$ is ergodic and has zero mean, the variance is simply equal to

$$\sigma^2[\overline{y}_k] = \langle \overline{y}_k^2 \rangle = I^2(\tau) \tag{5.85}$$

This quantity is a theoretical measure and is also referred to as *true variance*, as based on averaging an infinite number of samples. It is also denoted as $I^2(\tau)$, because it indicates that it is a measure of *instability* over the time interval τ.

For *stationary* frequency fluctuations, the true variance has the following limit values:

$$\lim_{\tau \to 0} I(\tau) = \sqrt{\langle y^2(t) \rangle}$$

$$\lim_{\tau \to \infty} I(\tau) = 0 \tag{5.86}$$

In other words, for $\tau \to 0$ we approach ideal instantaneous frequency measurement (yielding the root mean square value of $y(t)$) and for $\tau \to \infty$ stationary fluctuations tend to be completely averaged out.

Despite its mathematical simplicity, the true variance $I^2(\tau)$ is really not a useful tool for clock stability characterization, because its time-averaging does not converge for some common kinds of phase noise, such as flicker and random-walk frequency noise (see Section 5.9 for common types of clock noise). In particular, the limit value for $\tau \to \infty$ may approach infinity in such cases. Therefore, more suitable quantities for clock stability characterization, based on the concept of sample variance, were introduced beginning from 1966 by Allan [5.3] and others to cope with such convergence issues in most cases of practical interest.

5.8.3 M-Sample Variance of $y(t)$

The *M-sample variance* is based upon averaging an ensemble of a *finite* number M of samples $\overline{y}_k$ $(k = 1, 2, \ldots, M)$ defined as previously: each sample $\overline{y}_k$ is measured over an observation interval τ beginning at instant t_k. Sampling times are evenly spaced $T : t_{k+1} = t_k + T$. Thus, the dead time between measurements is $T - \tau$.

The M-sample variance may be defined in several possible ways, consisting in biased or unbiased estimates of the true variance (5.85) as described in [5.22]. Rutman suggested [5.2] as the main estimate of the average of M samples the definition

$$\sigma_y^2(M, T, \tau) = \frac{1}{M-1} \sum_{i=1}^{M} \left(\overline{y}_i - \frac{1}{M} \sum_{j=1}^{M} \overline{y}_j \right)^2 \tag{5.87}$$

This quantity is a random variable itself and its infinite time-average can be used as a measure of frequency stability over a time interval τ. In particular, for $\tau = T$ and white frequency noise (see Section 5.9 for noise types) the definition above is an unbiased estimate of the true variance, as

$$\left\langle \sigma_y^2(M, \tau, \tau) \right\rangle = I^2(\tau) \tag{5.88}$$

Equation (5.87) estimates the mean squared deviation of M samples of $\overline{y}_k$ from their mean value, as recommended by any good statistics textbook. There is a long story about why the denominator is $M - 1$ instead of the more obvious M. Actually, this denominator is found also for any stationary noises when T is much longer than the correlation time of $y(t)$ (i.e., when the samples $\overline{y}_k$ are independent)[12].

5.8.4   Allan Variance

To allow standard adoption of a unique time-domain measure that can be used unambiguously in all laboratories, the IEEE subcommittee on frequency stability recommended in 1971 [5.1][5.7] the use of the sample variance with $M = 2$ and adjacent samples with zero dead time (i.e., $T = \tau$), following the pioneering work of Allan in 1966 [5.3].

The resulting measure owes its name to its inventor and is thus known as *Allan variance*, defined as

$$\left\langle \sigma_y^2(2, \tau, \tau) \right\rangle = \left\langle \sum_{i=1}^{2} \left(\overline{y}_i - \frac{1}{2} \sum_{j=1}^{2} \overline{y}_j \right)^2 \right\rangle \tag{5.89}$$

---

[12] As noticed in the excellent book on numerical computing [5.35], which we definitely recommend, the $M - 1$ denominator *should* be changed to M if you ever are in the pleasant situation of measuring the variance of a distribution whose mean is known *a priori*, rather than being estimated from the data measured. By the way, the authors of [5.35] opportunely comment also that, if the difference between $M - 1$ and M ever matters to you, then you are probably up to no good anyway (for example, trying to substantiate a questionable hypothesis with marginal data).

With more compact notation, the definition of the Allan variance is usually written as

$$\sigma_y^2(\tau) = \frac{1}{2}\langle(\overline{y}_{k+1} - \overline{y}_k)^2\rangle \tag{5.90}$$

The Allan variance is also known as *two-sample variance*, since pairs of adjacent measurements are grouped together.

The Allan variance too is a theoretical measure, as based on averaging an infinite number of samples. However, it has much greater practical utility than $I^2(\tau)$, because it converges for all kinds of power-law noise (see Section 5.9 for noise types).

From Equation (5.82), the Allan variance definition can also be written as time-average of the second difference of phase or time error samples

$$\sigma_y^2(\tau) = \frac{1}{2\tau^2}\langle[x(t_k + 2\tau) - 2x(t_k + \tau) + x(t_k)]^2\rangle \tag{5.91}$$

To summarize, Allan variance computation is based on time-averaging combinations of *pairs* of frequency samples or of *triplets* of time samples.

Experimentally, only *estimates* of $\sigma_y^2(\tau)$ can be obtained from a finite number of samples $\overline{y}_k$, taken over a finite duration. Therefore, an inherent statistical uncertainty (usually plotted as error bars) exists when a finite number N of values of $\overline{y}_k$ is used to estimate $\sigma_y^2(\tau)$. A widely used estimator (adopted also by international telecommunications standard bodies, as shown in a later section) is

$$\sigma_y^2(\tau, N) = \frac{1}{2(N-1)}\sum_{i=1}^{N-1}(\overline{y}_{i+1} - \overline{y}_i)^2 \tag{5.92}$$

This quantity is itself a random variable, whose variance (i.e., the variance of the estimated variance) may be used to assess the error bars on the plot of $\sigma_y^2(\tau)$ versus τ [5.20][5.29]. For long-term measurements, the size of N is severely limited by the overall measurement duration. In general, along a plot of $\sigma_y^2(\tau)$ versus τ, error bars are negligible at the left (short τ) and large at the right (long τ).

5.8.5 Modified Allan Variance

The relatively poor discrimination capability of the Allan variance against white and flicker phase noise, as it will be shown in Section 5.10, prompted the development of the so-called *modified Allan variance* in 1981 [5.24][5.26][5.28], defined as

$$\text{Mod }\sigma_y^2(n\tau_0) = \frac{1}{2}\left\langle\left[\frac{1}{n}\sum_{i=1}^{n}(\overline{y}_{i+n} - \overline{y}_i)\right]^2\right\rangle$$

$$= \frac{1}{2n^2\tau_0^2}\left\langle\left[\frac{1}{n}\sum_{j=1}^{n}(x_{j+2n} - 2x_{j+n} + x_j)\right]^2\right\rangle \tag{5.93}$$

where τ_0 is the sampling period (previously denoted as T in the definition of the M-sample variance), the observation interval is given by $\tau = n\tau_0$ and the samples of the random time deviation $x(t_k)$ are simply denoted as x_k.

To summarize, modified Allan variance differs from basic Allan variance in the additional average over n adjacent measurements. From the definition above, moreover, it is evident that for $n = 1$ ($\tau = \tau_0$) the modified Allan variance coincides with the Allan variance.

5.8.6   Time Variance (TVAR)

The Allan variance and the modified Allan variance are measures of the stability of the fractional frequency and thus they are a-dimensional. More recently, an additional variance, closely related to the modified Allan variance, was introduced aiming at measuring directly time stability: the *Time variance*, better known with its abbreviated name TVAR [5.28]. It is defined as

$$\sigma_x^2(\tau) = \frac{\tau^2}{3}\text{Mod } \sigma_y^2(\tau) \tag{5.94}$$

with dimension [time^2].

The TVAR has been widely adopted by telecommunications international standards for the specification of timing interfaces.

5.8.7   Root Mean Square of the Time Interval Error (TIE$_{\text{rms}}$)

The root mean square value of the TIE is formally defined as

$$\text{TIE}_{\text{rms}}(t; \tau) = \sqrt{\text{E}\left\{[\text{TE}(t + \tau) - \text{TE}(t)]^2\right\}} \tag{5.95}$$

with dimension [time], where the operator E$\{\cdot\}$ (*expectation operator*) denotes *ensemble-averaging*[13]. Under the assumption of ergodic processes, the ensemble-average above is equal to infinite time-average and does not depend on the time t. Thus, this quantity can be denoted shortly as TIE$_{\text{rms}}(\tau)$.

Experimentally, a simple estimate of TIE$_{\text{rms}}(\tau)$ can be obtained from a finite number N of samples TE$_k$ = TE(t_k), taken over a finite duration, as

$$\text{TIE}_{\text{rms}}(n\tau_0) = \sqrt{\frac{1}{N - n}\sum_{i=1}^{N-n}(\text{TE}_{i+n} - \text{TE}_i)^2} \tag{5.96}$$

where $\tau = n\tau_0$, with $n = 1, 2, \ldots, N - 1$.

It is interesting to notice that, if in the timing signal model (5.4) the only non-ideal component is the random frequency deviation $\nu_a(t)$, then the TIE$_{\text{rms}}$ is directly linked to the classical variance of $y(t)$, as

$$\text{TIE}_{\text{rms}}(\tau) = I(\tau) \cdot \tau \tag{5.97}$$

5.8.8   Maximum Time Interval Error

Recalling the definition of the Time Interval Error TIE$_t(\tau)$ (cf. Equation (5.16)), the *Maximum Time Interval Error* function MTIE(τ, T) is defined as the maximum peak-to-peak

---

[13] The ensemble-averaging operation requires knowing the probability distribution of the ensemble. Under the common assumption of ergodic processes, the ensemble-average is equal to the infinite time-average (see for example [5.36]).

variation of TE in all the possible observation intervals τ[14] within a measurement period T, as

$$\text{MTIE}(\tau, T) = \max_{0 \le t_0 \le T-\tau} \left\{ \max_{t_0 \le t \le t_0+\tau} [\text{TE}(t)] - \min_{t_0 \le t \le t_0+\tau} [\text{TE}(t)] \right\} \qquad (5.98)$$

The sample plot in Figure 5.14, especially if compared to Figure 5.2, clarifies the meaning of MTIE(τ, T) and its relationship with TIE(τ).

It should be noted, however, that the standards in force specify the MTIE limits simply as a function of τ (or S), thus implicitly assuming

$$\text{MTIE}(\tau) = \lim_{T \to \infty} \text{MTIE}(\tau, T) \qquad (5.99)$$

In other words, to rigorously assert the MTIE compliance of a clock with standard specifications, one should verify that, for every τ, MTIE(τ) stays below the allowed limits *for all the device life*. As an extreme case, if after *years* of regular operation a clock should exhibit a fortuitous over-mask phase hit, this event would cancel all the past (honest) history in the resulting MTIE curve.

Now, if phase fluctuations were modelled ideally by a Gaussian probability distribution of the amplitudes (this assumption has been somehow verified experimentally, especially under white phase noise [5.20]), the maximum range spanned by TE might reach infinitely large values. Increasing the measurement period T allows to observe the tails of the distribution: they are less likely but, at least in principle, unlimited.

Therefore, the measured value MTIE(τ, T) depends in general not only on τ but also, to a smaller extent, on the overall period T during which the clock has been under test.

It is clear that a single measurement of MTIE(τ, T), on one realization of the TE(t) random process and based on one measurement period T, yields a random variable result of that particular experiment. Therefore, alone, it is not adequate for a rigorous characterization of the oscillator under test.

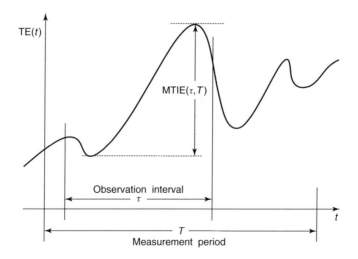

Figure 5.14. Definition of MTIE(τ, T)

---

[14] In early ITU-T Recommendations [5.37] and in ANSI standards [5.38], the observation interval τ is denoted with S.

In order to cope with these issues, ITU-T [5.39] redefined the quantity $MTIE(\tau, T)$ as a specified *percentile* β of the random variable X result of a single measurement of $MTIE(\tau, T)$, i.e. (cf. Equation (5.98))

$$X = \max_{0 \le t_0 \le T - \tau} \left\{ \max_{t_0 \le t \le t_0 + \tau} [TE(t)] - \min_{t_0 \le t \le t_0 + \tau} [TE(t)] \right\} \qquad (5.100)$$

Based on such a definition, therefore, a MTIE β-percentile mask

$$MTIE_{\beta - \text{perc}}(\tau) \le a_\beta(\tau) \qquad (5.101)$$

gives the limit $a_\beta(\tau)$ not to be exceeded in more than a $1 - \beta$ fraction of measurements, for any T.

ITU-T Rec. G.810 [5.39] suggests also an expression to estimate the β-percentile range from multiple range measures. Let $X_1, X_2, \ldots X_M$ be a set of single independent samples of $MTIE(\tau, T)$, measured according to Equation (5.100) on M measurement periods and sorted in ascending order, i.e. $X_1 \le X_2 \le \ldots \le X_M$. Let x_β be the β-percentile of the random variable X. Then, a confidence interval for x_β, expressed as the probability that x_β falls between the samples X_r and X_s (with $r < s$), is given by

$$P\left\{ X_r \le x_\beta \le X_s \right\} = \sum_{k=r}^{s-1} \frac{M!}{k!(M-k)!} \beta^k (1 - \beta)^{M-k} \qquad (5.102)$$

The MTIE has been widely adopted in telecommunications international standards for the specification of timing interfaces, even before TVAR since the 1980s. Actually, it was defined aiming directly at buffer size design in bit synchronizers. Recalling that described in Section 2.3.2, bit synchronization is accomplished by writing the bits of the asynchronous bit stream into an elastic store (buffer) at their own instantaneous arrival rate f_w and by reading them out with the frequency f_r of the equipment local clock. The elastic store absorbs limited random zero-mean frequency fluctuations between the write and read clocks. Nevertheless, larger fluctuations or any frequency offset $|f_w - f_r|$ will make the buffer empty or overflow sooner or later. If the buffer empties, some bytes are repeated in transmission; if the buffer overflows, some are lost. Such events are called *slips*. A larger buffer size allows reducing the slip rate, for any given clock accuracy.

Now, it is important to point out that in elastic stores the buffer-fill level is proportional to the TE between the input digital signal and the local equipment clock. Therefore, ensuring the clock compliance with timing specifications given in terms of MTIE guarantees that certain buffer thresholds are never exceeded. If MTIE limits are low enough, this may mean that no slips or pointer justifications should ever take place. This fact accounts for MTIE's appeal in supporting the design of equipment buffer size.

Papers [5.33][5.40][5.41] deal with specific topics about MTIE and its measurement. Moreover, a section in Chapter 7 is specifically devoted to MTIE practical measurement issues.

5.8.9   Standard Estimators of Stability Quantities Defined in ITU-T Recommendations and ETSI Standards

Among the several quantities defined in literature for characterizing time and frequency stability, the following five in the time domain have been considered by

ITU-T and ETSI standard bodies [5.39][5.42] for the specification of timing interface requirements:

- the *Allan Deviation* (ADEV) $\sigma_y(\tau)$, square root of the Allan Variance (AVAR) (5.90);

- the *Modified Allan Deviation* (MADEV) mod $\sigma_y(\tau)$, square root of the Modified Allan Variance (MAVAR) (5.93);

- the *Time Deviation* (TDEV) $\sigma_x(\tau)$, square root of the Time Variance (TVAR) (5.94);

- the *root mean square of Time Interval Error* (TIE$_{\mathrm{rms}}$) (5.95);

- the *Maximum Time Interval Error* (MTIE) (5.98).

Actually, while TIE$_{\mathrm{rms}}$ and MTIE have been defined based on the TE process, ADEV, MADEV and TDEV are theoretically defined in terms of the $x(t)$ process (5.11), i.e. of the random time deviation stripped of the deterministic components. Nevertheless, in order to overcome the complex problem of removing the frequency offset and drift from the TE measured data — and to achieve uniformity in the evaluation procedure of all the five standard stability quantities — the international telecommunication standards bodies simply recommend the use of TE(t) samples instead of $x(t)$ samples also for ADEV, MADEV and TDEV measurements. This choice is well justified based on the following:

- ADEV, MADEV and TDEV are not sensitive to any constant frequency offset in the TE data (they are based on a second difference operator, analogous in the discrete-time domain to the second derivative in the continuous-time domain);

- the frequency drift contribution is usually negligible in the observation intervals of interest in telecommunications ($\tau < 10^4$ s);

- the concern is not to rigorously characterize the noise types affecting the clock under test, but simply to limit the overall time deviations to the purposes of network design.

Therefore, to the practical purposes of telecommunications clock stability measurement, the processes TE(t) (5.13) and $x(t)$ (5.11) are considered synonymous. The Time Error samples TE(t_k) are referred to simply as $x(t_k)$. In conclusion, based on a sequence of N TE samples, defined as

$$x_k = x\big(t_0 + (k-1)\tau_0\big) \qquad \text{for } k = 1, 2, \ldots, N \tag{5.103}$$

i.e. measured with sampling period τ_0 over a measurement interval $T = (N-1)\tau_0$ and starting at an initial observation time t_0, the following five standard estimators have been defined by the ITU-T [5.39] and ETSI [5.42] bodies:

$$\text{ADEV}(\tau) = \sqrt{\frac{1}{2n^2\tau_0^2(N-2n)} \sum_{i=1}^{N-2n} \big(x_{i+2n} - 2x_{i+n} + x_i\big)^2}$$

$$\text{for } n = 1, 2, \ldots, \left\lfloor \frac{N-1}{2} \right\rfloor \tag{5.104}$$

$$\text{MADEV}(\tau) = \sqrt{\frac{1}{2n^4\tau_0^2(N-3n+1)} \sum_{j=1}^{N-3n+1} \left[\sum_{i=j}^{n+j-1} \left(x_{i+2n} - 2x_{i+n} + x_i \right) \right]^2}$$

$$\text{for } n = 1, 2, \dots, \left\lfloor \frac{N}{3} \right\rfloor \qquad (5.105)$$

$$\text{TDEV}(\tau) = \sqrt{\frac{1}{6n^2(N-3n+1)} \sum_{j=1}^{N-3n+1} \left[\sum_{i=j}^{n+j-1} \left(x_{i+2n} - 2x_{i+n} + x_i \right) \right]^2}$$

$$\text{for } n = 1, 2, \dots, \left\lfloor \frac{N}{3} \right\rfloor \qquad (5.106)$$

$$\text{TIE}_{\text{rms}}(\tau) = \sqrt{\frac{1}{N-n} \sum_{i=1}^{N-n} \left(x_{i+n} - x_i \right)^2} \qquad \text{for } n = 1, 2, \dots, N-1 \qquad (5.107)$$

$$\text{MTIE}(\tau) = \max_{1 \le k \le N-n} \left[\max_{k \le i \le k+n} x_i - \min_{k \le i \le k+n} x_i \right] \qquad \text{for } n = 1, 2, \dots, N-1 \qquad (5.108)$$

where $\tau = n\tau_0$ is the observation interval and $\lfloor z \rfloor$ denotes 'the greatest integer not exceeding z'.

Actually, most standard specifications are based on just two quantities among the five adopted: TVAR and MTIE. In Appendices 5A and 5B, efficient algorithms for fast computation of the TVAR and MTIE estimators are provided.

5.8.10 Translation from Frequency-Domain to Time-Domain Measures

Studying the translation relationships between stability quantities defined in the Fourier-frequency domain and in the time domain is interesting for several reasons [5.22]:

- relationships found provide a unified picture of frequency stability characterization;

- power-law noise types identified in the frequency domain (see Section 5.9) may be characterized by specific laws on the trends of time-domain quantities;

- stability measurements sometimes can be made only in one domain, due to equipment availability or capability; such relationships between quantities can then give an estimate of the performance in the other domain.

The translation between quantities in the two domains can be made by defining a suitable *transfer function* associated to each time-domain quantity [5.22]. Then, any time-domain quantity can be obtained by *integrating the noise spectral density filtered by the characteristic transfer function*, as it will be shown in the following. These relationships found are thus called *integral relationships*.

5.8.10.1   *True Variance of y(t)*

First, we notice that the true variance $I^2(\tau)$ (5.85) (assuming that $y(t)$ is zero-mean) can be written as

$$I^2(\tau) = \langle \bar{y}_k^2 \rangle = \left\langle \left[\frac{1}{\tau} \int_{t_k}^{t_k+\tau} y(t)dt \right]^2 \right\rangle$$

$$= \left\langle \left[\int_{-\infty}^{\infty} y(t)h_I(t_k - t)dt \right]^2 \right\rangle = \langle [y(t_k) * h_I(t_k)]^2 \rangle \quad (5.109)$$

where

$$h_I(t) = \begin{cases} 0 & t < -\tau \\ 1/\tau & -\tau \leq t \leq 0 \\ 0 & t > 0 \end{cases} \quad (5.110)$$

In other words, the true variance $I^2(\tau)$ can be expressed by convolving the frequency noise $y(t)$ by a function $h_I(t)$ that represents the basic measurement of one sample $\bar{y}_k$. The convolution integral above, therefore, represents the output of a hypothetical filter with impulse response $h_I(t)$ receiving an input signal $y(t)$. The true variance is the mean square value of this output signal.

Hence, the true variance can be also expressed in the frequency domain as the area under the spectral density of the signal output by this filter, as

$$I^2(\tau) = \int_0^{\infty} S_y(f)|H_I(f)|^2 df \quad (5.111)$$

where $S_y(f)$ is the one-sided PSD of $y(t)$ and $H_I(f)$ is the characteristic filter transfer function, i.e. the Fourier transform of $h_I(t)$, whose square magnitude is given by

$$|H_I(f)|^2 = \left(\frac{\sin \pi \tau f}{\pi \tau f} \right)^2 \quad (5.112)$$

The resulting integral relationship is then:

$$I^2(\tau) = \int_0^{\infty} S_y(f) \left(\frac{\sin \pi \tau f}{\pi \tau f} \right)^2 df \quad (5.113)$$

The impulse response $h_I(t)$ and the square magnitude of the transfer function $|H_I(f)|^2$ are plotted in Figures 5.15 and 5.16, respectively. Some limitations of the practical applicability of $I^2(\tau)$ do appear from the fact that the transfer function $H_I(f)$ is approximately equal to one for $\pi \tau f \ll 1$: the true variance is therefore very sensitive to low Fourier frequency components in $S_y(f)$. In particular, we note that $I^2(\tau) \rightarrow \infty$ if $S_y(f) \propto 1/f$, a very common kind of power-law noise (cf. Section 5.9).

Finally, please note that the above relationship between frequency-domain characterization $S_y(f)$ and time-domain characterization $I^2(\tau)$ *cannot* be reversed in a closed form, at least in most general cases.

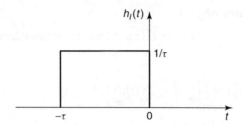

Figure 5.15. Impulse response $h_I(t)$ of the filter associated to the true variance

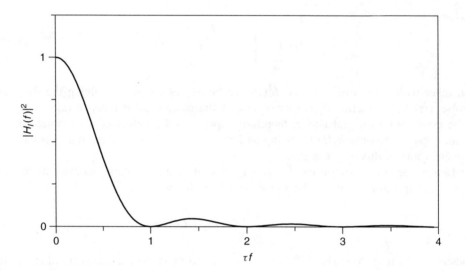

Figure 5.16. Square magnitude of the transfer function $H_I(f)$ associated with the true variance

5.8.10.2 *M-Sample Variance of* $y(t)$

Based on the M-sample variance definition given in Equation (5.87), for adjacent samples $(T = \tau)$ it results [5.22]:

$$\left\langle \sigma_y^2(M, \tau, \tau) \right\rangle = \frac{M}{M-1} \int_0^\infty S_y(f) \left(\frac{\sin \pi \tau f}{\pi \tau f} \right)^2 \left[1 - \left(\frac{\sin M \pi \tau f}{M \sin \pi \tau f} \right)^2 \right] df \quad (5.114)$$

The key point is that the transfer function now behaves as proportional to f^2 for $M \pi \tau f \ll 1$, thus ensuring convergence of the integral even for kinds of power-law noise such as $S_y(f) \propto f^{-1}$ and $S_y(f) \propto f^{-2}$ (cf. Section 5.9).

5.8.10.3 *Allan Variance*

By letting $M = 2$ in Equation (5.114), we obtain directly the expression for the two-sample Allan variance $\sigma_y^2(\tau)$, as

$$\sigma_y^2(\tau) = \int_0^\infty S_y(f) \frac{2 \sin^4 \pi \tau f}{(\pi \tau f)^2} df \quad (5.115)$$

The impulse response $h_A(t)$ and the square magnitude of the transfer function

$$|H_A(f)|^2 = \frac{2 \sin^4 \pi \tau f}{(\pi \tau f)^2} \tag{5.116}$$

associated to the Allan variance are plotted in Figures 5.17 and 5.18, respectively.

Once again, we note that the relationship cannot be inverted for the most general cases. Moreover, the integral converges for all kinds of power-law noise (cf. Section 5.9) as in the previous case.

5.8.10.4   Modified Allan Variance

Directly from the definition of the modified Allan variance (5.93), we can derive the time-domain measurement filter impulse response $h_{MA}(n, t)$, plotted in Figure 5.19 for example in the case $n = 6$.

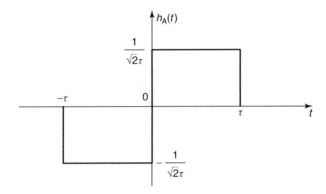

Figure 5.17.   Impulse response $h_A(t)$ of the filter associated with the Allan variance

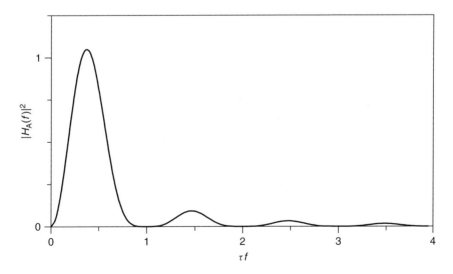

Figure 5.18.   Square magnitude of the transfer function $H_A(f)$ associated with the Allan variance

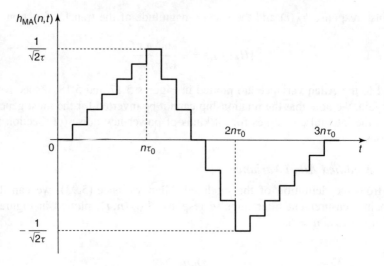

Figure 5.19. Impulse response $h_{MA}(n, t)$ of the filter associated with the modified Allan variance ($n = 6$)

The resulting integral relationship between the modified Allan variance and $S_y(f)$ results then

$$\text{Mod } \sigma_y^2(\tau) = \int_0^\infty S_y(f) \frac{2 \sin^6 \pi \tau f}{(n\pi\tau f)^2 \sin^2 \pi \dfrac{\tau}{n} f} df \tag{5.117}$$

The square magnitude of the characteristic transfer function $H_{MA}(n, f)$ is thus

$$|H_{MA}(n, f)|^2 = \frac{2 \sin^6 \pi \tau f}{(n\pi\tau f)^2 \sin^2 \pi \dfrac{\tau}{n} f} \tag{5.118}$$

The limit value of the transfer function $|H_{MA}(n, f)|^2$ for $n \to \infty$ and keeping $n\tau_0 = \tau$ constant is

$$\lim_{\substack{n \to \infty \\ n\tau_0 = \tau}} |H_{MA}(n, f)|^2 = 2 \frac{\sin^6 \pi \tau f}{(\pi \tau f)^4} \tag{5.119}$$

The square magnitude of the transfer function $|H_{MA}(n, f)|^2$ is plotted in Figure 5.20, for increasing values of the parameter n. It can be seen that the limit (5.119) is approached quickly for fairly low values of n.

5.8.10.5 Time Variance (TVAR)

Directly from the definition of TVAR (5.94) and from Equation (5.117), we can derive the following integral relationship between TVAR and $S_y(f)$

$$\sigma_x^2(\tau) = \int_0^\infty S_y(f) \frac{2 \sin^6 \pi \tau f}{3(n\pi f)^2 \sin^2 \pi \dfrac{\tau}{n} f} df \tag{5.120}$$

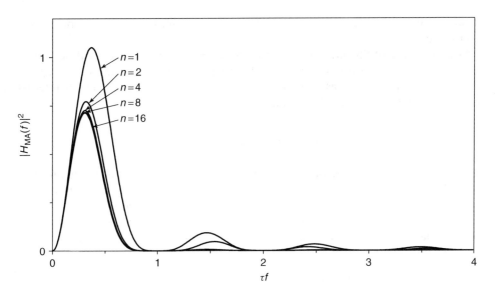

Figure 5.20. Square magnitude of the transfer function $H_{MA}(n, f)$ associated with the modified Allan variance for increasing values of the parameter n

5.8.10.6   Root Mean Square of the Time Interval Error (TIE$_{rms}$)

If the only non-ideal component in the timing signal model (5.4) is the random frequency deviation $v_a(t)$, then the TIE$_{rms}$ is directly linked to the classical variance of $y(t)$ by the relationship (5.97). Only under this assumption, the integral relationship between TIE$_{rms}$ and $S_y(f)$ can be derived directly from (5.113) and results

$$\text{TIE}_{rms}(\tau) = \sqrt{\int_0^\infty S_y(f) \left(\frac{\sin \pi \tau f}{\pi f}\right)^2 df} \qquad (5.121)$$

5.8.10.7   Maximum Time Interval Error (MTIE)

The Maximum Time Interval Error function MTIE(τ, T) is defined as the maximum peak-to-peak variation of TE in all the possible observation intervals τ within a measurement period T (cf. Equation (5.98)). Due to this peculiar nature of raw *peak* measure, it is not possible to translate to it from Fourier-frequency domain characterization by way of some integral relationship, based on a suitable transfer function, as we did for 'normal' stability quantities that are all based on some infinite time-averaging.

5.8.10.8   Transfer Functions as Tool to Define New Time-Domain Measures

Most time-domain quantities have been defined originally in the time domain, by defining a suitable processing of the time or frequency measurement samples. The shapes of the correspondent transfer functions have been derived from such definition in the time domain.

Nevertheless, the transfer function concept can be used also as a creative tool to define new quantities. According to this approach, a transfer function is first designed, depending on what kind of sensitivity is desired in the frequency domain. Then, the filter impulse response is derived and hence the time-domain measurement data processing is determined.

As an example of this alternative approach, the Hadamard variance will be presented in the next section. Also other variances have been defined according to this approach (high-pass variance, band-pass variance, etc.). For further details, the reader is referred to the fundamental paper [5.22] by Rutman.

5.8.11   Hadamard Variance

The *Hadamard variance* has been developed by Baugh [5.43] to achieve high-resolution spectral analysis of $y(t)$ from measurements of $\bar{y}_k$. In other words, the goal was to estimate the frequency-domain quantity $S_y(f)$ from time-domain data provided by digital counters.

With this aim, the Hadamard variance was defined by designing a corresponding transfer function that contains a narrow main lobe, well suited for spectral analysis. The resulting time-domain definition is

$$\langle \sigma_H^2(M, T, \tau) \rangle = \langle (\bar{y}_1 - \bar{y}_2 + \bar{y}_3 - \cdots - \bar{y}_M)^2 \rangle \tag{5.122}$$

The Hadamard variance is thus calculated from groups of M samples $\bar{y}_k$, with $k = 1, 2, \ldots, M$. Let us note, moreover, that the Allan variance $\sigma_y^2(\tau)$ is one-half times the Hadamard variance $\langle \sigma_H^2(M, T, \tau) \rangle$, with $M = 2$.

It has been shown that the Hadamard variance is related to $S_y(f)$ by the integral relationship

$$\langle \sigma_H^2(M, T, \tau) \rangle = \int_0^\infty S_y(f) \left(\frac{\sin \pi \tau f}{\pi \tau f} \right)^2 \left(\frac{\sin M \pi T f}{\cos \pi T f} \right)^2 df \tag{5.123}$$

The square magnitude of the characteristic transfer function $H_H(f)$ (i.e., the Fourier transform of the measurement sequence $h_H(t)$) is thus given by

$$|H_H(f)|^2 = \left(\frac{\sin \pi \tau f}{\pi \tau f} \right)^2 \left(\frac{\sin M \pi T f}{\cos \pi T f} \right)^2 \tag{5.124}$$

The measurement sequence $h_H(t)$ (for $M = 6$ and $T = \tau$) and the square magnitude of the transfer function $|H_H(f)|^2$ (for $M = 2, 4, 6$ and $T = \tau$) associated to the Hadamard variance are plotted in Figures 5.21 and 5.22, respectively. It can be seen that the transfer function exhibits a main lobe centred at the Fourier frequency $f_1 = 1/(2T)$. Its bandwidth decreases with increasing M.

Limitations and improvements for the practical use of the Hadamard variance are discussed in paper [5.22].

5.9   COMMON TYPES OF CLOCK NOISE

Experimental measurements on clocks may exhibit a variety of types of noise, either generated by physical processes intrinsic to the oscillator hardware or due to external phenomena. Such external phenomena include environmental perturbations, mechanical vibrations, residual ripples from the power supply, signal coupling via power supplies and ground paths, electromagnetic interference, etc.

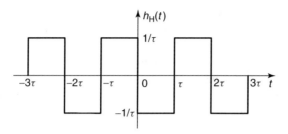

Figure 5.21. Impulse response $h_H(t)$ of the filter associated with the Hadamard variance ($M = 6$, $T = \tau$)

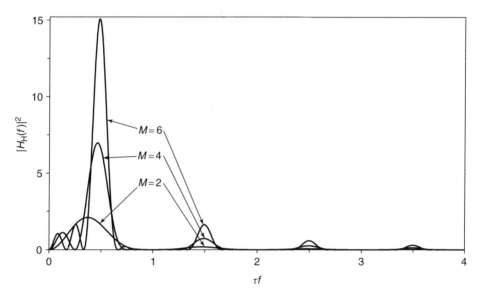

Figure 5.22. Square magnitude of the transfer function $H_H(f)$ associated with the Hadamard variance ($M = 2, 4, 6$ and $T = \tau$)

5.9.1   Power-Law Noise

In the frequency domain, the model most frequently used to represent the output phase noise measured on clocks is the so-called *power-law model*. In terms of the one-sided PSD of $x(t)$, such model is expressed by

$$S_x(f) = \begin{cases} \dfrac{1}{(2\pi)^2} \displaystyle\sum_{\alpha=-4}^{0} h_{\alpha+2} f^{\alpha} & 0 \le f \le f_h \\[2mm] 0 & f > f_h \end{cases} \tag{5.125}$$

where the h_{-2}, h_{-1}, h_0, h_{+1} and h_{+2} coefficients are device-dependent parameters[15] and f_h is an upper cut-off frequency (*clock hardware bandwidth*), mainly depending on

---

[15] The reason for the subscript $\alpha + 2$ (for $\alpha = -4, -3, -2, -1, 0$) is that, historically, the coefficients h_{-2}, h_{-1}, h_0, h_{+1} and h_{+2} have been used in the definition of the power-law model, originally in terms of $S_y(f)$ (cf. Equation (5.76)).

low-pass filtering in the oscillator and in its output buffer amplifier. This clock upper cut-off frequency has been measured in the range 10–100 kHz in precision frequency sources [5.44].

In practice, the measurement set-up introduces a further low-pass filtering on the clock output noise, with cut-off frequency $\hat{f}_h$ (*measurement hardware bandwidth*). The actual bandwidth of the $x(t)$ process measured is therefore limited by the smaller between f_h and $\hat{f}_h$. However, the measurement hardware bandwidth $\hat{f}_h$ of modern stability measurement set-ups, based on digital measurement of the TE, can be in the range of MHz and above: this fact usually ensures that all the clock phase noise components are fully captured in performing measurements.

The five noise types of the power-law model are: *White Phase Modulation* (WPM) for $\alpha = 0$, *Flicker Phase Modulation* (FPM) for $\alpha = -1$, *White Frequency Modulation* (WFM) for $\alpha = -2$, *Flicker Frequency Modulation* (FFM) for $\alpha = -3$ and *Random Walk Frequency Modulation* (RWFM) for $\alpha = -4$. All the stability quantities defined in the previous sections are sensitive, according to different laws, to the presence of these noises in the timing signal (see e.g. [5.2][5.22][5.31]).

According to this model in the frequency domain, when the PSD $S_x(f)$ is plotted on a log–log diagram, a broken line made of straight segments is approximately obtained, one per noise type and each having slope equal to the corresponding power α (see Figure 5.23).

In the time domain, on the other hand, the random realizations of the noise process of each single type exhibit characteristic trends, which could be even recognized at a glance by an experienced eye. To give an idea, Figures 5.24(a) through 5.24(e) show sample realizations of TE, each affected respectively by one of the five types of power-law noise. Each realization was obtained by numerical simulation of the power-law model (5.125), according to the value of α specified.

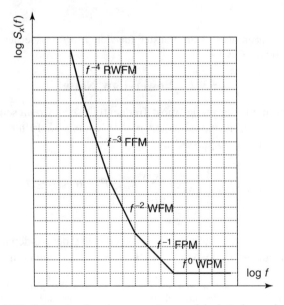

Figure 5.23. PSD $S_x(f)$ obeying the power-law model plotted on a log–log diagram

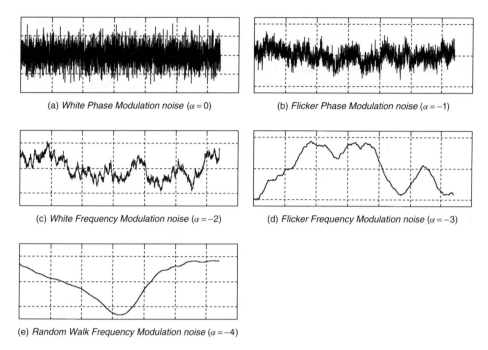

(a) *White Phase Modulation noise* ($\alpha = 0$)   (b) *Flicker Phase Modulation noise* ($\alpha = -1$)

(c) *White Frequency Modulation noise* ($\alpha = -2$)   (d) *Flicker Frequency Modulation noise* ($\alpha = -3$)

(e) *Random Walk Frequency Modulation noise* ($\alpha = -4$)

Figure 5.24.   Sample realizations of the TE random process featuring the five types of power-law noise. (Reproduced from [5.40], ©1997 IEEE, by permission of IEEE)

First, in order to simulate WPM ($\alpha = 0$) noise, two white and uniformly distributed pseudo-random sequences of length $N = 2^{13} = 8192$ were generated. Then, applying a well-known transformation formula [5.45], one white Gaussian pseudo-random sequence of the same length was obtained, thus approximating a WPM noise. Spectral shaping was accomplished by filtering repeatedly in the Fourier domain the WPM ($\alpha = 0$) noise sequence through integrators of fractional order $-\alpha/2$ [5.46], having transfer function $H_{-\alpha/2}(f) = K(j2\pi f)^{\alpha/2}$, to generate the FPM ($\alpha = -1$), WFM ($\alpha = -2$), FFM ($\alpha = -3$) and RWFM ($\alpha = -4$) noise sequences according to the power-law model (5.125).

It is evident how TE realizations become smoother as we proceed from WPM to FPM, WFM, FFM and RWFM noise. This is obvious, as increasingly more power is concentrated at lower frequencies ($S_x(f) \propto f^{-\alpha}$).

The WFM noise has spectrum proportional to $1/f^2$. By remembering a well-known property of power spectra, it is therefore integral of a WPM noise, whose spectrum is constant. Actually, looking at the simulation procedure outlined above, the WFM noise realization in Figure 5.24(c) is *the* integral of the WPM noise plotted in Figure 5.24(a). In the same way, the RWFM noise realization in Figure 5.24(e) is the integral of the WFM noise plotted in Figure 5.24(c).

Then, what are then the flicker noise realizations plotted in Figures 5.24(b) and 5.24(d)? They were obtained by filtering the noise realizations plotted in Figures 5.24(a) and 5.24(c) by a transfer function of the form $H(f) = K(j2\pi f)^{-1/2}$. In the Fourier-frequency domain, it comes easy to say that such a transfer function corresponds to a *one-half order integrator* (it is less obvious to define such an operator in the time domain) [5.46]. Actually, a visual inspection of the waveforms plotted in Figures 5.24(b) and 5.24(d) may lead to say that these processes are really something *in the middle* between the processes obtained by regular single and double integration.

These five power-law noise types may be due to different physical causes [5.25] and, on a particular oscillator, they may all be recognized or some may not. The main features and the supposed origin of each of the five types of power-law noise will be now summarized in brief.

- *Random Walk Frequency Modulation* (h_{-2}/f^4). Difficult to measure as close to the carrier (the ideal timing signal). It is mostly ascribed to environmental effects: if RWFM noise dominates, then we may suppose that frequent perturbations like mechanical shocks or temperature variations cause random shifts in the oscillation frequency.

- *Flicker Frequency Modulation* (h_{-1}/f^3). The causes of this type of noise are not fully understood, but they are mostly ascribed to the physical resonance mechanism of an active oscillator or to phenomena in the control electronic devices. FFM noise is commonly recognized in high-quality oscillators, but can be hidden by WFM or FPM noise in lower-quality oscillators.

- *White Frequency Modulation* (h_0/f^2). It is a type of noise commonly recognized in passive-resonator frequency standards, which are based on a slave oscillator (mostly a quartz) locked to a resonance of another device. Caesium-beam and rubidium standards feature a dominant WFM noise.

- *Flicker Phase Modulation* (h_1/f). Although it may be related to a physical resonance mechanism of the oscillator, it is mostly added by noisy electronics, especially in the output amplification stages and in the frequency multipliers.

- *White Phase Modulation* (h_2). It has little to do with the clock resonance mechanism, but it is mainly added by noisy electronics. In the past, this type of noise was often negligible in high-quality clocks, featuring very low-noise output stages. Nowadays, conversely, clocks based on digital control electronics (such as the Digital PLLs, DPLLs) became very common, especially in telecommunications. This fact made the WPM noise the most commonly found in telecommunications measurements. WPM noise in DPLLs is due to the quantization error in the phase-lock loop, which produces a broadband white noise in the output timing signal. Moreover, WPM is the test-bench background noise caused by the trigger and quantization errors of time counters in the digital measurement of TE.

5.9.2   Periodic Noise

Although the power-law model proved very general and suitable for describing most measurement results, yet other types of noise may result in experimental measurements. Periodic noises are quite common. They may be typically caused by:

- interference from 50/60 Hz AC power line;
- diurnal and seasonal variations of temperature, which can affect the output frequency of the oscillator under test or much more frequently the signal propagation speed, for example on long copper cable lines (*diurnal* and *annual wander*);

- sensitivity to acoustic or mechanical vibrations;

- intrinsic phenomena such as special frequency control algorithms in DPLLs.

Periodic noise is revealed in frequency-domain measurement results as a series of spikes (ideally, lines at discrete frequencies) in the noise power spectrum. In time-domain measurement results, as it will be better shown later, periodic noise appears as ripples on the measured quantity. Examples of real measurement results featuring periodic noise will be shown in Section 5.10.3.

5.9.3 Background White Phase Noise Due to Trigger and Quantization Errors

Recently, time-domain measurement techniques based on digital time counters have become very common. Actually, in telecommunications, all standard measures are based on such instrumentation. A digital time counter is an instrument able to measure the elapsed time between two events (*start* and *stop trigger events*). This is achieved by incrementing a counter register with a very high and stable reference frequency (as it will be explained in Chapter 7, tricks like Vernier interpolation allow to improve the measurement resolution).

A *trigger event* typically consists of detecting an electric signal $V(t)$ going over some threshold V_{thr} on the input channel. Actually, the time instant t_{tr} in which the threshold is exceeded is recognized with some error ε_{tr}, due to the uncertainty ΔV_{thr} in the detection of the threshold level, as shown in Figure 5.25. The resulting error ε_{tr} is then called *trigger error*. It can be reduced by increasing the $V(t)$ signal slope around the threshold level. For this reason, it is always advisable to perform measurements on square wave signals rather than on sine waves.

Moreover, the measure of the time interval between the start and stop trigger events, obtained by reading the counter register, is affected by some *quantization error*, depending on instrumentation resolution.

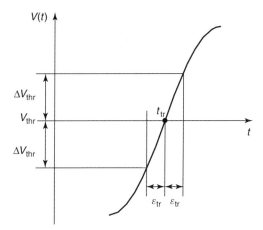

Figure 5.25. Error in recognizing the trigger-event time instant

Both trigger and quantization errors are revealed in measurement results as a background white noise. Therefore, WPM background noise is always experienced whenever time error is measured by a time counter, whereas WFM background noise is experienced if frequency measurements are carried out. Such background noise should be carefully evaluated before performing measurements on clocks.

Experimental Results

An example of real measurement results featuring background white noise due to trigger and quantization errors is provided in Figures 5.26 through 5.28. The measurement procedure is fully detailed in Appendix 7A. In short, the background noise of the measurement set-up was measured by splitting the 2.048 MHz reference timing signal and by feeding it directly into the time counter input ports. A sequence of $N = 96\,700$ samples of TE was acquired, with sampling period $\tau_0 \cong 37$ ms and over a measurement interval $T = 3600$ s. Then, ADEV and MADEV were computed with the standard estimators of Equations (5.104) and (5.105). The PSD $S_x(f)$ was computed (neglecting a multiplicative factor) through the Fast Fourier Transform (FFT) periodogram technique with triangular-shape data windowing, while the autocovariance function $C_x(\tau)$ was evaluated (again neglecting a multiplicative factor) as inverse transform of the PSD for $\tau \geq 0$ and mirrored copy for $\tau < 0$.

This background noise proves a pure WPM broadband noise: the ADEV and MADEV curves in Figure 5.26 are straight lines with slopes of τ^{-1} and $\tau^{-3/2}$ respectively, in perfect accordance with the theory (cf. Section 5.10.1), while the autocovariance $C_x(\tau)$ plotted in Figure 5.28 for $|\tau| \leq 10$ s features a $\delta(\tau)$ spike[16] at $\tau = 0$.

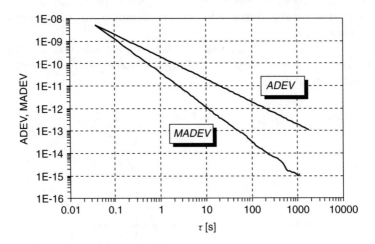

Figure 5.26. Measurement results featuring background WPM noise due to trigger and quantization errors — ADEV(τ) and MADEV(τ) ($N = 96\,700$, $\tau_0 \cong 37$ ms, $T = 3600$ s). (Reproduced from [5.32], ©1997 IEEE, by permission of IEEE)

---

[16] The spike has width really limited to the central sample $C_x(0)$. For the sake of precision, the first numerical values are $C_x(0) = 6.3 \times 10^{-6}$, $C_x(\pm\tau_0) = -6.8 \times 10^{-8}$, $C_x(\pm2\tau_0) = 8.3 \times 10^{-8}$, $C_x(\pm3\tau_0) = -3 \times 10^{-8}$, $C_x(\pm4\tau_0) = 3.2 \times 10^{-8}$, $C_x(\pm5\tau_0) = -1.8 \times 10^{-8}$, etc.

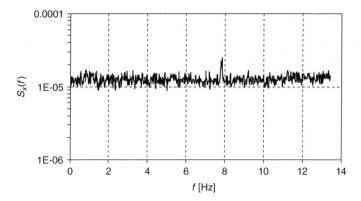

Figure 5.27. Measurement results featuring background WPM noise due to trigger and quantization errors — PSD estimate $S_x(f)$ ($N = 96\,700$, $\tau_0 \cong 37$ ms, $T = 3600$ s)

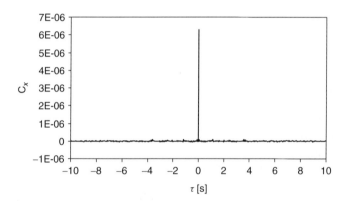

Figure 5.28. Measurement results featuring background WPM noise due to trigger and quantization errors — autocovariance function $C_x(\tau)$ ($N = 96\,700$, $\tau_0 \cong 37$ ms, $T = 3600$ s). (Reproduced from [5.32], ©1997 IEEE, by permission of IEEE)

5.10 BEHAVIOUR OF THE TIME-DOMAIN STABILITY QUANTITIES: AUTONOMOUS CLOCKS

The time-domain stability quantities defined in the previous sections reveal, with their particular behaviour, the presence of the different types of phase and frequency impairments on the signal under measurement. In this section, the effects of power-law noise, of frequency offset and drift and of periodic noise at the output of an autonomous clock are discussed. The characteristic trends of time-domain quantities for each type of such impairments are shown, thus allowing to properly interpret time-domain measurement results.

5.10.1 Power-Law Noise

Power-law noise has been previously defined in the frequency domain on the PSD $S_x(f)$ by the expression (5.125). Equivalent definitions can be given for other PSDs too, provided relationships such as those in Equation (5.76).

By considering separately each type of power-law noise, evaluation of the integral relationships provided in Equation (5.113) for the true variance, in Equation (5.114) for the M-sample variance, in Equation (5.115) for the Allan variance, in Equation (5.117) for the modified Allan variance, in Equation (5.120) for the Time variance and in Equation (5.121) for the TIE$_{\text{rms}}$ yields the corresponding time-domain expressions.

It is worthwhile noticing, at this point, that evaluation of those integrals features two critical points: for $f \to 0$ and for $f \to \infty$. Convergence for $f \to \infty$ is ensured by the upper cut-off frequency f_h in the model defined in Equation (5.125). Convergence for $f \to 0$, on the other hand, gets tougher as higher is the exponent of the Fourier frequency f at the denominator of Equation (5.125). For example, we noted that $I^2(\tau) \to \infty$ if $S_y(f) \propto f^{-1}$. The asymptotic behaviour for $f \to 0$ of the transfer function associated to other stability quantities ensures convergence even for $S_y(f) \propto f^{-4}$.

The results are summarized in the synoptical Tables 5.2 through 5.7, reporting for each power-law noise type the corresponding expressions of the true variance $I^2(\tau)$, of the M-sample variance $\langle \sigma_y^2(M, \tau, \tau) \rangle$, of the Allan variance $\sigma_y^2(\tau)$, of the modified Allan variance $\text{mod}\sigma_y^2(\tau)$, of the Time variance $\sigma_x^2(\tau)$ and of the root mean square of the Time Interval Error TIE$_{\text{rms}}(\tau)$.

An important remark is that including an upper cut-off frequency f_h in the model (5.125) is necessary, in order to ensure convergence of integrals for $f \to \infty$ also for WPM and FPM noise types. This upper cut-off frequency should always be specified, when due to the measurement instrumentation and not modelling the actual behaviour of the clock under test, because WPM and FPM noise measurement results depend on its value, as evident from Tables 5.2 through 5.7. In other words, measurements accomplished with test set-ups having different bandwidths may yield to different results. Moreover, results reported in tables above are valid only for $\tau \gg 1/(2\pi f_h)$, which is the most common case in practical measurements.

5.10.1.1   Expressions of the True Variance and of the M-Sample Variance

The expressions for the true variance reported in Table 5.2 can be derived by those for the M-sample variance in Table 5.3 by letting $T = \tau$ and evaluating the limit $M \to \infty$. Moreover, by inspection of the same tables, it is evident that these two variances depend, for FFM and RWFM noise, on the number M of samples and tend to infinite for $M \to \infty$.

Table 5.2   Expressions of the true variance for each power-law noise type

Noise Type	$S_x(f)$	$I^2(\tau)$
WPM	h_2	$\dfrac{1}{4\pi^2} h_2 f_h \cdot \tau^{-2}$ for $\tau \gg 1/(2\pi f_h)$
FPM	h_1/f	$\dfrac{h_1}{2\pi^2}[\gamma + \log(2\pi f_h \tau)] \cdot \tau^{-2}$ for $\tau \gg 1/(2\pi f_h)$
WFM	h_0/f^2	$\dfrac{h_0}{2} \cdot \tau^{-1}$
FFM	h_{-1}/f^3	$\displaystyle\lim_{M \to \infty} \dfrac{M \log M}{M - 1} h_{-1} = \infty$
RWFM	h_{-2}/f^4	$\displaystyle\lim_{M \to \infty} \dfrac{\pi^2}{3} M h_{-2} \cdot \tau = \infty$

$\gamma = 0.5772156649\ldots$ (Euler's constant)

Table 5.3 Expressions of the M-sample variance for each power-law noise

Noise Type	$S_x(f)$	$\langle \sigma_y^2(M, \tau, \tau) \rangle$
WPM	h_2	$2\dfrac{M + \delta_K(r-1)}{M4\pi^2} h_2 f_{\rm h} \cdot \tau^{-2}$ for $\tau \gg 1/(2\pi f_{\rm h})$
FPM	h_1/f	$\left\{ \gamma + \log(2\pi f_{\rm h}\tau) + \dfrac{1}{M(M-1)} \sum\limits_{n=1}^{M-1}(M-n)\log\left[\dfrac{n^2 r^2}{n^2 r^2 - 1}\right] \right\}$ $\cdot\dfrac{h_1}{2\pi^2} \cdot \tau^{-2}$ for $r \gg 1$ and $\tau \gg 1/(2\pi f_{\rm h})$
WFM	h_0/f^2	$\dfrac{h_0}{2} \cdot \tau^{-1}$ for $r \gg 1$
FFM	h_{-1}/f^3	$\left\{ \sum\limits_{n=1}^{M}(M-n)\left[-2(nr)^2\log(nr) + (nr+1)^2\log(nr+1)\right.\right.$ $\left.\left. + (nr-1)^2\log|nr-1| \right] \right\} \cdot \dfrac{h_{-1}}{M(M-1)}$
RWFM	h_{-2}/f^4	$\dfrac{\pi^2}{3}[r(M+1)-1]h_{-2} \cdot \tau$ for $r \geq 1$

$r = \dfrac{T}{\tau}, \ \delta_K(n) = \begin{cases} 1 & \text{for } n = 0 \\ 0 & \text{for } n \neq 0 \end{cases}, \ \gamma = 0.5772156649\ldots$ (Euler's constant)

Table 5.4 Expressions of the Allan variance for each power-law noise type

Noise Type	$S_x(f)$	$\sigma_y^2(\tau)$
WPM	h_2	$\dfrac{3}{4\pi^2}h_2 f_{\rm h} \cdot \tau^{-2}$ for $\tau \gg 1/(2\pi f_{\rm h})$
FPM	h_1/f	$\dfrac{h_1}{4\pi^2}\{3[\gamma + \log(2\pi f_{\rm h}\tau)] - \log 2\} \cdot \tau^{-2}$ for $\tau \gg 1/(2\pi f_{\rm h})$
WFM	h_0/f^2	$\dfrac{h_0}{2} \cdot \tau^{-1}$
FFM	h_{-1}/f^3	$2\log 2 \cdot h_{-1}$
RWFM	h_{-2}/f^4	$\dfrac{2\pi^2}{3}h_{-2} \cdot \tau$

$\gamma = 0.5772156649\ldots$ (Euler's constant)

Hence, true variance and M-sample variance are not suitable for clock stability characterization in presence of these types of noise.

5.10.1.2 *Expressions of the Allan Variance*

As far as the Allan variance is concerned, from Table 5.4 it is important to notice the following features.

- The Allan variance quantity is convergent for all kinds of power-law noise. Its convergence to a finite value for $f \to 0$, even for FFM and RWFM noise, is ensured by the asymptotic behaviour of the transfer function $H_A(f)$, in integral (5.115), which behaves as $\sim f^2$ for $f \to 0$.

- For all the five types of frequency-domain power-law noise ($\alpha = 0, -1, -2, -3, -4$), the Allan variance obeys in turn to a time-domain power law of the form

$$\sigma_y^2(\tau) = A_\mu \tau^\mu \tag{5.126}$$

(for $\mu = -2, -1, 0, +1$) with a slight modification for FPM noise because of the logarithm[17]. Therefore, a log–log plot of $\sigma_y^2(\tau)$ exhibits segments of straight lines, one per each noise type, whose slopes may be easily identified, as shown in the example diagram of Figure 5.29 (cf. Figure 5.23).

- Allan variance measurement results can thus be interpreted to identify the different types of power-law noise. Nevertheless, since both WPM and FPM noises yield very similar slopes, there is some ambiguity in noise identification whenever $\sigma_y^2(\tau) \propto \tau^{-2}$ is measured

5.10.1.3   Expressions of the Modified Allan Variance

As far as the modified Allan variance is concerned, on the other hand, the analytical evaluation of integral (5.117) proves quite cumbersome. As already mentioned, several works discussed in detail the properties of this quantity [5.24][5.26][5.28][5.29]. In particular, it has been studied the behaviour of the ratio

$$R(n) = \frac{\text{mod}\sigma_y^2(\tau)}{\sigma_y^2(\tau)} \tag{5.127}$$

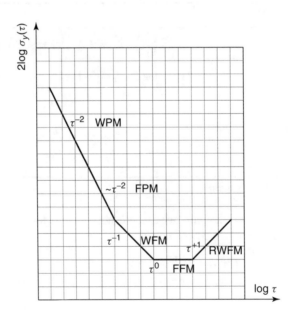

Figure 5.29.   Sample log–log plot of the Allan variance in presence of power-law noise

---

[17] Also the true variance and the M-sample variance exhibit a similar behaviour, but the fact that they do not converge for some types of power-law noise makes their time-domain power law less remarkable.

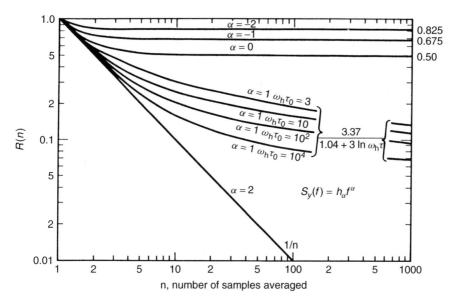

Figure 5.30.   Ratio $R(n)$ for all types of power-law noise. (Reproduced from [5.29], NIST. Not copyrightable)

where, once again, $n = \tau/\tau_0$. This ratio is plotted in Figure 5.30 (from [5.29], Section A6) versus n, for all types of power-law noise[18] and also for different values of the product $2\pi f_h \tau_0$, when relevant. The graph in Figure 5.30 was obtained directly from the basic definitions of the plain and modified Allan variances, by numerical computation. By inspection of this graph, we notice that:

- for WPM noise, the ratio $R(n)$ is $1/n$; therefore, for any given sampling period τ_0, while $\sigma_y^2(\tau)$ behaves as τ^{-2}, mod $\sigma_y^2(\tau)$ behaves as τ^{-3};

- for WFM, FFM and RWFM noises, the ratio $R(n)$ reaches quickly an asymptotic value for $n > 10$ (cf. Figure 5.20);

- for FPM noise, the value of the ratio $R(n)$ depends on the upper cut-off frequency f_h of the noise processes measured (as already pointed out, this cut-off frequency is the lower between the measurement hardware bandwidth and the clock hardware bandwidth); moreover, in this case, an asymptotic value is approached for at least $n > 100$.

In conclusion, from the results plotted in the diagram of Figure 5.30, from Equation (5.127) and from the expressions in Table 5.4, it is possible to derive the asymptotic expressions (i.e., strictly valid for $n \to \infty$ and keeping $n\tau_0 = \tau$ constant) reported in Table 5.5.

By inspection of these expressions, therefore, we notice that also the modified Allan variance, for all the five types of frequency-domain power-law noise, obeys in turn to

---

[18] Note that, in Figure 5.30, α denotes the exponent of the Fourier frequency in the PSD $S_y(f)$ rather than in $S_x(f)$ as we used elsewhere in this chapter. Hence, power-law noise types here are indexed by values of α from $\alpha = -2$ (RWFM) to $\alpha = 2$ (WPM).

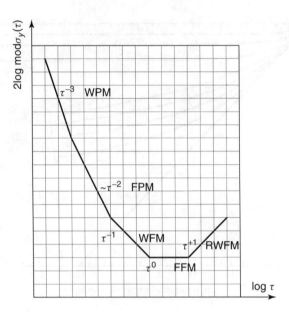

Figure 5.31.   Sample log–log plot of the modified Allan variance in presence of power-law noise

a time-domain power law of the form (5.126), as shown in the example diagram of Figure 5.31 (cf. Figure 5.29). Thus, modified Allan variance measurement results can be interpreted to identify the different types of power-law noise, but without the ambiguity between WPM and FPM noises as with Allan variance.

5.10.1.4   *Expressions of the Time Variance*

It is straightforward to extend all considerations about the modified Allan variance to the Time variance, based on its definition (5.94), and thus to derive the expressions reported in Table 5.6 from those in Table 5.5.

Moreover, as obvious, the Time variance obeys the same time-domain power law as the modified Allan variance (cf. Figure 5.31), having added +2 to the exponent μ.

5.10.1.5   *Expressions of the Root Mean Square of the Time Interval Error*

From the relationship between the root mean square of the Time Interval Error and the true variance in Equation (5.97), it is straightforward to derive the expressions reported in Table 5.7 from those in Table 5.2.

5.10.1.6   *Summary of Relationships between Frequency-Domain and Time-Domain Power Laws*

To reader's ease, in Table 5.8 we summarize the various values of the exponent μ (i.e., the slopes in log–log plots) in time-domain power laws such as Equation (5.126), for the stability quantities considered and for all types of power-law noise, resulting from expressions reported in Tables 5.2 through 5.7. In other words, for each type of power-law noise, having spectrum $S_x(f) = h_{\alpha+2}/f^{\alpha}$, it is specified the power τ^{μ} to which the stability quantity considered is proportional.

Table 5.5 Asymptotic expressions of the modified Allan variance for each power-law noise (strictly valid for $n \to \infty$ and keeping $n\tau_0 = \tau$ constant)

Noise Type	$S_x(f)$	$\text{mod}\,\sigma_y^2(\tau)$
WPM	h_2	$\dfrac{1}{n} \cdot \dfrac{3}{4\pi^2} h_2 f_h \cdot \tau^{-2}$ for $\tau \gg 1/(2\pi f_h)$
FPM	h_1/f	$\dfrac{3.37}{1.04 + 3\log(2\pi f_h \tau)} \cdot \dfrac{h_1}{4\pi^2} \{3[\gamma + \log(2\pi f_h \tau)] - \log 2\} \cdot \tau^{-2}$ for $\tau \gg 1/(2\pi f_h)$ and $n > 100$
WFM	h_0/f^2	$\dfrac{1}{2} \cdot \dfrac{h_0}{2} \cdot \tau^{-1}$ for $n > 10$
FFM	h_{-1}/f^3	$0.675 \cdot 2\log 2 \cdot h_{-1}$ for $n > 10$
RWFM	h_{-2}/f^4	$0.825 \cdot \dfrac{2\pi^2}{3} h_{-2} \cdot \tau$ for $n > 10$

$\tau = n\tau_0$, $\gamma = 0.5772156649\ldots$ (Euler's constant)

Table 5.6 Asymptotic expressions of the Time variance for each power-law noise (strictly valid for $n \to \infty$ and keeping $n\tau_0 = \tau$ constant)

Noise Type	$S_x(f)$	$\sigma_x^2(\tau)$
WPM	h_2	$\dfrac{1}{n} \cdot \dfrac{h_2 f_h}{4\pi^2}$ for $\tau \gg 1/(2\pi f_h)$
FPM	h_1/f	$\dfrac{3.37}{1.04 + 3\log(2\pi f_h \tau)} \cdot \dfrac{h_1}{12\pi^2} \{3[\gamma + \log(2\pi f_h \tau)] - \log 2\}$ for $\tau \gg 1/(2\pi f_h)$ and $n > 100$
WFM	h_0/f^2	$\dfrac{h_0}{12} \cdot \tau$ for $n > 10$
FFM	h_{-1}/f^3	$\dfrac{0.675 \cdot 2\log 2}{3} h_{-1} \cdot \tau^2$ for $n > 10$
RWFM	h_{-2}/f^4	$\dfrac{0.825 \cdot 2\pi^2}{9} h_{-2} \cdot \tau^3$ for $n > 10$

$\tau = n\tau_0$, $\gamma = 0.5772156649\ldots$ (Euler's constant)

Table 5.7 Expressions of the root mean square of the Time Interval Error for each power-law noise type

Noise Type	$S_x(f)$	$\text{TIE}_{\text{rms}}(\tau)$
WPM	h_2	$\sqrt{\dfrac{1}{4\pi^2} h_2 f_h}$ for $\tau \gg 1/(2\pi f_h)$
FPM	h_1/f	$\sqrt{\dfrac{h_1}{2\pi^2} [\gamma + \log(2\pi f_h \tau)]}$ for $\tau \gg 1/(2\pi f_h)$
WFM	h_0/f^2	$\sqrt{\dfrac{h_0}{2}\tau}$
FFM	h_{-1}/f^3	not convergent
RWFM	h_{-2}/f^4	not convergent

$\gamma = 0.5772156649\ldots$ (Euler's constant)

Table 5.8 Relationships between frequency-domain and time-domain power laws for stability quantities measured on autonomous clocks

Noise Type	$S_x(f)$	$I^2(\tau)$	$\langle \sigma_y^2(M, \tau, \tau) \rangle$	$\sigma_y^2(\tau)$	mod $\sigma_y^2(\tau)$	$\sigma_x^2(\tau)$	$\text{TIE}_{\text{rms}}(\tau)$
WPM	h_2	τ^{-2}	τ^{-2}	τ^{-2}	τ^{-3}	τ^{-1}	τ^0
FPM	h_1/f	τ^{-2}	τ^{-2}	τ^{-2}	τ^{-2}	τ^0	τ^0
WFM	h_0/f^2	τ^{-1}	τ^{-1}	τ^{-1}	τ^{-1}	τ^{+1}	$\tau^{+1/2}$
FFM	h_{-1}/f^3	—	—	τ^0	τ^0	τ^{+2}	—
RWFM	h_{-2}/f^4	—	—	τ^{+1}	τ^{+1}	τ^{+3}	—

— = Not convergent

5.10.2　Frequency Offset and Drift

As stated previously in Section 5.8.9, according to international standards, plain TE(t) samples are used to evaluate not only TIErms and MTIE, but also ADEV, MADEV and TDEV, which have been theoretically defined in terms of the $x(t)$ process (5.11). This fact well justifies the need of investigating the effects of frequency offset and drift on the behaviour of the various stability quantities.

If we assume that the reference clock is supplying the ideal time and that a constant frequency offset $\Delta\nu$ and a linear frequency drift D are the only non-ideal components in the timing signal under test, then, considering Equations (5.9), (5.10), (5.12) and (5.13), we have

$$\text{TE}(t) = \frac{\Delta\nu}{\nu_n}t + \frac{D}{2}t^2 \tag{5.128}$$

Now, substituting the TE(t) (5.128) in place of $x(t)$ within the definition of Allan variance (5.91), we get

$$\sigma_y^2(\tau) = \frac{1}{2\tau^2}\left\langle \left[D\tau^2\right]^2 \right\rangle = \frac{D^2}{2}\tau^2 \tag{5.129}$$

Therefore, the Allan variance is independent on a constant frequency offset $\Delta\nu$, but it reveals a linear frequency drift D according to a quadratic dependence on the observation interval τ. Note that no power-law noise yields a similar dependence of the Allan variance on τ.

On an analogous fashion, substituting the TE(t) (5.128) in place of $x(t)$ within the definition of modified Allan variance (5.93), we get

$$\text{mod}\sigma_y^2(\tau) = \frac{1}{2\tau^2}\left\langle \left[\frac{1}{n}\sum_{j=1}^{n}(\text{TE}_{j+2n} - 2\text{TE}_{j+n} + \text{TE}_j)\right]^2 \right\rangle$$

$$= \frac{1}{2\tau^2}\left\langle \left[\frac{1}{n}\sum_{j=1}^{n}D\tau^2\right]^2 \right\rangle = \frac{D^2}{2}\tau^2 \tag{5.130}$$

Therefore, the same considerations made for the Allan variance hold also for the modified Allan variance.

Table 5.9 Time-domain power laws for stability quantities in presence of constant frequency offset and linear frequency drift in the underlying TE data

Frequency Impairment	$\sigma_y^2(\tau)$	$\text{mod}\,\sigma_y^2(\tau)$	$\sigma_x^2(\tau)$	$\text{TIE}_{\text{rms}}(\tau)$
constant offset $v(t) = v_n + \Delta v$	not revealed	not revealed	not revealed	τ^{+1}
linear drift $v(t) = v_n + Dv_n t$	τ^{+2}	τ^{+2}	τ^{+4}	—

—= Not convergent

From the definition (5.94), we immediately get the expression of the Time variance in presence of a constant frequency offset Δv and a linear frequency drift D, as

$$\sigma_x^2(\tau) = \frac{D^2}{6}\tau^4 \qquad (5.131)$$

As far as the root mean square of the Time Interval Error is concerned, substituting the TE(t) (5.128) in place of $x(t)$ within the definition (5.95), we get

$$\text{TIE}_{\text{rms}}(t; \tau) = \sqrt{E\left\{\left[\frac{\Delta v}{v_n}\tau + Dt\tau + \frac{D^2}{2}\tau^2\right]^2\right\}} \qquad (5.132)$$

Therefore, the root mean square of the Time Interval Error is dependent both on the frequency offset Δv and on the frequency drift D. What's more, in presence of a frequency drift, it is even not a stationary quantity, being dependent not only on the observation interval τ, but also on the actual measurement time t.

Finally, to reader's ease, we complete previous Table 5.8 by summarizing in Table 5.9 the further results presented in this section.

5.10.3 Periodic Noise

Let us assume that the timing signal under measurement, with nominal frequency v_n, is frequency modulated by a sinusoidal signal of peak amplitude Δv_n and frequency f_m, i.e.

$$y(t) = \frac{\Delta v_n}{v_n}\sin 2\pi f_m t \qquad (5.133)$$

Then, in the frequency domain, the PSD will exhibit a discrete line at the modulation frequency f_m, as

$$S_y(t) = \frac{1}{2}\left(\frac{\Delta v_n}{v_n}\right)^2 \delta(f - f_m) \qquad (5.134)$$

In the time domain, considering for example the Allan variance, substitution of Equation (5.134) in Equation (5.115) yields

$$\sigma_y^2(\tau) = \left(\frac{\Delta v_n}{v_n}\right)^2 \frac{\sin^4 \pi \tau f_m}{(\pi \tau f_m)^2} \qquad (5.135)$$

Therefore, the effect of a sinusoidal frequency modulation is null whenever τ equals the modulation period $T_m = 1/f_m$ or one of its integer multiples, since the modulating signal is completely averaged out. The maximum effect, on the other hand, occurs when τ is near $T_m/2$ or one of its integer multiples.

Similar results can be obtained for other time-domain stability quantities. For example, for the modified Allan variance, we get

$$\text{mod}\sigma_y^2(\tau) = \left(\frac{\Delta\nu_n}{\nu_n}\right)^2 \left[2\frac{\sin^3 n\tau_0 f_m}{n\tau_0 f_m \sin \tau_0 f_m}\right]^2 \tag{5.136}$$

Seen from another point of view, this effect is, neither more nor less, the *aliasing* of the underlying periodic noise. Obviously, if we measure a periodic phenomenon at instants spaced an integer multiple of its period, we will always measure the same value. By varying the observation interval, therefore, we change the sensitivity to the periodic change in the quantity measured.

In practice, if we measure a time-domain stability quantity such as the Allan variance on a timing signal affected by sinusoidal frequency modulation, the resulting plot of $\sigma_y^2(\tau)$ will exhibit a *ripple* of period T_m. Measurement results often show such ripples superposed to power-law straight slopes.

Experimental Results

A first experimental example of periodic noise is provided in Figures 5.32 and 5.33, showing results measured on the SEC of a SDH Line Terminal Multiplexer STM-16 (LTM-16). Here and in the following of the book, we will denote the supplier of this equipment as supplier A.

The measurement procedure and set-up are fully detailed in Appendix 7A. In short, a digital time counter measured the TE between the output timing signal of the Clock Under Test (CUT) and its input reference (synchronized-clock configuration). A sequence of $N = 79\,000$ samples of TE was acquired, with sampling period $\tau_0 \cong 23$ ms and over a measurement interval $T = 1800$ s. Then, ADEV and MADEV were computed with the

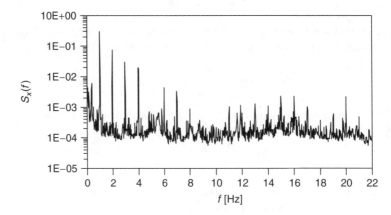

Figure 5.32. Periodic noise measured on the LTM-16 SEC from supplier A — PSD estimate $S_x(f)$ ($N = 79\,000$, $\tau_0 \cong 23$ ms, $T = 1800$ s). (Reproduced from [5.32], ©1997 IEEE, by permission of IEEE)

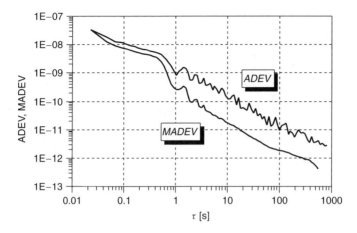

Figure 5.33. Periodic noise measured on the LTM-16 SEC from supplier A — ADEV(τ) and MADEV(τ) ($N = 79\,000$, $\tau_0 \cong 23$ ms, $T = 1800$ s). (Reproduced from [5.32], ©1997 IEEE, by permission of IEEE)

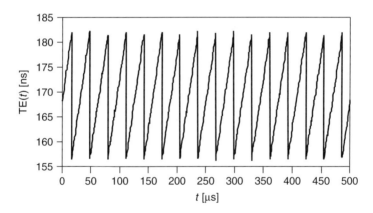

Figure 5.34. Periodic noise measured on the ADM-1 SEC from supplier B — first quarter of the TE sequence $\{x_i\}$ ($N = 4096$, $\tau_0 \cong 488$ ns, $T = 2$ ms). (Reproduced from [5.32], ©1997 IEEE, by permission of IEEE)

standard estimators of Equations (5.104) and (5.105), while the PSD $S_x(f)$ was computed (neglecting a multiplicative factor) through the Fast Fourier Transform (FFT) periodogram technique with triangular-shape data windowing.

In the frequency domain, the PSD in Figure 5.32 exhibits several discrete lines (spikes) at harmonic frequencies of about 1 Hz. In the time-domain analysis of ADEV and MADEV plots in Figure 5.33, on the other hand, the same periodic noise appears as a ripple.

A second experimental example of periodic noise is provided in Figures 5.34 through 5.36. These results were measured on the SEC of an early design of SDH Add-Drop Multiplexer STM1 (ADM1), according to the same procedure outlined above. Here and in the following of the book, we will denote the supplier of this equipment as

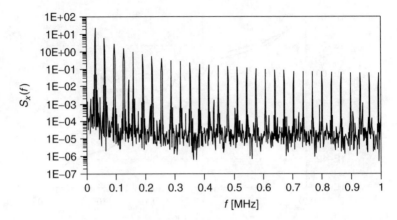

Figure 5.35. Periodic noise measured on the ADM-1 SEC from supplier B — PSD estimate $S_x(f)$ ($N = 4096$, $\tau_0 \cong 488$ ns, $T = 2$ ms). (Reproduced from [5.32], ©1997 IEEE, by permission of IEEE)

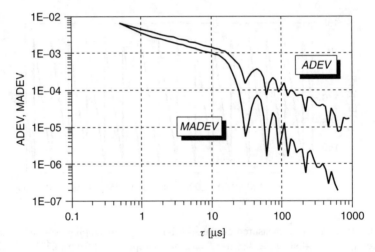

Figure 5.36. Periodic noise measured on the ADM-1 SEC from supplier B — ADEV(τ) and MADEV(τ) ($N = 4096$, $\tau_0 \cong 488$ ns, $T = 2$ ms)(Reproduced from [5.32], ©1997 IEEE, by permission of IEEE)

supplier B. This SEC was designed as DPLL, based on some numerical frequency-control algorithm producing high short-term noise. Aiming at studying the clock behaviour in the very short term, the maximum TE sampling rate was thus set, measuring the TE on *every* edge of the 2.048 MHz timing signals (thus $N = 4096$, $\tau_0 \cong 488$ ns, $T = 2$ ms).

Figure 5.34 zooms in on the first quarter (i.e., $t = 0 \div 500$ μs) of the acquired TE sequence. The plot reveals that this DPLL operates stepping alternatively between two discrete frequencies, controlled by a phase threshold mechanism. The frequency quantization error yields the saw-toothed waveform of 26-ns peak-to-peak amplitude and of period about 30–40 μs shown in the graph.

This is really a limit case, as the measured TE consists almost entirely of periodic noise. Hence, the PSD (shown in Figure 5.35) consists of a series of spikes and the ripples in the ADEV and MADEV plots (shown in Figure 5.36) are very wide. Finally, it is worthwhile noticing that such a short-term noise, when performing measurements over longer observation intervals, appears as a broadband WPM noise.

5.11 BEHAVIOUR OF THE TIME-DOMAIN STABILITY QUANTITIES: SLAVE CLOCKS

In the previous section, we described the behaviour of various time-domain stability quantities when measured on timing signals affected by power-law noise, by frequency offset and drift or by periodic noise. Such impairments model the most common non-idealities of autonomous clocks.

Now, let us suppose that the oscillator, affected for example by power-law noise, is working inside a PLL. When measurement is accomplished on the output of the slave clock, the noise generated by the internal oscillator is revealed as high-pass filtered (cf. Equation (5.75)). Thus, also the behaviour of stability quantities measured changes accordingly and is not the same anymore as reported in tables of the previous section.

Referring again to Figure 5.12, we consider three possible noise sources to investigate the behaviour of time-domain stability quantities at the output of a slave clock: the phase noise generated by the VCO ϕ_{VCO} [rad], the tension noise produced cumulatively by the phase detector and the loop filter V_{DF} [V] and the phase noise on the input signal ϕ_{in} [rad].

On the slave clock modelled in Figure 5.12, two possible measures can be envisaged. One consists of measuring the output phase noise $\phi_{out}(t)$, compared to the absolute time t (in practical measurements, this can be supplied by some 'good' external reference clock). The other consists of measuring the phase error $\phi_{out}(t) - \phi_{in}(t)$, based on $\phi_{in}(t)$ itself to establish the reference time. In ITU-T and ETSI standards [5.39][5.42], the former measurement scheme is called *independent-clock configuration*, the latter *synchronized-clock configuration*. The latter scheme, moreover, is of particular interest because it allows to observe directly the impairments added by the slave clock, without having to deal with frequency offsets between the clock under test and the reference clock.

For the sake of simplicity, we limit our analysis to the following cases, where only one noise source is considered active at a time:

(1) measurement in independent-clock configuration with only some noise $\phi_{VCO}(t)$ from the VCO;

(2) measurement in independent-clock configuration with only some noise $V_{DF}(t)$ from the phase detector and the loop filter;

(3) measurement in synchronized-clock configuration with only some phase noise $\phi_{in}(t)$ on the reference signal

In case (1), noise from the VCO is transferred to the measured quantity ϕ_{out} *high-pass filtered* according to the transfer function $H_B(s)$ of Equation (5.75). In case (2), noise from the phase detector and the loop filter is transferred to the measured quantity ϕ_{out}

low-pass filtered according to the transfer function $H_A(s)$ of Equation (5.74). In case (3), phase noise on the reference signal is transferred to the measured quantity $\phi_{\text{out}} - \phi_{\text{in}}$ *high-pass filtered* according to the transfer function

$$H_C(s) = \frac{\phi_{\text{out}}(s) - \phi_{\text{in}}(s)}{\phi_{\text{in}}(s)} = H(s) - 1 \qquad (5.137)$$

i.e., the same as in case (1) neglecting the sign.

The behaviour of time-domain stability quantities in the three measurement schemes above can be studied as we did for autonomous clocks, by evaluating the integral relationships provided in Equation (5.113) for the true variance, in Equation (5.114) for the M-sample variance, in Equation (5.115) for the Allan variance, in Equation (5.117) for the modified Allan variance, in Equation (5.120) for the Time variance and in Equation (5.121) for the TIErms. In this case, nevertheless, the noise spectrum $S_y(f)$ in those integral relationships must be replaced by the spectrum of noise signals $\phi_{\text{VCO}}(t)$, $V_{\text{DF}}(t)$ and $\phi_{\text{in}}(t)$, properly filtered as indicated in the previous paragraph.

In conclusion, the integral relationships are evaluated by letting

$$S_y(f) = \begin{cases} S_{\phi_{\text{VCO}}}(f)|H_B(s)|^2 & \text{in case (1)} \\ S_{V_{\text{DF}}}(f)|H_A(s)|^2 & \text{in case (2)} \\ S_{\phi_{\text{in}}}(f)|H_C(s)|^2 & \text{in case (3)} \end{cases} \qquad (5.138)$$

For example, we evaluated the integral relationships in the most common case of power-law noise in a second-order PLL, with closed-loop transfer function $H(s)$ characterized by bandwidth B. Examination of results allows to identify the asymptotic behaviour of the time-domain stability quantities, i.e. the corresponding power laws in the time domain as we did in the previous section for autonomous clocks.

For the sake of brevity, we report only the resulting slopes of the Allan variance and of the TIErms [5.47], in Tables 5.10 and 5.11 respectively, compared with the unfiltered slopes already reported in the previous section. Among power-law noise types, RWFM has not been evaluated for cases (1) and (3), because, on the one hand, this noise type is mostly negligible for $\tau \ll 1/B$ for all practical values of the PLL bandwidth and, on the other hand, it is filtered out as other noise types for $\tau \gg 1/B$. Moreover, in case (2), only computations for WPM and PPM were carried out, because

Table 5.10   Time-domain power laws for the Allan variance measured in slave clocks

Noise Type	Unfiltered	Cases (1) and (3)		Case (2)	
		$\tau \ll 1/B$	$\tau \gg 1/B$	$\tau \ll 1/B$	$\tau \gg 1/B$
WPM	τ^{-2}	τ^{-2}	τ^{-2}	τ^{-2}	τ^{-2}
FPM	τ^{-2}	τ^{-2}	τ^{-2}	τ^{0}	τ^{-2}
WFM	τ^{-1}	τ^{-1}	τ^{-2}	—	—
FFM	τ^{0}	τ^{0}	τ^{-2}	—	—

— = Not evaluated

Table 5.11 Time-domain power laws for the TIErms measured in slave clocks

Noise Type	Unfiltered	Cases (1) and (3)		Case (2)	
		$\tau \ll 1/B$	$\tau \gg 1/B$	$\tau \ll 1/B$	$\tau \gg 1/B$
WPM	τ^0	τ^0	τ^0	τ^{+1}	τ^0
FPM	τ^0	τ^0	τ^0	τ^{+2}	τ^{0**}
WFM	τ^{+1}	τ^{+1}	τ^0	—	—
FFM	*	τ^{+2}	τ^0	—	—

*Not convergent
**Depending on the loop filter
— = Not evaluated

these seem to be the only noise types encountered in practical PLL phase detectors and loop filters.

5.12 CHAINS OF SLAVE CLOCKS

In synchronization networks, chains of slave clocks are very common and are the basic architecture to transfer timing along synchronization trails (cf. the synchronization network reference chain). Hence the need to assess the performance of timing transfer along chains of slave clocks.

In this section, this complex topic is introduced and some basic result is provided. For further information, the reader may refer for example to the papers [5.48][5.49], which report studies carried out by numerical analysis and simulation under the simple hypothesis of linear behaviour of the slave clocks (i.e., the linear model of PLL is adopted). Moreover, the paper [5.50] studies random noise accumulation in chains of clocks by a time-domain state-space approach and by considering both linear and non-linear PLL models with additive noise sources.

To study the behaviour of clock chains, we present the simple models adopted by Carbonelli *et al.* in papers [5.48][5.49]. Although the assumption of linear behaviour is quite limiting, as it implies the assumption of small-amplitude phase noise, yet some indicative results can still be derived.

First, the model of clock chain is shown in Figure 5.37. A chain of K slave clocks is considered and the *synchronized-clock configuration* for relative TE measurement is adopted: the timing signal output by the Kth Slave Clock (SC) of the chain is compared to the input reference from a Reference Clock (RC). The timing signal at the input of the chain is affected by phase noise with PSD $S_{\varphi_{in}}(f)$ [rad^2/Hz]. On the other hand, the PSD of the phase noise at the output of the jth clock of the chain is denoted with $S_{\varphi out, j}(f)$ (for $j = 1, 2, \ldots, K$).

Under the assumption of clock linear behaviour, the model of the single slave clock of the chain can be derived directly from the model described in Section 5.5.5 and shown in Figure 5.12. Hence, the model for noise generation and filtering in the single slave clock is shown in Figure 5.38. In this figure, as before, φ_{in} and φ_{out} [rad] represent the phase noise on the input and output signals, φ_{VCO} [rad] is the phase noise generated by the VCO

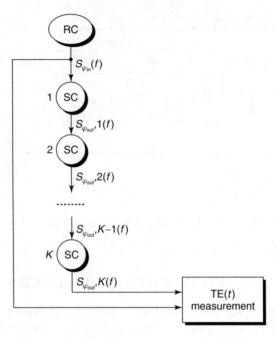

Figure 5.37. Model of chain of slave clocks with TE measurement in the synchronized-clock configuration

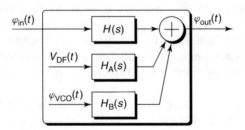

Figure 5.38. Model for noise generation and filtering in the single slave clock of the chain

and V_{DF} [V] is the tension noise produced cumulatively by the phase detector and the loop filter. Let $S_{\varphi_{in}}(f)$, $S_{\varphi_{out}}(f)$, $S_{\varphi_{VCO}}(f)$ and $S_{V_{DF}}(f)$ denote their PSDs, respectively. Finally, $H(s)$, $H_A(s)$ and $H_B(s)$ are the three transfer functions defined in Equations (5.73), (5.74) and (5.75).

Therefore, by the superposition principle, the PSD $S_{\varphi_{out}}(f)$ of the phase noise $\varphi_{out}(t)$ at the output of the clock is obtained as

$$S_{\varphi_{out}}(f) = S_{\varphi_{in}}(f)|H(f)|^2 + S_{V_{DF}}(f)|H_A(f)|^2 + S_{\varphi_{VCO}}(f)|H_B(f)|^2 \qquad (5.139)$$

where transfer functions in the Fourier-frequency domain have been used, by letting $s = j2\pi f$.

Now, under the further assumption that all clocks of the chains are characterized by the same parameter values (*homogeneous chain*), it is possible to evaluate the PSD $S_{\varphi_{out},K}(f)$

of the phase noise at the output of the last clock of the chain, as

$$S_{\varphi_{\text{out}},K}(f) = S_{\varphi_{\text{in}}}(f)|H(f)|^{2K} + \left[S_{V_{\text{DF}}}(f)|H_A(f)|^2 + S_{\varphi_{\text{VCO}}}(f)|H_B(f)|^2\right]$$

$$\cdot \left[\sum_{j=0}^{K-1}|H(f)|^{2j}\right]$$

$$= S_{\varphi_{\text{in}}}(f)|H(f)|^{2K} + \left[S_{V_{\text{DF}}}(f)|H_A(f)|^2 + S_{\varphi_{\text{VCO}}}(f)|H_B(f)|^2\right]$$

$$\cdot \left[\frac{1 - |H(f)|^{2K}}{1 - |H(f)|^2}\right] \tag{5.140}$$

Based on this expression and remembering the relationships (5.76), it is then possible to evaluate the trends of the time-domain stability quantities of interest at the output of the chain, by way of the integral relationships provided in Equation (5.113) for the true variance, in Equation (5.114) for the M-sample variance, in Equation (5.115) for the Allan variance, in Equation (5.117) for the modified Allan variance, in Equation (5.120) for the Time variance and in Equation (5.121) for the TIErms.

Nevertheless, analytical evaluation of such expressions by closed-form integration is unfeasible. Thus, paper [5.48] provides some results obtained by numerical integration, for realistic values of clock model parameters based on measurement results and manufacturer data sheets.

In particular, chains of both SEC- and SASE-type clocks were considered. Parameter values were chosen as follows: clock nominal frequency $\nu_n = 5$ MHz, PLL damping ratio $\zeta = 3$ and bandwidth $B = 1$ Hz for the SEC and $B = 1$ mHz for the SASE. Moreover, $\varphi_{\text{VCO}}(t)$ and $V_{\text{DF}}(t)$ were modelled by way of the following power-law PSDs:

$$\text{SEC}: \begin{cases} S_{V_{\text{DF}}}(f) = \dfrac{10^{-13}}{f} + 10^{-17} \quad [\text{V}^2/\text{Hz}] \\[2mm] S_{\varphi_{\text{VCO}}}(f) = \dfrac{10^{-5.5}}{f^3} + \dfrac{10^{-12.2}}{f^2} + \dfrac{10^{-10.3}}{f} + 10^{-15.5} \quad [\text{rad}^2/Hz] \end{cases}$$

$$\text{SASE}: \begin{cases} S_{V_{\text{DF}}}(f) = \dfrac{10^{-15}}{f} + 10^{-17} \quad [\text{V}^2/\text{Hz}] \\[2mm] S_{\varphi_{\text{VCO}}}(f) = \dfrac{10^{-12.2}}{f^3} + \dfrac{10^{-12.2}}{f^2} + \dfrac{10^{-13.15}}{f} + 10^{-15.5} \quad [\text{rad}^2/\text{Hz}] \end{cases} \tag{5.141}$$

with upper cut-off frequency $f_h = 10$ MHz. For such values of the model parameters and with null input noise $\varphi_{\text{in}}(t) = 0$, Carbonelli *et al.* [5.48] evaluated the TIErms measured in synchronized-clock configuration (there called RTIErms according to some old terminology), according to the scheme in Figure 5.37, separately with either φ_{VCO} or V_{DF} non-null at a time.

First, a chain of M SECs was considered. The graphs in Figure 5.39(a) (φ_{VCO} internal noise only) and 5.39(b) (V_{DF} internal noise only) show the numerical results obtained for values of M from 1 to 100 and observation interval 10^{-4} s $\leq \tau \leq 10^4$ s. Notice that the stability measure results independent on the number M of clocks in the chain for $\tau \ll 1/B$, while it exhibits increasing values with M for $\tau \gg 1/B$.

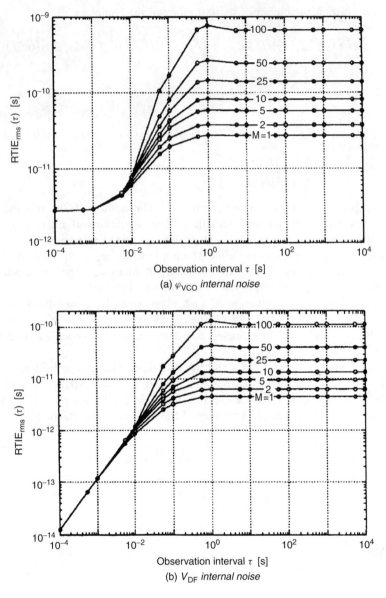

Figure 5.39. TIErms evaluated at the output of a chain of M SECs. (Reproduced from [5.48], ©1993 IEEE, by permission of IEEE)

Secondly, the effect of cascading N subchains, each made of $M = 25$ SECs and with one SASE between each subchain and the next one, was investigated. The graph in Figure 5.40 shows the results obtained for both φ_{VCO} and V_{DF} noises, for $N = 1, 2, 4$ and again for 10^{-4} s $\leq \tau \leq 10^4$ s. The VCO noise contribution proves to be dominant over the DF one, both in the short and in the long term, while no variations on N were found. Therefore, the heavy jitter filtering action of SASE clocks, due to their narrow bandwidth, allows to build long slave clock chains without significantly reducing the output stability.

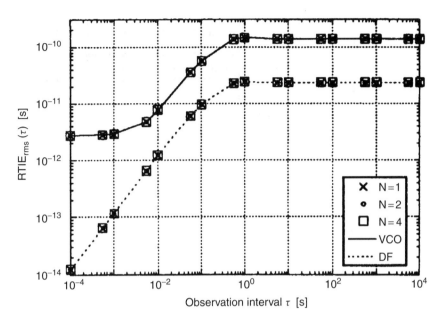

Figure 5.40. TIErms evaluated at the output of $N = 1, 2, 4$ cascaded subchains, each made of $M = 25$ SECs and with one SASE between each subchain and the next one. (Reproduced from [5.48], ©1993 IEEE, by permission of IEEE)

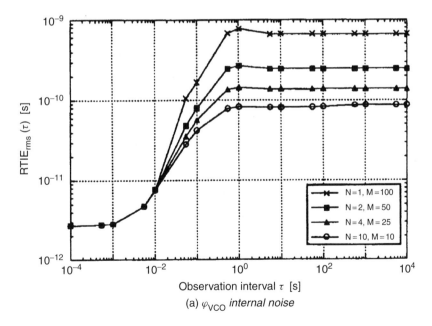

(a) φ_{VCO} *internal noise*

Figure 5.41. TIErms evaluated at the output of chains including a total of $M = 100$ SECs and $N = 1, 2, 4, 10$ SASE clocks. (Reproduced from [5.48], ©1993 IEEE, by permission of IEEE)

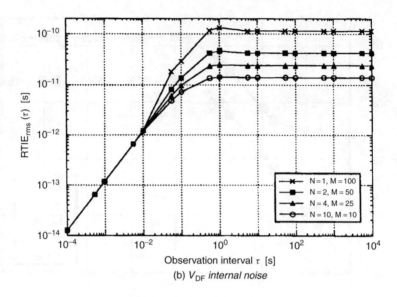

(b) V_{DF} internal noise

Figure 5.41. (*continued*)

Finally, referring to a worst-case chain including 100 SECs, four different combinations of subchains and SASE clocks were compared: $N = 1$ and $M = 100$, $N = 2$ and $M = 50$, $N = 4$ and $M = 25$, $N = 10$ and $M = 10$. The results are shown in Figures 5.41(a) (φ_{VCO} internal noise only) and 5.41(b) (V_{DF} internal noise only). Again, only the long-term stability results affected by the number of subchains: the bigger is the number of SECs cascaded without intervening SASE clocks, the higher is the TIErms measured. Therefore, the conclusion is that a combination with a greater number of shorter subchains should be preferred to one with fewer but longer subchains.

5.13 SUMMARY

A clock is a device able to supply a *timing signal*, i.e. a pseudo-periodic signal usable to control the timing of actions. A comprehensive model of the instantaneous frequency $v(t)$ supplied by actual clocks is provided in Equation (5.4).

The *stability* of a clock deals with the measurement of variations of its instantaneous frequency (or of the time generated) compared to the nominal value over a given observation interval. The *accuracy*, on the other hand, denotes the maximum frequency error Δv_{max}, compared to the nominal value, measured in general over the whole clock life.

An *autonomous clock* is a stand-alone device able to generate a timing signal, suitable for the measurement of time intervals, starting from some periodic physical phenomenon that runs independently of external influence. A *slave clock* is a device able to generate a timing signal, suitable for the measurement of time intervals, having phase (or much less frequently frequency) controlled by a reference timing signal at its input. The Phase-Locked Loop (PLL) and the Frequency-Locked Loop (FLL) are examples of slave clocks.

A *Phase-Locked Loop* is a control system, based on a negative feedback loop, used to automatically adjust the phase of a locally generated signal $\hat{s}(t)$ to the phase of an incoming signal $s(t)$. Its building blocks are a phase detector, a low-pass filter having transfer function $F(s)$ and a VCO. Second-order PLLs are the most common in practice, because they are able to track a constant frequency offset between the input signal and the local oscillator without residual steady-state phase error. Moreover, safe system stability can be easily ensured. The closed-loop transfer functions $H(s) = \phi_{out}/\phi_{in}$ and $H_A(s) = \phi_{out}/V_{DF}$ are low pass, that is the input phase noise and the internal tension noise produced by the phase detector and the loop filter are low-pass filtered to the output. Conversely, the transfer function $H_B(s) = \phi_{out}/\phi_{VCO}$ is high pass, that is the internal phase noise generated by the VCO is high-pass filtered to the output.

A slave clock may operate in the following three modes in a synchronization network:

- *locked mode*, when it is tracking the input frequency;

- *free-run mode*, when it has to work autonomously because the reference signal failed and the internal VCO is working with null control tension;

- *hold-over mode*, again when it has to work autonomously because the reference signal failed, but the last control tension value at the input of the VCO before the reference failure is maintained, so to hold the last output frequency value over.

The characterization of clock stability is usually carried out by characterizing, by means of suitable analytical tools, the random processes $\varphi(t)$, $x(t)$, TE(t), $y(t)$ or $\nu(t)$. Examples of stability measures in the *frequency domain* are the one-sided power spectral densities of the phase, time and frequency fluctuations, since they are functions of the Fourier frequency f. On the other hand, variances of the same fluctuations, averaged over a given observation interval, are examples of stability measures in the *time domain*, since they are functions of the observation interval τ (time).

Among the several quantities defined in the literature for characterizing time and frequency stability, the following five in the time domain have been considered by ITU-T and ETSI standard bodies: the Allan variance (AVAR), the modified Allan variance (MAVAR), the Time variance (TVAR), the root mean square of Time Interval Error (TIErms) and the Maximum Time Interval Error (MTIE). Moreover, five standard estimators have been defined for their practical evaluation.

In the frequency domain, the model most frequently used to represent the output phase noise measured on clocks is the so-called *power-law model* (5.125). The five noise types of this model are: White Phase Modulation (WPM), Flicker Phase Modulation (FPM), White Frequency Modulation (WFM), Flicker Frequency Modulation (FFM) and Random Walk Frequency Modulation (RWFM).

The time-domain stability quantities reveal, with their characteristic trends, the presence of the different types of power-law noise, of frequency offset and drift and of periodic noise on the signal under measurement.

5.14 REFERENCES

[5.1] J. A. Barnes, A. R. Chi, L. S. Cutler, D. J. Healey, D. B. Leeson, T. E. McGunigal, J. A. Mullen Jr., W. L. Smith, R. L. Sydnor, R. F. C. Vessot, G. M. R Winkler. Characterization of frequency stability. *IEEE Transactions on Instrumentation and Measurement* 1971; **IM-20**(2).

[5.2] J. Rutman, F. L. Walls. Characterization of frequency stability in precision frequency sources. *Proceedings of the IEEE*, vol. 79, no. 6, June 1991.

[5.3] D. W. Allan. Statistics of atomic frequency standards. *Proceedings of the IEEE*, vol. 54, no. 2, July 1966.

[5.4] F. L. Walls, D. W. Allan. Measurement of frequency stability. *Proceedings of the IEEE*, vol. 74, no. 1, Jan. 1986.

[5.5] J. A. Barnes. The measurement of linear frequency drift in oscillators. *Proceedings of the 15th Annual Precise Time and Time Interval (PTTI) Meeting*, 1983.

[5.6] R. L. Filler, J. R. Vig. Long term aging of oscillators. *IEEE Transactions on Ultrasonics and Ferroelectronics Frequency Control* 1993; **40**(4) 387–393.

[5.7] IEEE Std. 1139. *IEEE Standard Definitions of Physical Quantities for Fundamental Frequency and Time Metrology*. Revised version approved Oct. 20, 1988.

[5.8] International Organisation for Standardisation (ISO). *International Vocabulary of Basic and General Term in Metrology*. Geneva, 1993.

[5.9] D. W. Allan, N. Ashby, C. C. Hodge. *The Science of Timekeeping*. Application Note 1289, Hewlett-Packard Company, June 1997.

[5.10] A. J. Viterbi. *Principles of Coherent Communications*. New York: McGraw–Hill, 1966.

[5.11] W. C. Lindsey. *Synchronization Systems in Communications and Control*. Englewood Cliffs, NJ: Prentice Hall Inc., 1972.

[5.12] A. Blanchard. *Phase-Locked Loops*. New York: John Wiley & Sons, 1976.

[5.13] F. M. Gardner. *Phaselock Techniques*. New York: John Wiley & Sons, 1979.

[5.14] R. E. Best. *Phase Locked Loops*. New York: McGraw–Hill Book Company, 1984.

[5.15] P. Horowitz and W. Hill. The art of electronics, Sections 9.28 to 9.33. *Phase-Locked Loops*. Cambridge: Cambridge University Press, 1980.

[5.16] H. Meyr, G. Ascheid. *Synchronization in Digital Communications. Vol. 1: Phase-, Frequency-Locked Loops, and Amplitude Control*. New York: John Wiley & Sons, 1990.

[5.17] V. F. Kroupa. Noise properties of PLL systems. *IEEE Transactions on Communications* 1982; **COM-30**(10): 2244–2252.

[5.18] L. S. Cutler, C. L. Searle. Some aspects of the theory and measurement of frequency fluctuations in frequency standards. *Proceedings of the IEEE*, vol. 54, no. 2, Feb. 1966.

[5.19] J. A. Barnes. Atomic timekeeping and the statistics of precision signal generators. *Proceedings of the IEEE*, vol. 54, no. 2, Feb. 1966.

[5.20] P. Lesage, C. Audoin. Characterization of frequency stability: uncertainty due to the finite number of measurements. *IEEE Transactions on Instrumentation and Measurement* 1973; **IM-22**(2).

[5.21] W. C. Lindsey, C. M. Chie. Theory of oscillator instability based upon structure functions. *Proceedings of the IEEE*, vol. 64, no. 12, Dec. 1976.

[5.22] J. Rutman. Characterization of phase and frequency instabilities in precision frequency sources: fifteen years of progress. *Proceedings of the IEEE*, vol. 66, no. 9, Sept. 1978.

[5.23] P. Lesage, C. Audoin. Characterization and measurement of time and frequency stability. *Radio Science (American Geophysical Union)*, 1979; **14**(4): 521–539.

[5.24] D. W. Allan, J. A. Barnes. A modified Allan variance with increased oscillator characterization ability. *Proceedings of the 35th Annual Frequency Control Symposium*, 1981.

[5.25] D. A. Howe, D. W. Allan, J. A. Barnes. Properties of signal sources and measurement methods. *Proceedings of the 35th Annual Frequency Control Symposium*, 1981.

[5.26] P. Lesage, T. Ayi. Characterization of frequency stability: analysis of the modified Allan variance and properties of its estimate. *IEEE Transactions on Instrumentation and Measurement* 1984; **IM-33**(4).

[5.27] S. R. Stein. Frequency and time — their measurement and characterization. In *Precision Frequency Control*. Vol. 2, ch. 2, pp. 191–232. New York: Academic Press, 1985.

[5.28] L. G. Bernier. Theoretical analysis of the modified Allan variance. *Proceedings of the 41st Annual Symposium on Frequency Control*, 1987.

[5.29] D. B. Sullivan, D. W. Allan, D. A. Howe, F. L. Walls (eds). *Characterization of Clocks and Oscillators*. NIST Technical Note 1337, March 1990.

[5.30] D. W. Allan, M. A. Weiss, J. L. Jespersen. A frequency-domain view of time-domain characterization of clocks and time and frequency distribution systems. *Proceedings of the 45th Annual Symposium on Frequency Control*, 1991.

[5.31] M. Carbonelli, D. De Seta, D. Perucchini. Characterization of timing signals and clocks. *European Transactions on Telecommunications* 1996; **7**(1).

[5.32] S. Bregni. Clock stability characterization and measurement in telecommunications. *IEEE Transactions on Instrumentation and Measurement* 1997; **46**(6).

[5.33] S. Bregni. Measurement of maximum time interval error for telecommunications clock stability characterization. *IEEE Transactions on Instrumentation and Measurement* 1996; **IM-45**(5).

[5.34] S. M. Kay, S. L. Marple. Spectrum analysis—a modern perspective. *Proceedings of the IEEE*, vol. 69, no. 11, Nov. 1981.

[5.35] W. H. Press, B. P. Flannery, S. A. Teukolsky, W. T. Vettering. *Numerical Recipes in C—The Art of Scientific Computing*. Cambridge: Cambridge University Press. (Also available for other programming languages)

[5.36] A. Papoulis. *Probability, Random Variables and Stochastic Processes*. 2nd edition. New York: McGraw-Hill, 1984.

[5.37] ITU-T Recs. G.810 *Considerations on Timing and Synchronization Issues*; G.811 *Timing Requirements at the Outputs of Primary Reference Clocks Suitable for Plesiochronous Operation of International Digital Links*; G.812 *Timing Requirements at the Outputs of Slave Clocks Suitable for Plesiochronous Operation of International Digital Links*. Geneva, Blue Book, 1988.

[5.38] ANSI T1.101 *Telecommunications—Synchronization Interface Standard*. January 1997.

[5.39] ITU-T Rec. G.810 *Definitions and Terminology for Synchronisation Networks*. Geneva, August 1996.

[5.40] S. Bregni, F. Setti. Impact of the anti-aliasing prefiltering on the measurement of MTIE. *Proceedings of IEEE GLOBECOM '97*, Phoenix, AZ, USA, 3–8 November 1997.

[5.41] S. Bregni, S. Maccabruni. Fast computation of maximum time interval error by binary decomposition. *IEEE Transactions on Instrumentation and Measurement* 2000; **40**(6).

[5.42] EN 300 462 *Transmission and Multiplexing (TM); Generic Requirements for Synchronization Networks*; Part 1-1: *Definitions and Terminology for Synchronization Networks*; Part 2-1: *Synchronization Network Architecture*; Part 3-1: *The Control of Jitter and Wander within Synchronization Networks*; Part 4-1: *Timing Characteristics of Slave Clocks Suitable for Synchronization Supply to Synchronous Digital Hierarchy (SDH) and Plesiochronous Digital Hierarchy (PDH) Equipment*; Part 5-1: *Timing Characteristics of Slave Clocks Suitable for Operation in Synchronous Digital Hierarchy (SDH) Equipment*; Part 6-1: *Timing Characteristics of Primary Reference Clocks*.

[5.43] R. A. Baugh. Frequency modulation analysis with the hadamard variance. *Proceedings of the 25th Annual Symposium on Frequency Control*, April 1971, pp. 222–225.

[5.44] F. L. Walls, A. De Marchi. RF spectrum of a signal after frequency multiplication: measurement and comparison with a simple calculation. *IEEE Transactions on Instrumentation and Measurement* 1975; **24**(3).

[5.45] D. E. Knuth. *The Art of Computer Programming*. Vol. 2, p. 118. London: Addison-Wesley, 1981.

[5.46] J. A. Barnes, D. W. Allan. A statistical model of flicker noise. *Proceedings of the IEEE*, vol. 54, no. 2, pp. 176–178, Feb. 1966.

[5.47] S. Bregni, M. Carbonelli, D. De Seta, D. Perucchini. Impact of slave clock internal noise on Allan variance and root mean square time interval error measurement. *Proceedings of IEEE Instrumentation and Measurement Technology Conference*, Hamamatsu, Japan, 10–12 May 1994.

[5.48] M. Carbonelli, D. De Seta, D. Perucchini. Root mean square time interval error accumulation along slave clock chains. *Proceedings of IEEE ICC'93*, Geneva, Switzerland, May 1993.

[5.49] M. Carbonelli, D. De Seta, D. Perucchini. Jitter and Wander performance in synchronization distribution chains. *Proceedings of IEEE Instrumentation and Measurement Technology Conference*, Brussels, Belgium, 4–6 June 1996.

[5.50] G. M. Garner. Accumulation of random noise in a chain of slave clocks. *Proceedings of the 48th Annual Symposium on Frequency Control*, June 1994, pp. 798–811.

APPENDIX 5A
FAST COMPUTATION OF TVAR ESTIMATOR BY RECURSION ALGORITHM

In this Appendix, a fast recursive algorithm to evaluate the TVAR/TDEV standard estimator defined in Equation (5.106) is provided. For ease of understanding, the TVAR estimator is reported again here below:

$$
\text{TVAR}(\tau) = \frac{1}{6n^2(N - 3n + 1)} \sum_{j=1}^{N-3n+1} \left[\sum_{i=j}^{n+j-1} (x_{i+2n} - 2x_{i+n} + x_i) \right]^2
$$

$$
\text{for } n = 1, 2, \ldots, \left\lfloor \frac{N}{3} \right\rfloor \tag{5.142}
$$

Plain computation of this TVAR standard estimator requires execution of two nested summation loops, thus yielding a computational complexity in the order of N^2 for each value $\text{TVAR}(\tau)$ to compute.

To save computation time, the expression (5.142) can be written as follows:

$$
\text{TVAR}(\tau) = \frac{1}{6n^2(N - 3n + 1)} \sum_{j=1}^{N-3n+1} A_j^2(n) \tag{5.143}
$$

where

$$
A_j(n) = \sum_{i=j}^{n+j-1} (x_{i+2n} - 2x_{i+n} + x_i) \tag{5.144}
$$

with

$$
n = 1, 2, \ldots, \left\lfloor \frac{N}{3} \right\rfloor
$$
$$
j = 1, 2, \ldots, N - 3n + 1 \tag{5.145}
$$

The computational weight of this expression can be reduced by noting that the terms $A_j(n)$ can be evaluated recursively, as

$$
A_1(n) = \sum_{i=1}^{n} (x_{i+2n} - 2x_{i+n} + x_i)
$$

$$
A_{j+1}(n) = A_j(n) + (x_{3n+j} - 3x_{2n+j} + 3x_{n+j} - x_i) \tag{5.146}
$$

The first term $A_1(n)$ requires $3n - 1$ additions, but the next terms $A_j(n)$ for $j > 1$ can be calculated with only four further additions.

APPENDIX 5B
FAST COMPUTATION OF MTIE ESTIMATOR BY BINARY DECOMPOSITION

Plain computation of the MTIE standard estimator (5.108) proves cumbersome in most cases of practical interest, due to its heavy computational weight. In this Appendix, a fast algorithm based on binary decomposition to compute the MTIE estimator is described [5.41]. The computational weight of the binary decomposition algorithm is compared to that of the estimator plain calculation, showing that the number of operations needed is reduced to a term proportional to $N \log_2 N$ instead of N^2. A heavy computational saving is therefore achieved, thus making feasible MTIE evaluation based on even long sequences of Time Error (TE) samples.

5B.1   Plain Computation of MTIE Estimator

Let $N_T = T/\tau_0 + 1$ be the total number of available TE samples in the sequence $\{x_i\}$ and $N_\tau = \tau/\tau_0 + 1$ be the number of samples available in a window (observation interval) of span τ. Then, for *each* single value MTIE(τ, T) the following expression has to be computed (cf. Equation (5.108)):

$$\text{MTIE}(\tau, T) = \max_{j=1}^{N_T - N_\tau + 1} \left[\max_{i=j}^{N_\tau + j - 1} (x_i) - \min_{i=j}^{N_\tau + j - 1} (x_i) \right] \quad (5.147)$$

MTIE masks currently specified in standards span over a wide range of τ: four decades, namely from 10^{-1} s up to 10^3 s. For a long time this range was even wider, from a few milliseconds up to 10^5 s. Furthermore, more specific studies may require investigation over different wide ranges.

As pointed out in [5.33], the number of samples N_T to process may get easily to the order of 10^5 in most cases of practical interest, if we are interested in a somehow accurate characterization of the clock noise. It is obvious that the plain computation of the estimator (5.147) is unadvisable and quickly tends to be unmanageable, due to the number of operations nested in evaluation loops. Hence comes the need of contriving a suitable algorithm, effective in cutting down the computational weight of a plain implementation of the estimator (5.147).

5B.2   MTIE Computation by Binary Decomposition

The fast algorithm described in this Appendix is based on a binary decomposition of a TE sequence $\{x_i\}$ made of $N_T = 2^{k_{\text{MAX}}}$ samples in nested windows made of $N_\tau = 2^k$ samples ($k = 1, 2, 3, \ldots, k_{\text{MAX}}$). MTIE can then be evaluated recursively for each window size 2^k.

As the first step ($k = 1$), all the possible 2-points windows ($\tau = \tau_0$) are analysed in the TE sequence: for each of them, the maximum and minimum values are stored. Their difference is the MTIE(τ_0) measured in that window, and the maximum of the MTIE values of all the 2-points windows is the resulting MTIE(τ_0, T) of the whole sequence. At this first step, there is no computational saving yet compared to the plain computation of the standard estimator.

Then, as a second step ($k = 2$), all the possible 4-points windows ($\tau = 3\tau_0$) are considered. The maximum and minimum values of each of these windows can be obtained by comparing the maximum and minimum values of the two 2-points windows in which the 4-points window can be split. The difference between the maximum of the two maxima and the minimum of the two minima is the MTIE($3\tau_0$) measured in that 4-point window. The maximum of the MTIE values of all the 4-point windows is the resulting MTIE($3\tau_0$, T) of the whole sequence.

The next step ($k = 3$) is to consider all the possible 8-points windows ($\tau = 7\tau_0$), split in two 4-points windows. Then so on for increasing integer values of k. The computational saving of this algorithm, compared to the plain computation of the standard estimator, lies in avoiding the comparison of all the samples in the windows of size larger than 2. The price to pay is that we have to limit the evaluation of MTIE(τ,T) just to the $\log_2 N_T$ values corresponding to the windows made of $N_\tau = 2^k$ samples (this corresponds to a bit more than three MTIE values per decade on the τ axis, which may be considered adequate in most practical applications).

More formally, starting from the TE sequence vector $\mathbf{x}$ made of $N_T = 2^{k_{MAX}}$ TE samples x_i, two matrices $\mathbf{A}_M$ and $\mathbf{A}_m$ are built. Matrices are made of $N_T - 1$ columns (indexed by i) and $\log_2 N_T$ rows, indexed by k. The first $N_T - 2^k + 1$ elements of each kth row of the matrix $\mathbf{A}_M$ contain the maximum values of all the possible 2^k-points windows sliding from left to right along the TE sequence $\{x_i\}$. The matrix $\mathbf{A}_m$ contains, in an analogous fashion, the corresponding minimum values of the 2^k-points windows. Therefore, the set of all the possible 2^k-points windows in the whole TE sequence is completely described by the couple of vectors

$$\begin{aligned} \mathbf{a}_{M/k} &= \{a_{M/k,i}\} \\ \mathbf{a}_{m/k} &= \{a_{m/k,i}\} \end{aligned} \qquad i = 1, 2, \ldots, N_T - 2^k + 1 \qquad (5.148)$$

where $\mathbf{a}_{M/k}$ and $\mathbf{a}_{m/k}$ are the kth rows taken from the matrices $\mathbf{A}_M$ and $\mathbf{A}_m$ respectively.

The first row ($k = 1$) of matrices $\mathbf{A}_M$ and $\mathbf{A}_m$ is obtained directly by the TE sequence vector $\mathbf{x}$ as

$$\begin{aligned} a_{M/1,i} &= \max(x_i, x_{i+1}) \\ a_{m/1,i} &= \min(x_i, x_{i+1}) \end{aligned} \qquad (5.149)$$

for $i = 1, 2, \ldots, N_T - 1$. The next rows ($k > 1$), instead, are obtained recursively as

$$\begin{aligned} a_{M/k,i} &= \max(a_{k-1,i}, a_{k-1,i+p}) \\ a_{m/k,i} &= \min(a_{k-1,i}, a_{k-1,i+p}) \end{aligned} \qquad (5.150)$$

where $p = 2^{k-1}$, for $i = 1, 2, \ldots, N_T - 2^k + 1$.

Finally, the value MTIE(τ, T) for $\tau = (N_\tau - 1)\tau_0$ and $N_\tau = 2^k$ (here denoted as MTIE$_k$ for the sake of brevity) can be evaluated from the kth rows of the matrices $\mathbf{A}_M$ and $\mathbf{A}_m$ as

$$\text{MTIE}_k = \max_{i=1,\ldots,N_T-2^k+1} (a_{M/k,i} - a_{m/k,i}) \qquad (5.151)$$

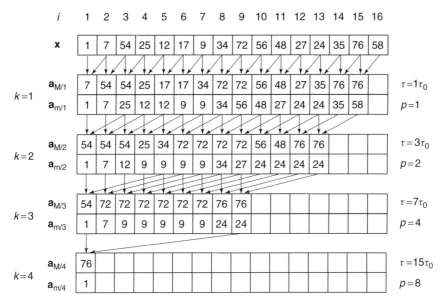

Figure 5B.1. Example of execution of the binary decomposition algorithm ($N_T = 16$). (Reproduced from [5.41], ©2000 IEEE, by permission of IEEE)

An example of the binary decomposition tree, applied on a TE sequence $\{x_i\}$ made of $N_T = 16$ samples ($k_{MAX} = 4$), is shown in Figure 5B.1, which depicts the four couples of vectors $\mathbf{a}_{M/k}$ and $\mathbf{a}_{m/k}$ (for $k = 1, 2, 3, 4$) built recursively starting from the TE vector $\mathbf{x}$.

5B.3   Computational Saving

The number of operations involved in the estimator plain computation and in the binary decomposition algorithm has been evaluated, in order to assess the resulting computational saving.

5B.3.1   Plain Computation of the Estimator

As far as a plain computation of the estimator (5.147) is concerned, three nested loops can be identified:

(1)   an external loop increasing the observation interval τ, executed one time per each single value $\mathrm{MTIE}(\tau, T)$ to compute;

(2)   a first internal loop executed, given τ, for each N_τ-points sliding window (the external max[·] function in Equation (5.147)), i.e. $N_T - N_\tau + 1$ times;

(3)   the most internal loop to find the maximum and minimum value in a set of N_τ samples, thus involving $2(N_\tau - 1)$ comparison test branches and a variable number of assignments according to the particular TE sequence (we neglect here the possibility to use a more efficient algorithm to extract the maximum and minimum values)

If we limit MTIE computation to one value per octave on the τ axis, as in the binary decomposition algorithm, then the first loop is executed $k_{MAX} = \log_2 N_T$ times, the second loop $N_T - 2^k + 1$ times ($k = 1, 2, \ldots, k_{MAX}$) and the third loop involves $2^{k+1} - 2$ branches. Thus, the computational weight results are approximately (from now on, N_T will be denoted simply as N for the sake of brevity):

$$\frac{4}{3}N^2 + \cdots \qquad \text{comparison test branches}$$

$$3\left(N \log_2 N - 2N\right) + \cdots \qquad \text{assignments (best case)}$$

$$\frac{2}{3}N^2 + \cdots \qquad \text{assignments (worst case)}$$

$$N \log_2 N - 2N + \cdots \qquad \text{additions}$$

(5.152)

Comparison test branches are the most time-consuming operations in most programming languages.

It is worthwhile noticing that MTIE plain computation turned out to be a N^2-problem because we decided to limit MTIE computation to one value per octave on the τ axis. If MTIE is computed for all the possible $N - 1$ values of τ, then the number of operations required becomes proportional to N^3 instead.

5B.3.2   Binary Decomposition Algorithm

As far as the binary decomposition algorithm is concerned, on the other hand, the following loops can be identified:

(1)   a first loop initializing the first row ($k = 1$) of matrices $\mathbf{A}_M$ and $\mathbf{A}_m$ and then computing $MTIE_1$, involving in particular $2(N - 1)$ comparison test branches;

(2)   a second main loop increasing the row index k ($k > 1$), executed $\log_2 N - 1$ times;

(3)   a loop, internal to the previous one, to compute the next rows ($k > 1$) of matrices $\mathbf{A}_M$ and $\mathbf{A}_m$ and to evaluate the corresponding $MTIE_k$, involving in particular $3(N - 2^k + 1)$ comparison test branches

Thus, the computational weight results are approximately:

$$3N \log_2 N - 7N + \cdots \qquad \text{comparison test branches}$$

$$2N + \cdots \qquad \text{assignments (best case)}$$

$$3\left(N \log_2 N - 2N\right) + \cdots \qquad \text{assignments (worst case)}$$

$$N \log_2 N - 2N + \cdots \qquad \text{additions}$$

(5.153)

5B.3.3   Comparison in Terms of Computational Weight

We notice that, in the binary decomposition algorithm, the number of comparison test branches and worst-case assignments needed has been reduced to a term proportional to $N \log_2 N$ instead of the N^2 involved in the plain computation of the estimator (5.147).

The graph of Figure 5B.2 compares, on a logarithmic scale, the number of comparison test branches needed by the two algorithms considered as a function of the total number of available TE samples N, for $2^1 \leq N \leq 2^{25}$ (to build this graph, all the lower order

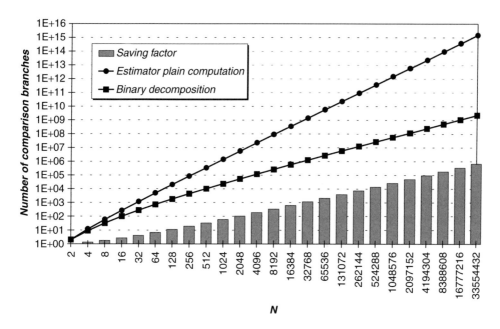

Figure 5B.2.   Number of comparison test branches required by the MTIE estimator plain computation and the binary decomposition algorithm as a function of the total number of TE samples N. (Reproduced from [5.41], ©2000 IEEE, by permission of IEEE)

terms not shown in (5.152) and (5.153) have been taken into account). Moreover, the ratio between the two numbers (i.e., the *computational saving factor*) is plotted as well for ease of comparison. It may be noticed that in the most common range $2^{14} \leq N \leq 2^{19}$ (i.e., $16\,384 \leq N \leq 524\,288$) the saving factor turns out to be in the remarkable order of 10^3 to 10^4.

6

PHYSICAL PRINCIPLES AND TECHNOLOGY OF CLOCKS

The oldest historical examples of clocks are mostly dated back to 3500 BC: Egyptian obelisks and sundials. Across several technical improvements, sundials (based on measuring the rotation of a shadow with the sun) and clepsydrae (based on measuring the level of water in a vessel with a regulated flow of water at the input or output) have been the clocks in use during the Middle Ages. On the other hand, early mechanical clocks date back to the thirteenth century, while the first pocket clocks were constructed in the fifteenth century, based on a spring mechanism. However, a true milestone was the invention of pendulum clocks, due to Galileo and Huygens (sixteenth and seventeenth century), and the introduction of the swing-wheel as an oscillating element. After that, until the introduction of electrical clocks and then of clocks based on quartz and atomic oscillators, mechanical clocks did not go through substantial changes, at least in the operation principle, until today. Some milestones in the progress of timekeeping are mentioned by Allan *et al.* in their outstanding review paper [6.1]:

3500 BC	Egyptian obelisks and sundials
2000–1500 BC	Mayan calendar
1900–1600 BC	Stonehenge site
1094	Sung Su's Chinese water clock perfected
1583	Galileo discovers pendulum period constancy
1656	Huygens' pendulum clock
1736	Harrison's maritime chronometer tested at sea
1918	Quartz-crystal oscillator developed
1948–1949	Lyons develops the first atomic clock (ammonia)
1955	Essen and Parry start keeping time with a caesium atomic clock
1978	First NAVSTAR GPS Block I satellite launched
1995	GPS announced fully operational

This chapter summarizes the physical principles of operation of high-precision clocks based on quartz oscillators and atomic frequency standards, suitable for operation in digital telecommunications networks. Finally, the characteristics and architecture of the Global Positioning System (GPS) are outlined.

6.1 CLOCKS

From a theoretical point of view, the operation principle of any kind of clock consists of a generator of oscillations and an automatic counter of such oscillations. With different ability, the oscillator can be based indeed on (pseudo-)periodic physical phenomena of *any* kind: the swinging of a pendulum or a wheel in mechanical clocks, the vibration of atoms in a crystal around their minimum-energy position in quartz clocks, the radiation associated with specific quantum atomic transitions in atomic clocks are just the best known examples due to their widest application, but not the only ones.

Allan *et al.*, in their charming paper [6.1], report that in 1982 the first millisecond pulsar was discovered. The arriving pulse stream, which left the star some 13 000 years ago, is exceedingly uniform with a period of only 1.55780645169838 ms. The authors of the paper are right to point out that it challenges the mind to think of a neutron star, with a mass of one or two times than our sun collapsed in a radius of only about 10 km, spinning 642 revolutions per second and flashing us with its electromagnetic beam at each rotation! Today, more than 30 celestial objects of this kind are known. Using GPS, it has been measured that these pulse signals have time stabilities that approach the best of atomic clocks (something like 10^{-14}!).

Before presenting the principle of operation of clocks, we begin by distinguishing between frequency sources and time sources. A *frequency source* supplies a (pseudo-)-periodic signal whose meaningful information is frequency, usable as reference for frequency synchronization. A *time source*, on the other hand, supplies a signal that carries absolute or relative time information (e.g., the Big Ben clock in London). The most remarkable time sources are the UTC sources. Being time integral of frequency, it is evident that frequency sources do not deliver, at least intrinsically, either UTC or any other time scale.

Moreover, we distinguish between primary and secondary frequency standards. A *primary frequency standard* is a source of standard frequency, that is a frequency signal usable as reference for frequency synchronization or for other metrology purposes, which does not need to be steered by any other external reference. Examples of primary frequency standards are the caesium-beam and hydrogen-MASER atomic standards. On the other hand, a *secondary frequency standard* is a source of standard frequency that can be, and usually is, steered by some primary standard source. Examples of secondary frequency standards are the high-quality quartz-crystal oscillators and the rubidium-gas-cell atomic standard.

6.2 QUARTZ-CRYSTAL OSCILLATORS

Quartz-crystal oscillators are based on piezoelectricity, discovered by Curie in 1880 (Nobel prize, 1903), but they became practically feasible upon the invention of the first electronic amplifier: the triode, by de Forest in 1907. They have been used in communications systems since 1920 and they were largely employed during World War II.

In modern telecommunications, they have pervaded practically all kinds of systems. For example, cheap crystal oscillators supply timing to electronic boards as well as to wristwatches. Stand-alone high-quality crystal oscillators are used as secondary frequency standards (e.g., in SASE). Moreover, they are the output frequency source in atomic standards (e.g., in PRC and SASE), where they are steered by the atomic resonance.

For a closer look on the physics and technology of quartz-crystal oscillators and on their performance the reader is referred to references [6.2]–[6.7] as well as to the wide bibliography cited by the articles of [6.2].

6.2.1   Plain Crystal Oscillators

As stated above, quartz oscillators are based on the piezoelectric effect: a mechanical strain in the crystal yields an electrical field and *vice versa*. A *Crystal Oscillator* (XO) is thus an electronic oscillator, where a quartz crystal is excited by a periodic electrical signal at the resonance frequency (in the range from 10 kHz to 1 GHz, but most commonly from 5 to 10 MHz).

The resonance frequency is determined mainly by the properties of the bulk material, but it is also strongly dependent on environmental conditions such as the temperature, humidity, etc. (for example, temperature affects crystal dimensions). The resulting resonator has a quality factor Q quite high (10^3 to 10^6 or even more) and, used in a positive feedback circuit, allows to generate a timing signal featuring an excellent short-term stability (in particular over observation intervals smaller than one second).

More in detail, the resonator is made of a plate of quartz crystal. Owing to the piezoelectric effect, mechanical deformations propagate in the bulk coupled with a variable electrical field. When the waves fulfil given boundary conditions, the crystal enters resonance. The perfection of the quartz lattice and its low cost make it the crystal of choice to build oscillators.

The design of a high-quality quartz crystal resonator is aimed at minimizing the effect of environmental perturbations. For example, the orientation of the crystal cut is chosen to reduce the influence of temperature on the oscillation frequency. In an advanced design (BVA[1] quartz oscillator), the electrodes are not even placed in direct contact on the faces of the vibrating plate, but on auxiliary plates at a distance of a few micrometers from the active one. This is for the aim of suppressing the mass loading of the main plate by the electrodes, as well as any polluting atom migration from the metal to the quartz.

The crystal resonator is inserted into a positive feedback loop, to stir up and maintain oscillation: in steady state at the resonance frequency, the closed-loop phase shift is 0 or π and the oscillation amplitude is constant. Examples of various crystal oscillator circuits, based on transistor or integrated-circuit amplifiers to achieve high loop gain, are provided for example in [6.9].

6.2.2   Voltage-Controlled Crystal Oscillators

From a functional point of view, a VCO is an oscillator whose output frequency is controlled, within given limits, by an external driving voltage. This device, mostly realized as a *Voltage-Controlled Crystal Oscillator* (VCXO), is the key component in PLLs, as explained in Chapter 5.

The VCXO is a quartz-crystal oscillator, in which a variable capacity, somewhere along the positive-feedback loop, makes possible the adjustment of the output frequency. Usually, this capacitor is made of two parts: one fixed capacitor and a varactor. The latter is regulated for oscillation-frequency fine tuning by the external DC voltage source.

---

[1] *'Boîtier à Vieillissement Amélioré'*, which in French language means 'packaging for improved ageing performance'.

6.2.3   Temperature-Compensated Crystal Oscillators

The main problem of a plain XO is the dependence of its natural frequency on ageing (around 10^{-7}/day in plain models) and on the temperature (in the order of 10^{-7}/°C or above).

To overcome the latter problem, *Temperature-Compensated Crystal Oscillators* (TCXOs) implement an automatic control on the oscillation frequency based on the measurement of the crystal temperature. Such a trick allows to achieve a frequency stability of 10^{-7} over a temperature interval from 0°C to 50°C. More sophisticated models, by way of digital control, achieve a frequency stability even in the order of 10^{-8} in the wider temperature interval from 0°C to 70°C.

6.2.4   Oven-Controlled Crystal Oscillators

Far better than compensating temperature variations with a feedback control is to insulate the oscillator thermally and to make it work in a constant-temperature closed environment. Such clocks are called *Oven-Controlled Crystal Oscillators* (OCXOs).

In OCXOs, the resonator and the other temperature-sensitive elements are placed in a controlled oven whose temperature is set as closely as possible to a point where the resonator frequency does not depend on temperature, so to minimize the effect of residual temperature variations.

Frequency stability values exceeding 10^{-9}/day are thus achieved. The aforementioned BVA quartz oscillators are state-of-the-art OCXOs, invented at the University of Besançon (France) and manufactured first by Oscilloquartz SA (Neuchâtel, Switzerland). Owing to a double-oven temperature control and to the auxiliary-plates crystal excitation, they achieve a frequency stability even in the order of 10^{-11}/day, thus approaching the performance of some atomic secondary frequency standards (rubidium clocks).

6.2.5   Performance and Characteristics of Crystal Oscillators

Some typical performance data and characteristics of quartz oscillators available on the market are summarized in Tables 6.1 (XO and TCXO) and 6.2 (OCXO). For further data,

Table 6.1   Typical performance data and characteristics of quartz oscillators available on the market (XO and TCXO)

	XO	TCXO
Short-term stability $\sigma_y(\tau = 1$ s$)$		1×10^{-9}
Linear frequency drift D	$>1 \times 10^{-6}$/year	5×10^{-7}/year
Frequency accuracy (1 year)	2×10^{-6} to 1×10^{-5}	2×10^{-6}
Temperature sensitivity	$>1 \times 10^{-7}$/°C	5×10^{-8} to 5×10^{-7} (−55°C to 85°C)
Warm-up time		10 s (to 1×10^{-6})
Lifetime (performance guaranteed)	10 years to 20 years	>5 years

Table 6.2 Typical performance data and characteristics of quartz oscillators available on the market (OCXO)

	Miniature Single Oven OCXO	Double Oven OCXO	Double Oven BVA OCXO
Short-term stability $\sigma_y(\tau = 1 \text{ s})$		1×10^{-11} to 5×10^{-11}	1×10^{-13} to 5×10^{-13}
Linear frequency drift D	2×10^{-8}/year to 4×10^{-7}/year	1×10^{-8}/year to 1×10^{-7}/year	1×10^{-9}/year to 4×10^{-9}/year
Frequency accuracy (1 year)	2.5×10^{-8} to 9×10^{-7}	1.1×10^{-8} to 1.1×10^{-7}	1.7×10^{-9} to 4.8×10^{-9}
Temperature sensitivity	5×10^{-9} to 4×10^{-7} ($-30°C$ to $60°C$)	2×10^{-10} to 8×10^{-9} ($-30°C$ to $60°C$)	1×10^{-10} to 2×10^{-10} ($-30°C$ to $60°C$)
Warm-up time		30 min (to 1×10^{-9})	2 hours (to 1×10^{-10})
Lifetime (performance guaranteed)	10 years to 20 years	10 years to 20 years	10 years to 20 years

the reader is referred for example to References [6.5][6.6], as well as to manufacturer data sheets.

In Tables 6.1 and 6.2, clock performance is expressed in terms of the short-term frequency stability over one second (i.e., the Allan variance for $\tau = 1$ s), of the linear frequency drift D (expressed in parts $\Delta v/v_n$ per year), of the frequency accuracy expected over one year ($\Delta v_{max}/v_n$) and of the temperature sensitivity (fractional frequency variation per Celsius degree or over a temperature range). Moreover, the warm-up time is expressed as the time needed to achieve the frequency accuracy specified between brackets.

6.3 ATOMIC FREQUENCY STANDARDS

In contrast to quartz oscillators, atomic frequency standards are based on the intrinsic properties of atoms. The reference frequency is determined by fundamental constants such as the energy gap between two quantum levels of certain atoms. In our present state of knowledge, we are allowed to postulate that such atomic properties are the same everywhere in the universe, in space and in time, within known relativistic effects.

An atomic clock, therefore, uses as frequency reference the oscillation of an electromagnetic signal associated with a quantum transition between two energy levels in certain atoms. The quantic bundle of electromagnetic energy is called a *photon* and is equal to the difference in energy between these two levels in one atom. Practical atomic frequency standards are currently based on exploiting the properties of three elements: *hydrogen* (in the hydrogen-MASER frequency standard), *rubidium* (in the rubidium-gas-cell frequency standard) and *caesium* (in the caesium-beam frequency standard). The principles of operation of these three standards will be thus outlined in the following of this section.

For a closer look on the physics and technology of atomic frequency standards and on their performance, the reader is referred to references [6.2] — [6.5][6.8] as well as to the wide bibliography cited by the articles of [6.2].

6.3.1 Physical Principle of Operation

Quantum Physics asserts that an atom can be only at quantum, discrete energy levels. Let E_1 and E_2 be the energy of two quantum levels 1 and 2, with $E_2 > E_1$. Then, the atom can transit between these two levels by absorption or emission of a photon of electromagnetic radiation at frequency ν_0. The value of the frequency ν_0 is determined by the energy conservation principle, according to the Bohr's law

$$E_2 - E_1 = h\nu_0 \tag{6.1}$$

where $h \cong 6.625 \times 10^{-34}$ J $\cdot$ s is the Planck's constant.

The frequency ν_0 is thus determined by fundamental physical constants, with only a small uncertainty $\Delta\nu$ obeying the Heisenberg's relation

$$\Delta\nu\Delta t \geq 1 \tag{6.2}$$

where Δt is the duration of observation of the atomic transition. To keep the frequency uncertainty width $\Delta\nu$ small, Δt is aimed to be as long as possible. The trick in making atomic clocks is harnessing the frequency of these photons, observing the phenomenon for as long as possible while producing minimum perturbations on the natural atomic resonance.

In practical atomic frequency standards, the energy levels 1 and 2 are determined by the interaction between the magnetic moment of the unpaired electron of the atom in the ground state and the magnetic moment of the nucleus (*hyperfine interaction*). This interaction is weak and thus leads to a small energy gap $E_2 - E_1$, which yields the resonance frequency ν_0 to lie in the microwave range, making practically feasible the electronic system for realizing the atomic resonator.

Unfortunately, another consequence of being small the energy gap $E_2 - E_1$ is that the two levels are almost equally populated at thermal equilibrium. This equilibrium has to be broken in order to make the atomic transition observable, for example by measuring the absorption or emission of some detectable amount of energy. Two methods are used to achieve this goal.

The first method is based on the fact that the atom exhibits opposite magnetic moments whether it is in one or the other energy level. Hence, atoms of the two populations can be separated by deflection through an inhomogeneous magnetic field. This method is used in caesium-beam and hydrogen-MASER frequency standards.

The second method is based on optical pumping, in an analogous fashion to what is done in Erbium-Doped Fibre optical Amplifiers (EDFA). An optical radiation of appropriate wavelength is used to excite (pump) atoms for example from level E_2 to another level at higher energy E_3. Spontaneous decay from level E_3 down to levels E_1 and E_2 happens very fast and the net result is that the population of one of the two levels E_1 and E_2 is increased. This principle is applied in rubidium and optically pumped caesium-beam frequency standards.

Based on the concepts above, the principle of operation of atomic frequency standards is based on locking the resonance frequency ν_0 by maximizing the number of atomic transitions between the two quantum energy levels 1 and 2. A feedback system is used to control the frequency that is synthesized from a quartz VCXO and that is used to probe the atomic resonator.

In *passive atomic frequency standards* (e.g., the caesium-beam frequency standard), the microwave signal is applied to the atoms in order to excite transitions. Some feedback system aims at maximizing the number of atomic transitions, by adjusting the frequency synthesized from the VCXO accordingly. Resonance is thus achieved when the frequency of the exciting microwave signal is precisely v_0.

In *active atomic frequency standards*, a device (e.g., the hydrogen-MASER active cavity) generates a self-sustained oscillation. An electronic system then phase-locks a VCXO to the oscillation generated. Since the VCXO operates in a frequency range in the order of 1 to 100 MHz, the principle of heterodyne detection must be used to lock to the microwave frequency v_0.

6.3.2 Caesium-Beam Frequency Standard

The caesium-beam frequency standard is based on the transition of atoms of the isotope 133 of caesium (^{133}Cs) in the ground state between the hyperfine levels characterized by magnetic moment $F = 4$ (level 1, with energy E_1) and $F = 3$ (level 2, with energy E_2).

The scheme of principle of a basic caesium-beam clock is shown in Figure 6.1. An oven with a few grams of the isotope 133 of caesium (^{133}Cs) effuses a beam of caesium atoms, uniformly distributed among 16 quantum energy levels, into a vacuum chamber. Then, the non-homogeneous magnetic field in the magnet 1 (*polarizer*) deflects the level-2 atoms through the resonant cavity (Ramsey resonator), where they are irradiated by a microwave field. When the condition (6.1) is fulfilled, the microwave signal has precisely frequency $v_0 = 9.192631770$ GHz and stimulates atom transition to level 1. Then, the non-homogeneous magnetic field in the magnet 2 (*analyser*) deflects only the level-1 atoms to a hot-wire ionization detector. Here, the output current is proportional to the incoming ^{133}Cs atom flux and hence to the transition probability from level 2 to level 1.

A resonance peak is exhibited by the output current when the frequency of the probing microwave field is swept across the value v_0. The spectral line width around v_0 is typically in the order of 100 Hz, due to the flight time of atoms through the cavity and according to Equation (6.2).

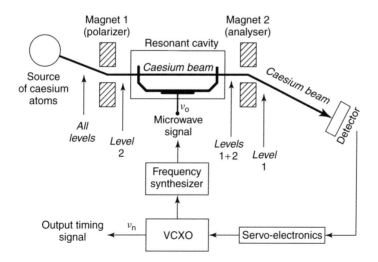

Figure 6.1. Scheme of principle of a caesium-beam frequency standard

The signal output by the detector, finally, is used to steer a quartz VCXO, from which the microwave radiation and the output timing signal are synthesized, aiming at maximizing the number of transitions. Therefore, the excellent short-term stability of the quartz oscillator is coupled with the long-term stability of the steering atomic resonator.

The value of ν_0 of unperturbed ^{133}Cs atoms has been used to define the *second* in the International System (SI), as 9 192 631 770 times the resonance oscillation period $1/\nu_0$[2]. References cited in the bibliography list some causes of frequency unaccuracy in practical caesium clocks, which limit to what extent the SI second can be practically realised with this frequency standard.

6.3.3   Hydrogen-MASER Frequency Standard

The principle of operation of a hydrogen-MASER (Microwave Amplification by Stimulated Emission of Radiation) frequency standard is based on the stimulated emission of electromagnetic radiation at the frequency ν_0, corresponding to the transition of hydrogen atoms between the states having magnetic moment $F = 1$ (level 2) and $F = 0$ (level 1).

As shown in the scheme of principle in Figure 6.2, a beam of hydrogen atoms is selected by the polarizer magnet, which produces a magnetic field of intensity 1 T. Atoms at the upper energy level E_2 ($F = 1$) are thus injected into a storage bulb, surrounded by a high-Q microwave resonant cavity tuned to the $\nu_0 = 1.42040575177$ GHz atomic resonance frequency. The resonant cavity is exposed to an internal homogeneous magnetic field of intensity 10^{-7} T, in order to shield the interaction region from the ambient magnetic field.

In the bulb, atoms bounce around and decay to the lower energy level E_1 ($F = 0$), by emitting radiation at the frequency ν_0. This emission is stimulated by the microwave field produced by the atoms themselves (active MASER) or by an external probing field (passive MASER). The coating of the inner surface is studied to have the highest number of atoms elastic collisions as possible (Teflon® coating is used). An average interaction

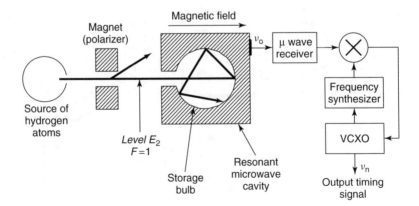

Figure 6.2.   Scheme of principle of an active hydrogen-MASER frequency standard

---

[2] In 1967, it was agreed by the 13th General Conference of Weights and Measures (Conférence Générale des Poids et Mesures) that: *'The second is the duration of 9 192 631 770 periods of the radiation corresponding to the transition between the two hyperfine levels of the ground state of the cesium-133 atom'.*

time of the atoms with the microwave field in the order of 1 s is thus achieved. Hydrogen atoms are continuously pumped away from the cavity, to maintain a constant pressure not higher than 10^{-5} Pa.

In the *active hydrogen MASER*, the cavity volume is big enough to make possible self-sustained oscillation. Therefore, the microwave radiation at frequency ν_0 emitted by the cavity is detected by an antenna and used to lock a quartz VCXO, which generates the output timing signal, by way of frequency multiplication and mixing (heterodyne detection). The radiation power detectable by the antenna outside the cavity is extremely small, in the order of something like -100 dBm to -110 dBm (i.e., 100 fW to 10 fW).

The short-term and medium-term stability of active hydrogen MASERs is the best today achievable in practical clocks. Their long-term stability, on the other hand, is severely affected by mechanical shocks and temperature variations, because the resonance frequency of the cavity depends on its physical dimensions. Therefore, very effective mechanical and thermal shields must surround the cavity. This is the main reason for the uncomfortable size and expensive price of active hydrogen MASER clocks.

In the *passive hydrogen MASER*, the cavity volume is smaller. A microwave-probing field must thus be applied to stimulate emission, which cannot self-sustain. The servo-mechanism to control the output frequency is similar to that used in caesium-beam clocks. A passive MASER is, in general, less sensitive to environmental variations than active MASERs, but it features a lower short-term stability.

6.3.4 Rubidium-Gas-Cell Frequency Standard

The rubidium-gas-cell frequency standard is based on the transition of atoms of the isotope 87 of rubidium (^{87}Rb) between the base levels characterized by magnetic moments $F = 1$ (level E_1) and $F = 2$ (level E_2). Atom state selection and transition detection are achieved by optical pumping.

The scheme of principle of a rubidium frequency standard is shown in Figure 6.3. Light from a lamp filled with ^{87}Rb is filtered through a cell (hyperfine filter) containing ^{85}Rb vapour, before it excites ^{87}Rb atoms in a cell (absorption cell) filled with buffer gas inside a resonant microwave cavity. The purpose of the buffer gas (a mixture of inert gases) is to

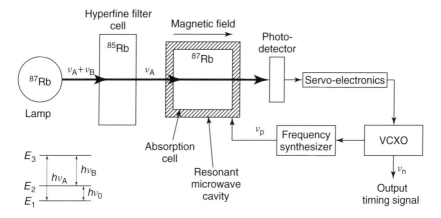

Figure 6.3. Scheme of principle of a rubidium-gas-cell frequency standard

increase the interaction time of the ^{87}Rb atoms with the microwave field in the resonator, by causing lots of elastic collisions of the ^{87}Rb atoms with the buffer atoms before an inelastic collision with the cell walls.

The hyperfine filter cell allows only the light spectrum component ν_A, which optically pumps atoms at the lower level E_1 ($F = 1$), to reach the absorption cell. Here, this level is thus depopulated as ^{87}Rb atoms absorb the light radiation, are pumped to the upper level E_3 and then decay down to both the base levels E_1 and E_2 (see the energy level diagram at the left bottom of Figure 6.3). If the level E_1 is depopulated, the cell then becomes transparent to radiation ν_A. Nevertheless, microwave probing radiation at frequency $\nu_p \cong 6.3846826128$ GHz is applied to the atoms, to excite their transition to level E_1 again. Hence, this level is populated back and optical absorption can start again.

Table 6.3 Typical performance data and characteristics of commonly available atomic frequency standards

	Rubidium standard	Caesium-beam standard	Hydrogen MASER standard
Short-term stability $\sigma_y(\tau)$	for $\tau = 1$ s: 5×10^{-12} to 5×10^{-11}	for $\tau = 100$ s: $<10^{-13}$ (prim. std.)[a] to 3×10^{-12}	for $\tau = 1$ s: 2×10^{-13} (active) 2×10^{-12} (passive)
Linear frequency drift D	5×10^{-11}/year to 5×10^{-10}/year	0	$<10^{-13}$/year to 5×10^{-12}/year
Frequency accuracy (1 year)	1×10^{-10} to 1×10^{-9} [b]	1×10^{-14} (prim. std.) to 7×10^{-12}	10^{-12}
Temperature sensitivity	1×10^{-12}/°C to 1×10^{-11}/°C	1×10^{-14}/°C to 1×10^{-13}/°C	1×10^{-14}/°C
Sensitivity to magnetic fields (per 10^{-4} T)	5×10^{-12} to 2×10^{-11}	$<1 \times 10^{-14}$ to 1×10^{-12}	3×10^{-14} (active) 1×10^{-14} (passive)
Resonance frequency ν_0	6.3846826128 GHz	9.192631770 GHz[c]	1.42040575177 GHz
Warm-up time	2 min to 30 min (to 5×10^{-10})	30 min (to 3×10^{-12})	24 hours (to 1×10^{-12})
Lifetime (performance guaranteed)	up to 10 years	3 years to 10 years	3 years
Basic wearout mechanism	Rb lamp and cavity life (15 years)	Cs beam tube (3 to 10 years)	Ion pumps and H_2 source depletion, cavity resonance shift (7 years)
Portability and intended location	space, air, ground	ground and laboratory oriented	Definitely ground and laboratory oriented
Typical volume	$<10^{-3}$ m^3 to 10^{-2} m^3	$<10^{-2}$ m^3	0.5 m^3 (active)
Typical weight	0.5 kg to 2 kg	10 kg to 30 kg	>50 kg
Power consumption	10 W to 50 W	25 W to 50 W	70 W to 100 W

[a]Caesium-beam clocks such as those operated in national primary standard laboratories, under constant environmental conditions.
[b]Being the rubidium a secondary frequency standard, these figures refer to a rubidium clock properly calibrated at the beginning.
[c]By standard definition of *second*.

In resonance, the frequency of the applied microwave radiation ν_p is exactly ν_0, because the signal at the output of the photo-detector exhibits a minimum by sweeping ν_p around ν_0. The VCXO is thus steered accordingly by the servo-electronics to keep light absorption at its maximum.

The centre frequency of the generated resonance line may deviate considerably (10^{-9} in fractional units) from the theoretical value ν_0 of unperturbed rubidium atoms, due to environmental causes of several kinds and to the physics itself of this device. For this reason, rubidium clocks are not suited as primary frequency standards, but need to be calibrated against caesium-beam or hydrogen-MASER clocks. For this reason, they are considered secondary frequency standards. However, their short-term stability is usually better than for cesium-beam standards. After laboratory calibration, rubidium clocks can *transport* frequency accuracy in the order of 10^{-11}.

6.3.5 Performance and Characteristics of Atomic Frequency Standards

Some typical performance data and characteristics of atomic frequency standards are summarized in Table 6.3. For further data, the reader is referred for example to References [6.2]–[6.8] as well as to manufacturer data sheets, from where most of the data reported below were taken.

In Table 6.3, as for quartz oscillators, clock performance is expressed in terms of the short-term frequency stability over τ seconds (i.e., the Allan variance), of the linear frequency drift D (expressed in parts $\Delta\nu/\nu_n$ per year), of the frequency accuracy expected over one year ($\Delta\nu_{max}/\nu_n$) and of the temperature sensitivity (fractional frequency variation per Celsius degree or over a temperature range). Moreover, the warm-up time is expressed as the time needed to achieve the frequency accuracy specified between brackets. Other typical characteristics are summarized as well.

6.4 THE GLOBAL POSITIONING SYSTEM

The Global Positioning System (GPS) is not a clock, but rather a complex system of clocks and satellites. It is treated in this section as it is indeed a sort of *super-clock* in the sky, available to everyone equipped with inexpensive receivers. GPS receivers are often used as additional reference of SASE clocks, especially in wide-area synchronization networks to face the issue of the timing transfer over long distances. Moreover, GPS receivers are commonly used to time-synchronize BTSs in mobile telephone cellular networks based on CDMA.

GPS is a satellite radio system providing continuously and in real time three-dimensional position, velocity and time information to suitably equipped land, sea and airborne users anywhere on Earth. Born essentially as a navigation and positioning tool, it is used also as pure time reference to disseminate precise time, time intervals and frequency.

The NAVSTAR (NAVigation Satellite Timing And Ranging) system, developed, funded and operated by the US Dept of Defense (DoD), is the first GPS system made available to civilian users. It has been designed, since 1973, to take the place of the older navigation system LORAN-C. The constellation of satellites was completed in 1994. On 17 July

1995, the US Air Force announced the GPS fully operational. The Russian system, called GLONASS, is very similar to NAVSTAR GPS. Both systems consist of three sets of facilities: the *space*, the *control* and the *user* segments.

This section provides a very general overview on GPS, addressing the NAVSTAR system in particular. To know more on GPS and its applications, starting points may be for example References [6.10][6.11]. Clear and thorough overviews, moreover, are available on the Internet, for example at [6.12]–[6.16]. Most information provided in this section has been derived from there.

Still on the World Wide Web, official pages on NAVSTAR are provided by the US Air Force [6.17] and by the US Coast Guard [6.18]. The official page on GLONASS is provided by the Russian Federation Ministry of Defense [6.19]. These official pages may not look very appealing for basic users, but they provide lots of specific technical information, such as signal specifications and nearly real-time data, satellite status reports, ephemeris data, almanacs, etc.

6.4.1   How Does GPS Work

GPS is based on a set of 24 orbiting satellites (space segment), each carrying two atomic clocks on board and continuously monitored by ground control stations (control segment). Each satellite broadcasts spread-spectrum signals bearing unique pseudo-random codes with very long periods.

Since at any time and at any place on Earth from five to eight satellites can be seen, *triangulation* principle is used to determine the receiver's position. In short, precise distance measurements (range measurements) to four satellites yield a set of four equations, which can be solved for four unknowns: latitude, longitude, altitude and GPS-system time.

Distance from the GPS receiver to each observable satellite is determined by measuring the time of flight of the signal received from that satellite. This measurement is carried out by determining the relative phase error between the received pseudo-random code and one generated locally (measurement from one extra satellite removes the ambiguity of the local time alignment). Since the signal speed is known almost exactly (i.e., the light speed in vacuum accounting for some additional delays crossing the neutral and ionized parts of the atmosphere), the receiver's computer can estimate accurately its distance from each satellite in view.

Triangulation of range measurements to determine the receiver's position is based on the assumption that satellite positions are known exactly. Actually, GPS-satellite orbits are determined with highest precision and GPS receivers have almanacs stored in their memory, with the positions of each satellite listed moment by moment (current almanac data are broadcasted by satellites at periodic times). Moreover, minor variations in satellite orbits are measured by the GPS control stations to compute correction data that are uploaded to satellites once or twice a day.

6.4.2   Space Segment

The Space Segment is made of a constellation of satellites, named Space Vehicles (SV), equipped with caesium and rubidium clocks and broadcasting spread-spectrum signals carrying pseudo-random codes. The NAVSTAR nominal constellation is made of 24 satellites, with a minimum of 21 operating for at least 98% of the time, but there are often more than 24 operational satellites in orbit, as new ones are launched to replace older satellites.

The 24 satellites are placed in near circular orbits with a period of about 12 hours. The satellite orbits repeat almost the same ground track once per day, as the Earth rotates beneath and after the satellites. There are six orbital planes, with nominally four satellites on each, equally spaced 60° apart and inclined at about 55° with respect to the equatorial plane. This constellation allows to see from five to eight GPS satellites from any point on Earth and at any time. The nominal data of the NAVSTAR GPS constellation are thus summarized as follows:

- 24 satellites, with a minimum of 21 operating for at least 98% of the time;

- six orbital planes, equally spaced 60° apart and inclined at about 55° with respect to the equatorial plane;

- nearly circular orbits, with nominal height 20 183 km above Earth and period 11h58′;

- about 5 h visibility window for each satellite.

There are three types of NAVSTAR GPS satellite launched (as of 2000) [6.15][6.17].

The *Block I* SVs, built by Rockwell International as development prototypes, were launched on the Atlas-Centaur between 1978 and 1985 from the Vanderberg Air Force Base in California. These SVs supported most of the system testing and are now non-operational. Out of a total of 11 Block I SVs launched, only one is still occasionally switched on. The main difference between these and later-generation satellites is that there was no ability to degrade purposely the positioning signal accuracy (selective availability, see Section 6.4.5).

The *Block IIA* SVs were also built by Rockwell International. The first of 28 Block IIA SVs was launched on 14 February 1989 from Cape Canaveral, Florida, and was set 'healthy' for global use on 15 April 1989. A Block IIA SV weighs about 930 kg (in orbit), is 5.1 m in size, is equipped with two caesium and two rubidium clocks, is designed for a lifetime of 7.5 years and is launched by a Delta rocket.

Significant Block IIA SV enhancements to the Signal-In-Space (SIS) interface have been:

- radiation hardened electronics to improve SIS reliability;

- capacity to store 180-days worth of Navigation Message data (see Section 6.4.5), compared to only 3.5-days worth of storage in the Block I SV;

- full Selective-Availability (SA) and Anti-Spoofing (AS) capabilities (see Section 6.4.5) to provide for SIS security;

- automatic detection of certain error conditions and switching to non-standard code transmission to protect users from tracking a faulty SV.

In 1989, 21 additional satellites (*Block IIR* SV) were procured from Lockheed Martin, formerly General Electric (*R* stays for 'replenishment'). The Block IIR SVs present the same SIS interface to the User Segment. In addition, they are capable of autonomously navigating themselves and generating their own Navigation Message data, maintaining full SIS accuracy for at least 180 days without Control Segment support. Additional differences between the Block IIR and IIA SVs are: satellite-to-satellite communication ability, accuracy improvement, additional radiation hardening, reprogrammable microprocessor, two atomic clocks on at all times (hot backup).

Block IIR satellites were launched starting from 1996, in order to maintain full constellation. A further follow-on satellite category has also been planned: the *Block IIF* SV.

We conclude with a few notes on the Russian system. The GLONASS constellation is composed of 24 satellites, on three orbital planes. The satellites operate in circular 19 100 km orbits with inclination angle of 64.8° and period of 11h15'. Each satellite transmits on two L frequency groups (the L1 group centred on 1609 MHz and the L2 group centred on 1251 MHz), on a unique pair of frequencies. Unlike NAVSTAR, both the GLONASS signals carry a precise code and a coarse/acquisition code, where the former is encrypted for military use and the latter is available for civilian use.

6.4.3   NAVSTAR Control Segment

The NAVSTAR Control Segment (also known as Operational Control Segment) is composed of all the GPS facilities based on the ground, monitoring and controlling the satellites. The Control Segment consists of five Monitor Stations (MS), a Master Control Station (MCS) and uplink Ground Antennas (GA) [6.17].

The *Monitor Stations* are located at Colorado Springs, Hawaii, Ascension Island, Diego Garcia and Kwajalein (Marshall Islands), as shown in the pictorial map in Figure 6.4. They track the signals from all GPS satellites in view, collecting ranging data from each satellite (approximately every 1.5 s). The raw data are then smoothed, using ionospheric and meteorological information, and sent in real time to the MCS for processing. MSs are unmanned installations operating under remote control of the MCS.

The *Master Control Station* is located at Schriever Air Force Base (formerly Falcon Air Force Base), in Colorado. It is the central processing facility for the control segment

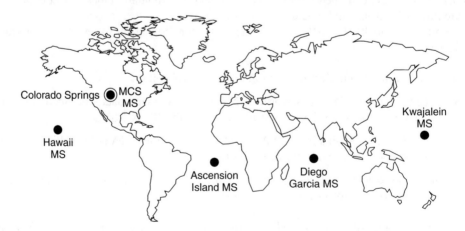

Figure 6.4.   Location of NAVSTAR GPS Monitor Stations

network and it is manned 24 hours per day, 7 days a week. It computes precise orbital data (ephemeris) and clock correction parameters for each satellite. It then uploads the ephemeris and the clock data, with other maintenance commands, to each satellite once or twice per day by the uplink Ground Antennas.

The *Ground Antennas* are unmanned installations operating under remote control of the MCS. Their function is to enable the MCS to command and control the on-orbit NAVSTAR satellites. The GAs are co-located with the MSs at Ascension Island, Diego Garcia and Kwajalein. There is no GA at the MCS or in Hawaii. The Pre-launch Compatibility Station (PCS), located at Cape Canaveral Air Force Base, is used primarily for checkout of the SVs prior to launch, but can also be used as backup GA in the event of a failure in one of the overseas GAs.

6.4.4   User Segment

The GPS User Segment consists of the GPS receivers and the user community. A GPS receiver can estimate in real-time its three-dimensional position, velocity and time information from the signals received from at least four satellites in view, as explained in the previous sections.

The receiver antenna is usually designed omni-directional with a gain of 3 dB: the signal coming from below the antenna ground plane is mostly ignored. The antenna is connected to the receiver front-end amplifier by a normal coaxial cable.

GPS receivers come in many different sizes, shapes and price ranges, according to their performance and the target application they are intended for. There are two main types of GPS receivers: those that track multiple GPS satellites simultaneously and those that sequence between satellites.

Sequencing receivers use a single channel to detect the C/A signal and move from one satellite to the next while gathering data. They have fewer circuits and thus are less expensive and consume less power. Unfortunately, sequencing interrupts signal measurement and thus can limit the overall accuracy.

Parallel receivers (also known as continuous receivers) can monitor several satellites simultaneously. They are valuable for mapping, surveying and scientific purposes. Parallel detection of satellite signals is useful not only because measurement interruptions are avoided, but also because comparing channels to each other allows to identify possible channel biases.

6.4.5   NAVSTAR GPS Signals

Each NAVSTAR GPS satellite broadcasts a pair of microwave (L-band) carrier signals. The reason for transmitting on two frequencies is being able to compensate errors introduced by ionospheric refraction. The two signals are generated from a standard frequency of $f_0 = 10.23$ MHz and are the following:

- L1, at frequency $154 f_0 = 1575.42$ MHz ($\lambda \cong 19$ cm);
- L2, at frequency $120 f_0 = 1227.60$ MHz ($\lambda \cong 24$ cm).

These two sine-wave carriers convey, by Binary Phase Shift Keying (BPSK) modulation, the following three binary codes.

- The *C/A (Coarse/Acquisition) code*, which modulates the L1 carrier phase and is a 1-MHz pseudo-random noise code, with repetition period equal to 1023 bits (code length). This period length corresponds to about 1 ms in time or to 300 km of ambiguity in the range measurement. Obviously, each satellite transmits a unique C/A code to be identifiable. This C/A code is the basis of the Standard Positioning Service (SPS) for civil public use (see Section 6.4.6).

- The *P (Precise) code*, which modulates the phase of both the L1 and L2 carriers and is a *very* long (in the order of 10^{14} bits) 10.23-MHz pseudo-random noise code. This code has a period of about 267 days. Each satellite uses only one-week segment of the P code. The segment used is unique to each satellite and is reset every week. In the so-called *Anti-Spoofing* (AS) mode of operation, the P code is encrypted. This P code is the basis of the Precise Positioning Service (PPS), the use of which is restricted to authorized users equipped with a classified AS Module (see Section 6.4.6).

- The *Navigation Message* is a 50-Hz data signal carrying satellite orbits data (ephemeris), clock corrections, almanacs and other system parameters. It is modulo-2 added to both the C/A and P codes.

The C/A code and the P code modulate the L1 signal in quadrature, so that they do not interfere with each other. Moreover, the C/A code is amplified to be 3 or 6 dB stronger than the P code.

Selective Availability (SA) is the intentional degradation of the SPS signals, controlled by the US DoD, to limit accuracy for users not belonging to US military and government agencies. The SA is a changing bias, different for each satellite, with low-frequency components (few hours of period). The potential positioning accuracy of the C/A code equal to about 30 m is then reduced to 100 m (two standard deviations of error). The United States decided to 'discontinue the use of SA by 2006 with an annual assessment of its continued use beginning this year' (Statement by the US President issued on May 1, 2000 [6.20]).

6.4.6   NAVSTAR Standard and Precise Positioning Services

The *Standard Positioning Service* (SPS) is the positioning and timing service provided by the C/A code transmitted on the L1 frequency. It is therefore the service provided to a user with a basic C/A code receiver, not able to decrypt the P code.

The SPS is available to civil users worldwide without charge or restrictions. The SPS predictable accuracy is not lower than:

- 100 m in horizontal position (for 95% of the time);
- 156 m in vertical position (for 95% of the time);
- 300 m in horizontal position (for 99.99% of the time);
- 500 m in vertical position (for 99.99% of the time);
- 340 ns in UTC timing (see Section 6.4.8) (for 95% of the time).

We remark again that the US DoD implemented Selective Availability (SA) to intentionally degrade the horizontal positioning accuracy from 30 m to 100 m and Anti-Spoofing (AS) to deny the P code to civil users.

The GPS accuracy figures above are from the 1999 Federal Radionavigation Plan. The 95%-accuracy figures express the value of *two standard deviations (2σ) of error* from the actual antenna position to an ensemble of position estimates made under specified (conservative) satellite visibility conditions. In particular, the 95% horizontal accuracy is the equivalent of twice the radial error standard deviation, while the 95% vertical and timing accuracy is the value of twice the standard deviation of vertical or time error. Receiver manufacturers may adopt other accuracy measures [6.12]:

- the *Root-Mean-Square (RMS) error*, i.e. the value of one standard deviation (σ) of the error, corresponding to 68% confidence;
- the *Circular Error Probable (CEP)*, i.e. the value of the radius of a circle centred on the actual position and that contains 50% of the two-dimensions position estimates;
- the *Spherical Error Probable (SEP)*, i.e. the spherical equivalent of CEP.

It is worthwhile noticing that some receiver vendors specify accuracy in RMS or CEP or even without SA, thus making their receivers appear more accurate than those from competitors that use more conservative error measures.

The *Precise Positioning Service* (PPS) is the positioning and timing service available to authorized users provided with cryptographic keys and equipment. US and Allied military, certain US Government Agencies and selected civil users specifically approved by the US Government can use PPS. The PPS consists of the SPS signal plus the P code transmitted on both L1 and L2 frequencies. Its predictable accuracy figures are 22 m in horizontal position, 27.7 m in vertical position and 200 ns in UTC timing (see Section 6.4.8) for 95% of the time.

Most scientific geodetic GPS receivers are capable of advanced measures on both frequencies L1 and L2 (e.g., cross-correlation to estimate the difference between their flight-times), which allow to attain PPS precision even without being able to decode the P code.

6.4.7 Differential GPS

Differential GPS (DGPS) techniques are based on correcting bias errors at one location with bias errors that are measured at a near, known position. The most precise technique consists of correcting individual range measurements before computing the navigation solution. Applying a simple position correction from the reference receiver to the remote receiver is less effective [6.12].

DGPS techniques are able to effectively remove the SA and other bias error, provided that differential corrections are computed at the reference station and applied to the remote receiver at an update rate that is less than the correlation time of SA. Suggested DGPS update rates are less than 20 s.

If the reference and the remote receivers are relatively close (say, closer than 100 km apart), DGPS accuracy in the order of 1 to 10 m is attainable based only on SPS signals. Differential carrier GPS, based on comparing the carrier phases detected at two locations within about 30 km to each other, allows to attain relative positioning accuracy even in the order of less than 10^{-2} m.

6.4.8   Use of GPS for Time and Frequency Dissemination

GPS can be used also as pure time reference to disseminate precise time (GPS system time), time intervals and frequency with excellent accuracy. It is available worldwide, with no charge, to any user equipped with relatively inexpensive equipment. For this reason, GPS receivers are often used as additional reference at the input of SASE clocks in synchronization networks, or to time-synchronize BTSs in mobile telephone cellular networks based on CDMA. This section provides a few notes, summarized from paper [6.14], on how GPS can be used as timing source.

GPS system time is closely related to UTC. GPS time is measured from the zero point at midnight, 5 January, 1980. Controlled by UTC, GPS time is not corrected with leap seconds and thus was ahead of UTC by six seconds in 1990. With this exception, GPS time is steered to within 1 μs of UTC with the difference reported in the GPS Navigation Message with a precision of 90 ns.

Making a GPS receiver operate as a time source is simple. It is enough that the receiver tracks one satellite on one or both the L-band carriers. SV time (i.e., the time as given by the individual satellite's clock) can be recovered by noting the time alignment of the pseudo-random code necessary to lock on the spread-spectrum carrier. Time intervals at the receiver can be reproduced by the 1-ms repetition rate of the C/A pseudo-random code. Time tagging of each millisecond in the SV time is accomplished by reading the Week-Number and Time-Of-Week fields in the Navigation Message. This also solves the ambiguity of the code period in recovering the SV time by noting the time alignment of the pseudo-random code.

Obviously, the receiver time scale so obtained must be corrected by subtracting some estimate of the total path transmission delay from the satellite to the receiver. The path delay is computed based on the range measurement, divided by the speed of light, and on several additional delay estimates (ionospheric delay, hardware delays, etc.). Hence, the relationship between the receiver clock and the GPS time can be established.

Actually, to set precise time, the three-dimensional position of the receiver must be known from an independent source to exploit the full potential of GPS: a position error of 1 m yields about 3 ns of time error and about 1 mHz of frequency error, due to the variation in the signal propagation delay. Moreover, SV time at the receiver must be corrected for errors of the SV clock with respect to the GPS time and for periodic relativistic effects. The SV clock error is transmitted in each data frame as a set of polynomial coefficients, based on uploaded MCS control data, while the relativistic correction is computed from the SV orbital parameters normally used for SV position determination.

Now, time or time intervals can be produced by the receiver. Receivers at different locations, therefore, can be time-synchronized to GPS time. If synchronization to UTC is required, the UTC correction in the Navigation Message can be applied.

The accuracy of GPS time signals is related to the ability of the receiver to accurately track the received C/A code. Accuracies in the order of 100 ns are possible with undegraded GPS signals (i.e., with no SA) and correct receiver position.

Beyond the use for time synchronization, a GPS receiver can be used also as pure frequency source. Frequency synchronization can be established by steering a local oscillator using integrated code-phase measurements or by directly measuring the SV carrier frequency. In the latter case, carrier frequency tracking is mostly accomplished by phase-locking to the de-spread carrier. The transmitted carrier signals are controlled by the on-board atomic clocks and thus carry a very stable and accurate frequency. Obviously, the

Doppler shift that results from SV orbital moving must be properly corrected. Frequency accuracy in the order of 10^{-12} is possible with undegraded signals (i.e., with no SA).

The *common-mode time transfer* is a technique useful for improving the accuracy of GPS time synchronization. It is analogous to DGPS. If two receivers track a satellite at the same time, we can assume that the path-delay corrections, the SV-time error and the data-message error are the same for both received signals. Therefore, by comparing one receiver to a reference clock such as a national PRC, the error in the SV time measured at that reference location can be transmitted to the other location and then used to adjust the clock there. Tests over paths of thousands of kilometres showed a time transfer accuracy within 25 ns.

We conclude this section with a few remarks on the impact of SA. The figures reported above (time accuracy in the order of 100 ns and frequency accuracy in the order of 10^{-12}) were experienced in the 1980s with Block I satellites, with no SA capabilities. Today, the SPS provided by Block II satellites, implementing SA, allows predictable time accuracy in the order of 340 ns (for 95% of the time). Frequency control accuracy is degraded to 10^{-10}. Fortunately, the careful use of common-mode techniques can substantially reduce SA impact on time transfer, when needed.

6.5 SUMMARY

From a theoretical point of view, the operation principle of clocks consists of a generator of oscillations and in an automatic counter of such oscillations.

A frequency source supplies a (pseudo)periodic signal whose meaningful information is frequency, usable as reference for frequency synchronization. A time source, on the other hand, supplies a signal that carries absolute (e.g., UTC) or relative time information.

A primary frequency standard is a source of standard frequency, that is a signal usable as reference for frequency synchronization, which does not need to be steered by any other external reference. On the other hand, a secondary frequency standard is a source of standard frequency that can be, and usually is, steered by some primary standard source.

Quartz oscillators are based on the piezoelectric effect: a mechanical strain in the crystal yields an electrical field and *vice versa*. An XO is an electronic oscillator, where a quartz crystal is excited by a periodic electrical signal at the resonance frequency. The crystal resonator is inserted into a positive feedback loop, to stir up and maintain oscillation. The VCXO is a quartz-crystal oscillator, where a variable capacity makes possible the fine adjustment of the output frequency. TCXOs implement an automatic control on the oscillation frequency based on the measurement of the crystal temperature. In OCXOs, the resonator and the other temperature-sensitive elements are placed in a controlled oven, whose temperature is set as closely as possible to a point where the resonator frequency does not depend on temperature.

An atomic clock uses as frequency reference the oscillation of an electromagnetic signal associated with a quantum transition between two energy levels in an atom. In practical atomic clocks, the energy levels are determined by the hyperfine interaction and are characterized by different magnetic moments. Atomic frequency standards are based on locking the resonance frequency by maximizing the number of atomic transitions between the two quantum energy levels. A feedback system is used to control the frequency used to probe the atomic resonator.

The caesium-beam frequency standard is based on the transition of ^{133}Cs atoms between two hyperfine levels. Hydrogen-MASER frequency standards are based on the stimulated emission of electromagnetic radiation at the frequency that corresponds to the transition of H atoms between two hyperfine levels. The rubidium-gas-cell frequency standard is based on the transition of ^{87}Rb atoms between two hyperfine levels and on optical pumping to achieve atom state selection and transition detection.

GPS is a satellite system providing continuously and in real-time a three-dimensional position, velocity and time information to suitably equipped users anywhere on Earth. The NAVSTAR system, operated by the US DoD, is the first GPS system made available to civilian users. The Russian GLONASS system is very similar to NAVSTAR. Both systems consist of three sets of facilities: the space, the control and the user segments.

GPS is based on a set of 24 orbiting satellites (space segment), each carrying two atomic clocks on board and continuously monitored by ground control stations (control segment) that compute precise orbital data and clock correction parameters for each satellite. Triangulation principle is used to determine the receiver's position, based on range measurements to at least four satellites.

GPS can be used also as pure time reference to disseminate precise time, time intervals and frequency with excellent accuracy. It is available worldwide, with no charge, to any user equipped with relatively inexpensive equipment. With undegraded GPS signals, time accuracy in the order of 100 ns and frequency accuracy in the order of 10^{-12} are possible. With SA, the SPS allows predictable time accuracy in the order of 340 ns (for 95% of the time) and frequency control accuracy in the order of 10^{-10}. Careful use of common-mode techniques can substantially reduce SA impact on time transfer, when needed.

6.6   REFERENCES

[6.1] D. W. Allan, N. Ashby, C. C. Hodge. *The Science of Timekeeping*. Application Note 1289, Hewlett-Packard Company, June 1997.

[6.2] Special issue on time and frequency. *Proceedings of the IEEE*, vol. 79, no. 7, July 1991.

[6.3] E. A. Gerber and A. Ballato (eds). *Precision Frequency Control*. New York: Academic Press, 1985.

[6.4] *Precision Time and Frequency Handbook*. Irvine, CA, USA: Ball Corp., Efratom Time and Frequency Products, 1993.

[6.5] *Handbook Selection and Use of Precise Frequency and Time Systems*. Geneva: ITU-R, 1997.

[6.6] J. Vig. *Quartz Crystal Resonators and Oscillators — A Tutorial*. Tech. Report SLCET-TR-88-1, US Army Electronics Technology and Devices Lab., Fort Monmouth NJ, July 1992.

[6.7] J. R. Norton, J. M. Cloeren, P. G. Sulzer. Brief history of the development of ultra-precise oscillators for ground and space applications. *Proceedings of the 1996 IEEE International Frequency Control Symposium*, pp. 47–57.

[6.8] Jacques Vanier, Claude Audoin. *The Quantum Physics of Atomic Frequency Standards*, vol. 1. Bristol: Adam Hilger, 1989.

[6.9] P. Horowitz, W. Hill. The art of electronics, Secs 4.11 to 4.17. *Oscillators*. Cambridge: Cambridge University Press, 1980.

[6.10] I. A. Getting. The global positioning system. *IEEE Spectrum*, December 1993, pp. 36–47.

[6.11] E. Kaplan. *Understanding GPS: Principles and Applications*. London: Artech House, 1996.

[6.12] P. H. Dana. Department of Geography, The University of Colorado at Boulder. *Global Positioning System Overview*.
http://www.colorado.edu/geography/gcraft/notes/gps/gps_f.html.

[6.13] J. T. Beadles. *Introduction to GPS Applications*. http://ares.redsword.com/gps/.

[6.14] P. H. Dana, B. Penrod. *The Role of GPS in Precise Time and Frequency Dissemination* (reprinted from GPS World, July/August 1990). http://www.bancomm.com/cgprole.htm.

[6.15] Quality Engineering and Survey Technology (QUEST) Ltd, used by Mercator, Inc. (a.k.a. Mercator GPS Systems) with the permission of QUEST. *The GPS Tutor*. http://www.mercat.com/QUEST/gpstutor.htm.

[6.16] Trimble Navigation Limited. *All About GPS*. http://www.trimble.com/gps/index.htm.

[6.17] U.S. Air Force, Los Angeles Air Force Base. *NAVSTAR Global Positioning System Joint Program Office*. http://gps.losangeles.af.mil/index.html.

[6.18] U.S. Coast Guard. *USCG Navigation Center GPS PAGE*.
http://www.navcen.uscg.mil/gps/.

[6.19] Russian Federation Ministry of Defense, Coordination Scientific Information Center. *Global Navigation Satellite System, GLONASS*. http://mx.iki.rssi.ru/SFCSIC/english.html.

[6.20] White House Press Release. *Statement by the President Regarding the United States' Decision to Stop Degrading Global Positioning System Accuracy*. 1 May, 2000.

7

TIME AND FREQUENCY MEASUREMENT TECHNIQUES IN TELECOMMUNICATIONS

Several experimental set-ups have been contrived in the last few decades to measure the accuracy and stability of oscillators in both time and frequency domains [7.1]–[7.10]. In the past, measurement techniques in the frequency domain have been the most common, being based on very low-noise analogue electronics. Nowadays, the availability of high-resolution time counters has made digital time-domain measurements to be more widely used. In particular, this became the standard choice in telecommunications.

The goal of this chapter is to describe experimental methods suitable for time and frequency measurements in telecommunications. To begin, several basic concepts are introduced: for example, how to estimate frequency offset and drift of a clock, to assess the statistical confidence of Allan variance estimates and to separate the noise of the Clock Under Test (CUT) from that of the Reference Clock (RC). Moreover, the impact of the measurement configuration and of the TE sampling period on the measured quantities is discussed. A brief survey on some of the most common devices and instruments used for time and frequency measurements is also provided.

Then, the main techniques for measuring time and frequency stability are overviewed, from direct digital measurement to heterodyne and homodyne methods for sensitivity enhancement. Several test set-ups are described.

Finally, clock stability measurement in telecommunications is examined. Its distinctive features are discussed. The standard technique, based on the acquisition of sequences of TE samples, is detailed. Moreover, MTIE measurement is addressed in particular, due to its peculiar issues. Other measurements on equipment clocks and on network interfaces, such as jitter and phase transients measurements, are described.

7.1 BASIC CONCEPTS

In this section, the basic concepts of experimental measurement of time and frequency accuracy and stability are introduced.

Beginning from the simple direct measurement of the timing-signal power spectral density, the quantities recommended by IEEE and ITU-T for characterizing time and frequency fluctuations are summarized. Then, the hierarchy of time and frequency measurement techniques is outlined. Guidelines for estimating the frequency offset and drift of a clock, for assessing the statistical confidence of Allan variance estimates and for separating the contribution of the Clock Under Test (CUT) from that of the Reference Clock (RC) in measurement results are provided.

Finally, two key issues are the impact of the measurement configuration (independent versus synchronized clocks) and of the TE sampling period on the trends of the stability quantities. They have to be definitely considered when performing stability measurement, in order to draw meaningful conclusions from the measurement results. Therefore, these two issues are addressed in the last subsections.

For further details on all these topics, the reader is addressed to the Bibliography cited in the various sections. In particular, references [7.2]–[7.4] can be recommended, because of both the wide set of contents and the tutorial approach.

7.1.1 Measuring the Power Spectral Density of the Timing Signal (RF Power Spectrum)

Recalling that stated in Section 5.7.1, the most straightforward and intuitive way to characterize the accuracy and stability of the timing signal $s(t)$ [V] supplied by a clock is to evaluate directly its two-sided PSD $S_s^{\mathrm{RF}}(f)$ (Radio-Frequency power spectrum), or more precisely its base-band equivalent[1].

The RF spectrum $S_s^{\mathrm{RF}}(f)$ is a continuous function, proportional to the timing-signal power delivered by the oscillator to a matched load in a bandwidth unit centred at the Fourier frequency f. It is measured in [W/Hz] or [V^2/Hz] units.

The spectrum $S_s^{\mathrm{RF}}(f)$ can be measured directly on the chronosignal $s(t)$ by means of an *RF-spectrum analyser*, based on a narrow band-pass filter followed by a bolometer, as shown in Figure 7.1 [7.4]. The band-pass filter has transfer function $H(f - f_0)$ non-null only in a small interval around the Fourier frequency f_0 of interest. To measure $S_s^{\mathrm{RF}}(f)$ across some range of Fourier frequencies, f_0 can be swept accordingly.

The variance $\sigma_{\hat{s}(t)}^2$ of the band-pass filtered signal is given by the Parseval's theorem as

$$\sigma_{\hat{s}(t)}^2(f_0) = \int_{-\infty}^{\infty} |H(f - f_0)|^2 S_s^{\mathrm{RF}}(f) df \qquad (7.1)$$

Therefore, if the filter pass-bandwidth is narrow enough (i.e., in the pass-band $S_s^{\mathrm{RF}}(f)$ can be approximated as constant), then it is possible to invert Equation (7.1), to get

$$S_s^{\mathrm{RF}}(f_0) = \frac{\sigma_{\hat{s}(t)}^2(f_0)}{\displaystyle\int_{-\infty}^{\infty} |H(f - f_0)|^2 df} \qquad (7.2)$$

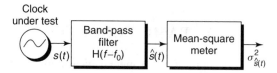

Figure 7.1. RF-spectrum analyser

---

[1] We remark once again that f indicates the Fourier frequency: positive values of f denote frequencies above the central frequency ν_{n}, negative values denote frequencies below ν_{n}.

In Section 5.7.1, it has been already pointed out that the RF-spectrum is definitely not a good tool to characterize clock frequency stability: given $S_s^{\text{RF}}(f)$, it is not possible to determine unambiguously whether the power at various Fourier frequencies is the result of amplitude rather than phase fluctuations in the timing signal $s(t)$.

7.1.2 Quantities Recommended by IEEE for Frequency Stability Measurement

By the mid-1960s, the IEEE convened a technical committee to recommend a set of standard measures for clock frequency stability, to cope with the little uniformity among manufacturers, metrologists and application engineers that was making it very difficult to compare different characterizations.

The recommendations of the IEEE committee (1971) are reported in [7.11][7.12]. To summarize, the following standard basic quantities were defined:

- the random fractional frequency deviation $y(t)$, defined as in Equation (5.11);
- the random time deviation $x(t)$, defined as in Equation (5.11);
- the instantaneous frequency $v(t)$, defined as in Equation (5.3).

Moreover, the IEEE committee recommended the following two standard measures of frequency stability:

- the one-sided PSD of $y(t)$, denoted as $S_y(f)$ [Hz^{-1}] (cf. Section 5.7.2);
- the two-sample Allan variance $\sigma_y^2(\tau)$ of the random fractional frequency deviation $y(t)$ (cf. Section 5.8.4).

7.1.3 Quantities Defined in International Standards and Their Estimators

By the mid-1990s, the international standard bodies ETSI and ITU-T faced the issue of recommending a uniform set of stability quantities, suitable for the specification of timing interface requirements in digital telecommunications networks.

Among the several quantities defined in the literature for characterizing time and frequency stability, the following five in the time domain have been considered by ETSI and ITU-T [7.13][7.14] for the specification of timing interface requirements:

- the Allan Deviation (ADEV) $\sigma_y(\tau)$,
- the Modified Allan Deviation (MADEV) mod $\sigma_y(\tau)$,
- the Time Deviation (TDEV) $\sigma_x(\tau)$,
- the root mean square of Time Interval Error (TIErms),
- the Maximum Time Interval Error (MTIE).

All five have been widely described and discussed in Chapter 5.

Moreover, ETSI and ITU-T defined five standard estimators of the quantities above, to allow uniform stability characterization and unambiguous comparison of measurement results. Recalling from Section 5.8.9, stability measurement begins with the acquisition of a sequence of TE samples, defined as

$$x_k = x\,(t_0 + (k-1)\tau_0) \quad \text{for } k = 1, 2, \ldots, N \tag{7.3}$$

The sequence is measured with sampling period τ_0 over a measurement interval $T = (N-1)\tau_0$ and starting at an initial observation time t_0. Then, the following standard estimators are recommended to compute the five stability quantities, as already reported in Section 5.8.9:

$$\mathrm{ADEV}(\tau) = \sqrt{\frac{1}{2n^2\tau_0^2(N-2n)} \sum_{i=1}^{N-2n} (x_{i+2n} - 2x_{i+n} + x_i)^2}$$

$$\text{for } n = 1, 2, \ldots, \left\lfloor \frac{N-1}{2} \right\rfloor \tag{7.4}$$

$$\mathrm{MADEV}(\tau) = \sqrt{\frac{1}{2n^4\tau_0^2(N-3n+1)} \sum_{j=1}^{N-3n+1} \left[\sum_{i=j}^{n+j-1} (x_{i+2n} - 2x_{i+n} + x_i)^2 \right]^2}$$

$$\text{for } n = 1, 2, \ldots, \left\lfloor \frac{N}{3} \right\rfloor \tag{7.5}$$

$$\mathrm{TDEV}(\tau) = \sqrt{\frac{1}{6n^2(N-3n+1)} \sum_{j=1}^{N-3n+1} \left[\sum_{i=j}^{n+j-1} (x_{i+2n} - 2x_{i+n} + x_i) \right]^2}$$

$$\text{for } n = 1, 2, \ldots, \left\lfloor \frac{N}{3} \right\rfloor \tag{7.6}$$

$$\mathrm{TIE}_{\mathrm{rms}}(\tau) = \sqrt{\frac{1}{N-n} \sum_{i=1}^{N-n} (x_{i+n} - x_i)^2} \quad \text{for } n = 1, 2, \ldots, N-1 \tag{7.7}$$

$$\mathrm{MTIE}(\tau) = \max_{1 \le k \le N-n} \left[\max_{k \le i \le k+n} x_i - \min_{k \le i \le k+n} x_i \right] \quad \text{for } n = 1, 2, \ldots, N-1 \tag{7.8}$$

where $\tau = n\tau_0$ is the observation interval and $\lfloor z \rfloor$ denotes 'the greatest integer not exceeding z'.

7.1.4 Hierarchy of Time and Frequency Measurement Techniques

It has been suggested that the techniques for time and frequency measurement can be organized in a hierarchy [7.4][7.5][7.15], with the measurement of the total phase $\Phi(t)$ at its summit:

(1) measure of the total phase $\Phi(t)$ or of the total Time $T(t)$;

(2) measure of time intervals (fluctuations) $\mathrm{TIE}(\tau)$, $\mathrm{TE}(t)$, $x(t)$;

(3) measure of the instantaneous frequency $v(t)$, $y(t)$;

(4) measure of frequency variations (fluctuations) $\delta y(t)$.

For a system capable of measuring for instance time intervals, frequency and frequency fluctuations can be easily deduced from the results, but absolute time cannot. More precisely, having measured some quantity at a given level of the hierarchy, lower level quantities can be obtained by *differentiating* one or more times the samples measured (or making simple differences between couples of them). For example, a sequence of instantaneous frequency samples $\{y_k\}$ can be obtained from a sequence of samples of random time deviation $\{x_k\}$ by applying the first-difference operator defined in Equation (5.83).

On the other hand, higher level quantities can be obtained by *integrating* one or more times the samples measured. Obviously, there is the ambiguity of the value of the integration constant (for example, the initial time deviation x_0 when integrating $\{y_k\}$ to obtain $\{x_k\}$).

Hence, gaps in sequences of measurement samples produce less deleterious effects on lower level than on higher level quantities. When computing the derivative, data gaps yield troubles only for times equal or shorter than the gap length. On the contrary, when integrating data, gaps must be filled with some data interpolation. From that point on, the error made propagates forever.

For this reason, it should always be preferred to measure quantities at the highest possible level of the hierarchy.

7.1.5   Estimating of the Frequency Offset and Linear Drift

Frequency offset and drift can be revealed only in the independent-clock configuration, where the CUT output is compared with the output of an independent RC, and not in the synchronized-clock configuration, where the time error between the input and the output of a slave CUT is measured.

Sometimes, it may be of interest to separate the deterministic from the random part in the TE data measured, i.e. the frequency offset and drift contribution from the pure $x(t)$ process. For example, one may be interested in evaluating the stability quantities ADEV, MADEV and TDEV according to the rigorous, theoretical definition, which has been provided in terms of the $x(t)$ process and not of TE(t), as it is customary in telecommunications (cf. Section 5.8.9). Therefore, in this subsection, we will not consider synonymous $x(t)$ and TE(t), as we did in several other cases when computing standard stability quantities.

In most practical cases, it is sufficient to take into account only the first two deterministic terms in the TE data, corresponding to the frequency offset and linear drift. Hence, the Time Error process is modelled as

$$\begin{aligned}
\text{TE}(t) &= x_0 + \frac{\Delta v}{v_n}t + \frac{1}{2}Dt^2 + \frac{\varphi(t)}{2\pi v_n} \\
&= x_0 + \frac{\Delta v}{v_n}t + \frac{1}{2}Dt^2 + x(t)
\end{aligned} \tag{7.9}$$

The first term x_0 is called *synchronization error* and the second, due to a constant frequency offset Δv, *syntonization error*.

Before proceeding further, it is opportune to remark at this point that the various Allan variances are not influenced in any way by synchronization and syntonization errors x_0 and $\Delta \nu$, because they are based on the second difference of TE data. A linear frequency drift, on the contrary, causes a quadratic ramp in TE(t) (third term in Equation (7.9)) and a trend of AVAR and MAVAR proportional to τ^2 (cf. Section 5.10.2).

The variance of the Allan variances evaluated in the presence of a significant linear frequency drift would be very small where the τ^2 trend dominates, further demonstrating that deterministic behaviour has been improperly described in statistical terms. Unfortunately, it is difficult to estimate the clock deterministic behaviour without introducing a bias in the noise at Fourier frequencies comparable to the inverse of the measurement interval. What's more, the optimum estimator depends on the type of dominating noise affecting the TE data.

Nevertheless, Stein [7.4] suggests the following simple practical procedure for computing the Allan variances, in presence of a significant linear frequency drift D that obscures the noise we are interested to characterize:

(1)   start from a sequence {TE$_k$} of Time Error samples, measured with sampling period τ_0;

(2)   evaluate the second difference {z_k} of the sequence {TE$_k$} (first difference of frequency samples {y_k}) on a period $\tau = n\tau_0$, obtaining then

$$z_k = D + \frac{x_{k+2n} - 2x_{k+n} + x_k}{n^2 \tau_0^2} \tag{7.10}$$

(3)   use the simple average of {z_k} to estimate the mean D;

(4)   subtract the estimated mean D from {z_k};

(5)   integrate twice, to obtain the TE data with drift removed;

(6)   further analysis, such as the evaluation of the Allan variance, may then proceed.

A similar procedure can be envisaged to estimate the frequency offset, by averaging the first difference of the TE sequence.

We notice that the average evaluated at step (3) is the optimum estimator of D only if the random process {z_k} is white, i.e. if {TE$_k$} has spectrum proportional to $1/f^4$. In other words, the period τ chosen to compute the second difference should be long enough so that the predominant TE noise process is random-walk frequency modulation.

Fortunately, a maximum-likelihood estimate of the parameters for some typical cases has shown that the mean second difference of phase $\langle\{z_k\}\rangle$, as done above, is still a good estimator of linear frequency drift D also for other noise types, in the sense that it introduces negligible bias in the Allan variance as evaluated at step (6). Therefore, the procedure suggested above can be recommended also when it is not the theoretical optimum.

7.1.6   Confidence of the Allan Variance Estimate

The Allan variance σ_y^2 is defined as *infinite-time* average of the square difference of two consecutive frequency samples (cf. Equation (5.90)). In practice, nevertheless, we can have only *finite* sequences of N measurement samples x_k. From that, by evaluating the first difference, we can get $N-1$ frequency samples y_k (or, more precisely, $\bar{y}_k$ (5.81)). Hence, it is possible to compute only an *estimate* s_y^2 of the Allan variance σ_y^2.

The estimate s_y^2 is a random variable. Its variability depends, among the other things, on the length N of the data sequence collected. To model its distribution, the so-called χ^2 (read: *chi-squared*) distribution with l degrees of freedom is used [7.4].

To clarify the rationale of using this distribution, let us suppose we have only three TE samples x_1, x_2, x_3. From them, we can calculate two adjacent frequency samples y_1, y_2 and one Allan variance value $(y_1 - y_2)^2/2$. Now, let us assume, as verified in common cases, that $(y_1 - y_2)$ is normally distributed (i.e., with Gaussian probability density) with zero mean. Its square $(y_1 - y_2)^2$ is *not* normally distributed: while the Gaussian distribution is symmetrical around zero, with negative values being as likely as positive ones, the square is always positive.

The distribution of the square of one normally distributed random variable is called χ^2 *distribution with one degree of freedom*. In general, if we consider the sum of M independent random variables, each being square of a normally distributed variable, the resulting distribution is the χ^2 *distribution with M degrees of freedom*.

By estimating the Allan variance by means of the standard estimator (7.4), $N - 2n$ *non-independent* squared random variables are summed, because each frequency sample is taken into computation several times to evaluate different Allan-variance samples. Nevertheless, the resulting distribution of the estimate s_y^2 is still χ^2-type. The problem, in this case, is to assess the number l of degrees of freedom, which can assume a fractional value, depending on the number N of available samples and on the type of underlying noise, which determines the correlation between adjacent samples.

More precisely, the estimate s_y^2 of the Allan variance is a random variable distributed as χ^2, defined as [7.4]

$$\chi^2 = l \frac{s_y^2}{\sigma_y^2} \tag{7.11}$$

where l is the number of degrees of freedom (not necessarily an integer) and σ_y^2 is the true value of the Allan variance, which we are interested to know but that we can only estimate imperfectly. The probability density of χ^2 is given by [7.4]

$$p(\chi^2) = \frac{1}{2^{\frac{l}{2}} \Gamma\left(\frac{l}{2}\right)} (\chi^2)^{\frac{l}{2}-1} e^{-\frac{\chi^2}{2}} \tag{7.12}$$

where $\Gamma(t)$ is the so-called *gamma function*, defined by the integral

$$\Gamma(t) = \int_0^\infty x^{t-1} e^{-x} dx \tag{7.13}$$

A typical shape of the distribution $p(\chi^2)$ is shown in Figure 7.2.

Knowing the exact χ^2 distribution of the Allan-variance estimate s_y^2 computed is useful to determine its *confidence interval*. For example, let us suppose to have obtained some estimate $s_y^2(n\tau_0)$, based on a sequence of N TE samples. Then, it is possible to infer the

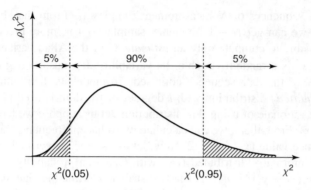

Figure 7.2. Determination of the confidence interval of the Allan-variance estimate by means of the χ^2 distribution

number of degrees of freedom l.[2] At this point, it is fully known the χ^2 distribution of the estimate. Hence, by referring to numerical tables of the χ^2 distribution, it is possible to determine for example the interval $\chi^2(0.05) < \chi^2 < \chi^2(0.95)$ where the 90% of the distribution area falls, as shown in Figure 7.2. Obviously, the above interval 5%–95% is arbitrary and a specific problem may dictate a different choice.

Now, it is straightforward to obtain the confidence interval

$$\frac{s_y^2 l}{\chi^2(0.95)} < \sigma_y^2 < \frac{s_y^2 l}{\chi^2(0.05)} \tag{7.14}$$

where, with probability 90%, the true value σ_y^2 falls. The value s_y^2 is a *point* estimate. The interval (7.14), on the other hand, is an estimate of the *most probable interval* (i.e., the interval where 90% of the estimates fall).

An alternative approach to the method based on the χ^2 distribution consists of assessing the variance of the Allan variance [7.16]. Nevertheless, this latter approach is less versatile, since it yields only symmetric error intervals ($\pm\sigma$).

7.1.7   Separating the Variances of the Clock under Test and of the Reference Clock

Up to now, we understood to measure the Time Error between the Clock Under Test (CUT) and a Reference Clock (RC) supposed ideal. Actually, the stability quantities measured include mixed noise contributions from both clocks. Being able to distinguish these contributions is a real issue, in order to draw meaningful results.

In particular, when the TE is measured between the input and the output of a slave clock (*synchronized-clock configuration* [7.13][7.14], cf. also Section 7.1.8), the measurement results do include both the noises from the CUT and from the RC, differently filtered to

---

[2] In order to be able to assess the number of degrees of freedom l, empirical equations have been found, by both analytical and Monte-Carlo simulation techniques, which are accurate to 1% for WPM, WFM and RWFM noise types [7.4]. This tolerance is reported somewhat larger for FPM and FFM noises. Numerical tables are available, for example in [7.4], which give the estimated number of degrees of freedom l for selected values of the number of samples and of the number of intervals averaged.

the output. As far as the contribution from the CUT is concerned, phase noise generated by the internal oscillator is high-pass filtered to the output, while phase noise from other internal sources (such as the phase detector) is transferred to the output low-pass filtered. The noise from the RC, on the other hand, is transferred low-pass filtered to the output and high-pass filtered to the input–output TE measured.

In Section 4.6.3, we already discussed quite in detail the issue of distinguishing noise contributions from the CUT and the RC, depending on the slave-CUT bandwidth f_c. To summarize, there is not a simple and unambiguous rule to distinguish these noise contributions. Nevertheless, the different noise transfer properties for $f < f_c$ and for $f > f_c$ allowed us to devise some empirical guidelines for assessing noise contributions at different Fourier frequencies.

The case of the measurement in the *independent-clock configuration* [7.13][7.14] (cf. also Section 7.1.8), on the other hand, deserves some more detailed treatise. Obviously, the simplest case to cope with is when it is known *a priori* that the RC is much less noisy than the CUT. In this case, all the measured noise can be ascribed to the latter. If, on the contrary, the presumed quality of the two clocks is comparable, it is customary to share out the measured noise in equal parts between the two clocks (this Solomonic practice is typical for measurements on primary frequency standards such as caesium clocks).

Otherwise, the individual contributions can be distinguished by comparing three or more devices [7.4][7.17]. Let us compare, for example, the output frequencies of three clocks. Then, let us denote by $\sigma_{ij}^2, \sigma_{jk}^2, \sigma_{ik}^2$ the three possible joint (Allan) variances and by $\sigma_i^2, \sigma_j^2, \sigma_k^2$ the individual clock variances. Under the assumption that the three clocks are independent, the joint variances are given by the sum of the individual variances, as

$$\begin{cases} \sigma_{ij}^2 = \sigma_i^2 + \sigma_j^2 \\ \sigma_{jk}^2 = \sigma_j^2 + \sigma_k^2 \\ \sigma_{ik}^2 = \sigma_i^2 + \sigma_k^2 \end{cases} \tag{7.15}$$

and it is thus straightforward to obtain

$$\begin{cases} \sigma_i^2 = \dfrac{\sigma_{ij}^2 + \sigma_{ik}^2 - \sigma_{jk}^2}{2} \\[2mm] \sigma_j^2 = \dfrac{\sigma_{jk}^2 + \sigma_{ij}^2 - \sigma_{ik}^2}{2} \\[2mm] \sigma_k^2 = \dfrac{\sigma_{jk}^2 + \sigma_{ik}^2 - \sigma_{ij}^2}{2} \end{cases} \tag{7.16}$$

This method works best if the three clocks are comparable in performance. Caution must be exercised, because Equation (7.16) may yield a negative result, despite the fact that variances are positive definite. Such an absurd result may indicate that the confidence intervals of the three variance estimates are too wide and that collecting more data is required.

7.1.8 Measuring Frequency Stability in the Independent and in the Synchronized Clock Configuration

Two standard configurations for measuring clock stability have been defined by international bodies ITU-T [7.13] and ETSI [7.14]: the synchronized-clock measurement configuration and the independent-clock measurement configuration.

According to ITU-T Rec. G.810 [7.13] and to ETSI EN 462 [7.14], when the two timing signals compared in the measurement of Time Error (or equivalently of frequency or phase error) are traceable to a common master clock, the configuration is referred to as *synchronized-clock measurement configuration*. Two cases of practical interest are shown in Figure 7.3: in both examples, the Time Error $T(t) - T_{ref}(t)$ between the output and the input of the slave CUT is measured.

It is worthwhile remarking that the TE measured in synchronized-clock configuration is unaffected by the frequency offset and drift of the common master clock. This measure, therefore, is of particular interest because it allows observing directly the impairments added by the slave CUT (internal phase noise).

Again according to ITU-T Rec. G.810 [7.13] and to ETSI EN 462 [7.14], any test set-up where there is no common master clock controlling the timing signals between which the TE is measured is referred to as *independent-clock measurement configuration*. Two examples are shown in Figure 7.4: in the former case (Figure 7.4(a), the Time Error $T(t) - T_{ref}(t)$ between the output of the free-running CUT and the output of an independent reference clock is measured; in the latter (Figure 7.4(b), the CUT is slaved to an independent master.

It is worthwhile remarking that the TE measured in independent-clock configuration, besides being dependent on internal CUT phase noise, is affected also by any frequency offset or drift of all clocks involved in the measurement.

As it has been widely discussed in Sections 5.10 and 5.11, the measurement configuration substantially affects the behaviour of the stability quantities. For example, by

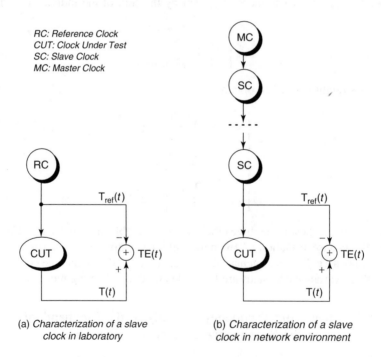

(a) *Characterization of a slave clock in laboratory*

(b) *Characterization of a slave clock in network environment*

Figure 7.3. Examples of TE measurement in the synchronized-clock measurement configuration. (Adapted from Figure 1/G.810 [7.13] and from Figure 1/EN 300 462-1-1 [7.14], © ETSI 1998, by permission of ITU and ETSI)

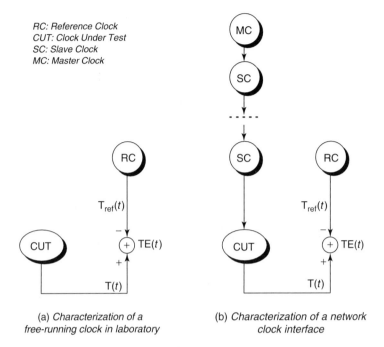

RC: Reference Clock
CUT: Clock Under Test
SC: Slave Clock
MC: Master Clock

(a) Characterization of a
free-running clock in laboratory

(b) Characterization of a network
clock interface

Figure 7.4. Examples of TE measurement in the independent-clock measurement configuration. (Adapted from Figure 2/G.810 [7.13] and from Figure 2/EN 300 462-1-1 [7.14], © ETSI 1998, by permission of ITU and ETSI)

application of the theory presented in those sections, the graph in Figure 7.5 [7.18] compares the plots of the Allan deviation evaluated by numerical computation of the analytical relationships for the two measurement configurations on a clock with loop bandwidth $B = 1$ Hz. In this figure, the measurement configurations of Figures 7.3(a) and 7.4(a) have been indicated with 'free-run mode' and 'locked mode' respectively.

We notice that the curves coincide in the short term (i.e., for observation intervals τ much shorter than the PLL time constant), but exhibit a substantially different trend in the long term, showing that the CUT internal noise is high-pass filtered in the synchronized-clock measurement configuration.

Experimental Results

An example of real measurement results in both the independent- and synchronized-clock configurations is provided in Figures 7.6 and 7.7 [7.18]. We studied the behaviour of a SASE clock, deployed in a Public Switched Telephone Network (PSTN) office building, equipped with an OCXO oscillator and with bandwidth set to $B = 5$ mHz. Here and in the following of the book, we will denote the supplier of this SASE as supplier C. No special environmental precautions were taken, aiming at observing its behaviour in real field conditions.

The measurement procedure is fully detailed in Appendix 7A. The configurations defined in Figures 7.3(a) and 7.4(a) were adopted. Two TE sequences were measured: the former in the synchronized-clock configuration ($N = 90\,000$, $\tau_0 \cong 60$ ms, $T = 5400$ s)

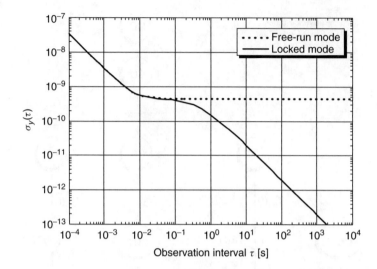

Figure 7.5. Comparison of the Allan deviation curves evaluated analytically in locked (cf. Figure 7.3(a)) and free-run (cf. Figure 7.4(a)) modes at the output of a clock with loop bandwidth $B = 1$ Hz. (Reproduced from [7.18], © 1994 IEEE, by permission of IEEE)

and the latter in the independent-clock configuration ($N = 60\,000$, $\tau_0 \cong 60$ ms, $T = 3600$ s). Then, ADEV, MADEV, TDEV and TIErms were computed with the standard estimators of Equations (7.4) through (7.7).

The dotted curves in Figures 7.6 and 7.7 were evaluated based on the former sequence, while the solid ones on the latter. Moreover, the upper two curves in Figure 7.6 depict ADEV results, while the lower two show those of MADEV. Similarly, Figure 7.7 shows TDEV and TIErms results. It is evident the agreement of these measurement results with the theoretical model of the graph in Figure 7.5: the stability curves, although measured in different days, match approximately in the short term but separate for longer observation intervals.

Unfortunately, the measurement background noise in these results is not negligible as compared to the intrinsic clock noise, mainly in the short term, being the clock we studied a very high-quality oscillator (OCXO). Nevertheless, to the purpose of verifying the expected long-term differences in the stability quantity behaviour between independent- and synchronized-clock configurations, the impact of the measurement noise has not been an issue.

7.1.9   Impact of the TE Sampling Period on the Behaviour of the Stability Quantities

An issue to consider carefully when performing stability measurements, in order to draw meaningful conclusions from the measurement data, is the possible impact of the TE sampling period on the behaviour of the stability quantities. Such dependence, for some of the standard stability quantities, has been pointed out somehow in several reference papers and in Chapter 5, which studied in depth the behaviour of the stability quantities analytically and under various aspects.

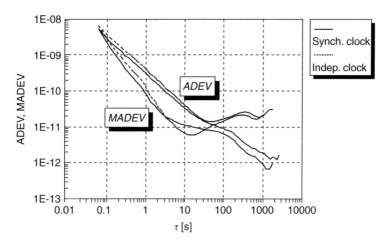

Figure 7.6. Comparison of the ADEV and MADEV curves measured in synchronized-clock (cf. Figure 7.3(a)) and independent-clock (cf. Figure 7.4(a)) configurations at the output of the SASE clock from supplier C. (Reproduced from [7.18], © 1994 IEEE, by permission of IEEE)

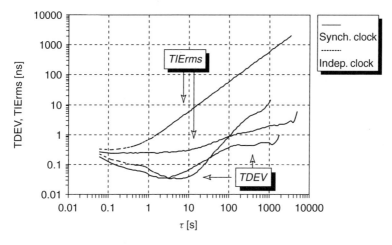

Figure 7.7. Comparison of the TDEV and TIErms curves measured in synchronized-clock (cf. Figure 7.3(a)) and independent-clock (cf. Figure 7.4(a)) configurations at the output of the SASE clock from supplier C. (Reproduced from [7.18], © 1994 IEEE, by permission of IEEE)

In order to further clarify the impact that the TE sampling period may have in practical measurements, the paper [7.19] reported a systematic simulation study together with some experimental results. This section is mainly based on that paper.

For all the stability quantities, the sampling period τ_0 used to acquire the TE sequence constrains both the minimum and the maximum values of the observation interval τ: for a fixed number of samples N, the shorter τ_0 the shorter is the minimum observation interval τ, the longer τ_0 the longer is the maximum τ. In other words, choosing τ_0 implies a trade-off between the resolution and the duration of the measurement.

In practical applications, τ might even range from 10^{-3} s up to 10^6 s. Aiming at characterizing the overall behaviour of the stability quantities on such a whole range of τ, one should set $\tau_0 = 10^{-3}$ s, thus increasing enormously both the need of storage capability and the computation time (e.g., for TDEV one should collect $N = 3 \times 10^9$ samples x_i).

In order to overcome storing and processing billions of samples, one might operate on subranges of the whole range of τ as follows:

(1)   for each subrange, choose τ_0 equal to the minimum τ of the subrange and collect TE measurement data;

(2)   for each subrange, evaluate the stability quantities based on the data collected;

(3)   *juxtapose* the resulting curves.

Unfortunately, it has been shown that while ADEV, MTIE and TIErms do not feature any significantly different behaviour as the sampling period is varied for all the power-law noise types considered, the behaviour of MADEV and TDEV is substantially dependent on the chosen measurement sampling period in the observation intervals where WPM or FPM noises dominate. In such cases, the above procedure fails when juxtaposing the curves of different subranges (...extremes do not meet!).

In the following subsections, the main results of the theoretical analyses reported in different reference papers and in Chapter 5 are first summarized. Then, a systematic simulation study, under the presence of all the clock noises obeying the so-called power-law model and for all the stability quantities, is reported. Finally, some measurements on a real clock are provided, with the aim at pointing out the impact that the choice of τ_0 can have on practical measurements in telecommunications applications.

7.1.9.1   Summary of Theoretical Analysis Results: Allan Variance and TIErms

The AVAR($n\tau_0$) and the TIErms($n\tau_0$) are substantially independent on the values chosen for n and τ_0, their product $\tau = n\tau_0$ being kept constant.

For comparison purposes among experimental results, therefore, it is sufficient to specify τ and, in case, the measurement hardware bandwidth $\hat{f}_h$, if smaller than the natural cut-off frequency f_h of the clock output noise. In fact, if $\hat{f}_h < f_h$, the measurement set-up cuts off the high-frequency components of the clock phase noise. Otherwise, if $\hat{f}_h > f_h$, the actual value of $\hat{f}_h$ does not affect substantially the total phase noise power measured.

7.1.9.2   Summary of Theoretical Analysis Results: Maximum Time Interval Error

On the other hand, MTIE($n\tau_0$), due to its peculiar nature of rough peak measure, can be hardly treated with analytical tools. Anyhow, it can be shown empirically that its value depends slightly on the values chosen for n and τ_0, their product τ being kept constant, since the peak-to-peak deviation of a set of n TE samples, over an observation interval τ, tends to be greater for larger n (keeping constant $\tau = n\tau_0$, i.e. by decreasing the sampling period τ_0).

For the sake of precision, this is especially true if a broadband noise such as WPM dominates and the noise bandwidth is large enough. Anyway, the resulting change in the MTIE values is usually not substantial, provided that n is large enough.

7.1.9.3 Summary of Theoretical Analysis Results: Modified Allan Variance and Time Variance

Finally, MAVAR($n\tau_0$) and TVAR($n\tau_0$), closely related, depend substantially on the parameters n and τ_0, their product τ being kept constant, mainly in the presence of WPM and FPM noise types.

This property results from the treatise in Chapter 5 and, for example, from reference papers [7.20]–[7.22]. To summarize, the reason for this behaviour lies in the fact that the definition of MAVAR is based on the first difference of the fractional frequency $y(t)$ averaged (unlike AVAR) over n adjacent samples (cf. Equation (5.93)). If $n = 1$, then AVAR($n\tau_0$) = MAVAR($n\tau_0$). Otherwise, the algorithm averages over n adjacent samples smoothing data fast fluctuations. Hence, the impact of the parameter n is the more noticeable the higher the TE noise power at high frequencies (i.e., especially with WPM or FPM noise).

More in detail, the reader is referred again to Section 5.10.1. By studying the behaviour of $R(n) =$ MAVAR($n\tau_0$)/AVAR($n\tau_0$) in Figure 5.30, the following conclusions can be drawn (cf. the comments already made in that section):

- if the noise type is WPM, the ratio $R(n)$ is equal to $1/n$; to give an example, this means that MAVAR($10n, \tau_0$) = $(1/10) \cdot$ MAVAR($n, 10\tau_0$) with this type of noise;

- if the noise type is FPM, the ratio $R(n)$ becomes smaller of a factor 10 as n increases in the interval $1 < n < 100$, but it changes very slowly for $n > 100$ (though, $\lim_{n \to \infty} R(n) = 0$);

- if the noise type is WFM, the ratio $R(n)$ keeps practically unchanged for $n > 10$;

- if the noise type is FFM or RWFM, the ratio $R(n)$ does not change significantly with n greater than few units.

From the remarks above, we note that choosing n large enough (with FPM, WFM, FFM and RWFM noises) limits effectively the impact of the sampling period on MAVAR.

Therefore, the final notice is that, for comparison purposes among experimental results in terms of MAVAR(τ) and TVAR(τ), it is not sufficient to specify τ and in case the measurement hardware bandwidth $\hat{f}_h$ (if $\hat{f}_h < f_h$), as stated for AVAR(τ) and TIErms(τ), but it is necessary to specify also n or τ_0.

7.1.9.4 Simulation Results

In order to make evident the impact that the TE sampling period may have in practical measurements, the paper [7.19] reported a systematic simulation study, for all the stability quantities, in the presence of all the power-law noise types.

First, in order to simulate WPM noise, two white and uniformly distributed pseudo-random sequences of length $N = 4096$ were generated. Then, applying a well-known transformation formula [7.23], one white Gaussian pseudo-random sequence was obtained, thus simulating a WPM noise. The amplitude was adjusted to set the sequence standard deviation $\sigma = 1$ ns. Finally, spectral shaping was accomplished by filtering the WPM noise sequence through a half-order integrator [7.24] with transfer function $H_{1/2}(f) = 1/\sqrt{j2\pi f}$, to generate the FPM noise sequence. Repeatedly filtering through $H_{1/2}(f)$ yielded WFM, FFM and RWFM noise sequences.

The stability quantities ADEV, MADEV, TDEV, TIErms and MTIE were evaluated basing on each of these five noise sequences, according to the standard estimators of Equations (7.4)–(7.8). Then, the same computation was carried out on sequences obtained by repeated decimation of the above five sequences, yielding sequences of length $N_1 = N/2$, $N_2 = N/4$ and $N_3 = N/8$. This procedure is clearly equivalent to sampling the same segment of a given noise process with the sampling periods τ_0, $2\tau_0$, $4\tau_0$ and $8\tau_0$, respectively.

The simulation results shown in [7.19] confirm the theoretical findings summarized above. ADEV and TIErms do not feature any different behaviour as the sampling period is varied, for all the power-law noise types considered. MTIE, on the other hand, exhibits a hardly noticeable change with the sampling period if the noise type is WPM. Conversely,

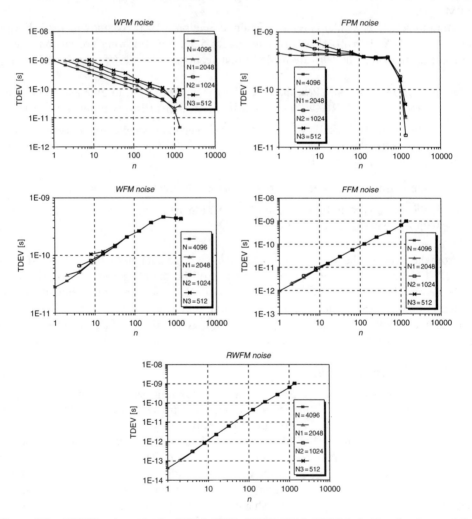

Figure 7.8. TDEV$(n\tau_0)$ results for WPM, FPM, WFM, FFM, RWFM simulated noises and for different sampling periods τ_0, $2\tau_0$, $4\tau_0$, $8\tau_0$ (sequence lengths $N = 4096$, $N_1 = 2048$, $N_2 = 1024$, $N_3 = 512$). (Reproduced from [7.19] by permission of AEI)

MADEV and TDEV exhibit substantial quantitative differences if WPM and FPM noises dominate.

Among those results, the graph plots of TDEV($n\tau_0$) for each power-law noise type and for the different sampling periods τ_0, $2\tau_0$, $4\tau_0$ and $8\tau_0$ (sequences of length $N = 4096$, $N_1 = 2048$, $N_2 = 1024$ and $N_3 = 512$, respectively) are shown in Figure 7.8. We leave the reader to verify the excellent agreement of these simulation results with the comments to the behaviour of the ratio $R(n)$ previously drawn in Section 7.1.9.3 (notice also the poor statistical confidence at the right ending part of some curves, where the stability quantities are computed averaging too small a number of samples).

7.1.9.5   *Experimental Results*

In order to support the conclusions stemming from both the theoretical analysis and the simulations with experimental evidence, measurements were carried out on the clock of a digital switching exchange, worldwide deployed in Public Switched Telephone Networks (PSTN). Here and in the following of the book, we will denote the supplier of this equipment as supplier D. The exchange was equipped with an OCXO clock with closed-loop bandwidth $B = 5$ mHz.

The measurement procedure is fully detailed in Appendix 7A. The synchronized-clock measurement configuration was adopted, according to Figure 7.3(a). First, a TE sequence of $N = 500\,000$ samples was acquired, with sampling period $\tau_0 \cong 7.5$ ms and over a measurement interval $T = 3600$ ms. Then, sample decimation was accomplished, as done in the simulation work, yielding a set of other six sequences with sampling period equal to respectively $2\tau_0 = 15$ ms, $5\tau_0 = 37.5$ ms, $10\tau_0 = 75$ ms, $20\tau_0 = 150$ ms, $50\tau_0 = 375$ ms, $100\tau_0 = 750$ ms and thus of length $N_1 = 250\,000$, $N_2 = 100\,000$, $N_3 = 50\,000$, $N_4 = 25\,000$, $N_5 = 10\,000$, $N_6 = 5000$. Based on these decimated sequences, six curves of MADEV(τ) and TDEV(τ) were evaluated with estimators (7.5)(7.6) and plotted in the graphs shown in Figures 7.9 and 7.10, respectively.

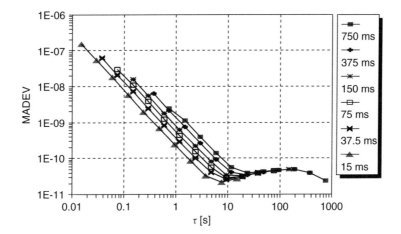

Figure 7.9.   MADEV(τ) curves measured on a PSTN digital switch clock (supplier D) for different sampling periods $2\tau_0 = 15$ ms, $5\tau_0 = 37.5$ ms, $10\tau_0 = 75$ ms, $20\tau_0 = 150$ ms, $50\tau_0 = 375$ ms, $100\tau_0 = 750$ ms   ($T = 3600$ s   and   respectively $N_1 = 250\,000$, $N_2 = 100\,000$, $N_3 = 50\,000$, $N_4 = 25\,000$, $N_5 = 10\,000$, $N_6 = 5000$)

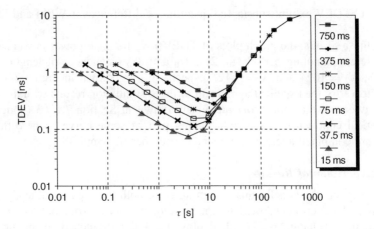

Figure 7.10.   TDEV(τ) curves measured on a PSTN digital switch clock (supplier D) for different sampling periods $2\tau_0 = 15$ ms, $5\tau_0 = 37.5$ ms, $10\tau_0 = 75$ ms, $20\tau_0 = 150$ ms, $50\tau_0 = 375$ ms, $100\tau_0 = 750$ ms ($T = 3600$ s and respectively $N_1 = 250\,000$, $N_2 = 100\,000$, $N_3 = 50\,000$, $N_4 = 25\,000$, $N_5 = 10\,000$, $N_6 = 5000$). (Reproduced from [7.19] by permission of AEI)

In all cases, the condition $2\pi f_h \tau_0 \gg 1$ is well satisfied. Moreover, the range of τ over which MADEV(τ) and TDEV(τ) have been evaluated fits almost entirely within the interval $\tau < 1/B = 200$ s. Therefore, in this range of τ, the noise generated by the internal oscillator is visible in the measurement results not affected significantly by the closed-loop filtering action.

These measurement results confirm the conclusions stemming from the theoretical analysis and the simulations. The slopes of MADEV(τ) and TDEV(τ) curves show the dominant presence in the timing signal under test of WPM and FFM noises. In particular, the WPM noise appears to be dominant for $\tau < 10$ s: for any given observation interval in that region, MADEV(τ) and TDEV(τ) take increasing values as the sampling period is increased. No such a dependence is recognizable for $\tau > 10$ s, where FFM noise dominates instead.

FFM noise is usually not recognizable in synchronized-clock measurement results, owing to its low-frequency nature, because the noise generated by the internal oscillator of the slave CUT is high-pass filtered by the closed loop to the output. In this case, nevertheless, the closed-loop time constant was very high and out of the range of τ under measurement.

It clearly appears, now, what kind of results we would have obtained should we have applied blindly the procedure outlined at the beginning of this section.

7.2   INSTRUMENTATION FOR TIME AND FREQUENCY MEASUREMENT IN TELECOMMUNICATIONS

The aim of this section is to provide a brief survey on some of the most common devices and instruments used for measuring clock stability in telecommunications: mixers, frequency synthesizers, spectrum analysers and time counters. The references cited do provide more details.

7.2.1 Low-Noise Mixer

In radio-frequency electronics, a circuit that forms the product of two analogue waveforms is used in a variety of applications and, depending on the context, is called modulator, mixer, synchronous detector, phase detector, etc.

Generally speaking, a *mixer* is a device that accepts two analogue pseudo-periodic signals at its inputs and forms an output signal with components at the sum and difference frequencies [7.25]. From the basic trigonometric relationship

$$\cos 2\pi v_1 t \cdot \cos 2\pi v_2 t = \tfrac{1}{2} \cos 2\pi \, (v_1 + v_2)\, t + \tfrac{1}{2} \cos 2\pi \, (v_1 - v_2)\, t \tag{7.17}$$

it is clear that a *four-quadrant multiplier* (i.e., a multiplier that forms the product of two signals according to their sign) is indeed a mixer.

Nevertheless, it is not necessary to form an accurate analogue product of two input signals in order to mix them. Actually, any non-linear combination of two signals produces sum and difference frequencies. For example, let us consider the square of the sum of two input signals, as

$$(\cos \omega_1 t + \cos \omega_2 t)^2$$
$$= 1 + \tfrac{1}{2} \cos 2\omega_1 t + \tfrac{1}{2} \cos 2\omega_2 t + \cos \, (\omega_1 + \omega_2)\, t + \cos \, (\omega_1 - \omega_2)\, t \tag{7.18}$$

A non-linear input–output law of this kind may be obtained, for instance, by applying two small signals to a forward-biased diode.

In this case, we notice that harmonics of the two input signals are transferred to the output, besides the sum and difference frequencies. The term *balanced mixer* is used to denote a circuit that transfers to the output only the sum and difference frequencies, but not the input signals neither their harmonics. The four-quadrant multiplier is a balanced mixer, whereas the squarer non-linear diode is not.

Among the most common techniques used to make mixers, there are for example simple non-linear transistor or diode circuits, dual-gate FETs with one signal applied to each gate, multiplier integrated circuits, balanced mixers constructed from transformers and arrays of diodes [7.25]. The advent of Schottky diodes made it possible to construct very-low-noise mixers, suitable for time and frequency stability measurement.

Finally, we remark that the oscillators that generate the signals sent to mixer inputs should be provided with well-buffered outputs, to be able to isolate perfectly the possible coupling between the two mixer inputs. Too good results, in fact, might be due to a tight coupling between the oscillators, via signal injection through their output ports. Moreover, it is advisable to operate near the maximum input levels, for best signal-to-noise ratio [7.3].

7.2.2 Frequency Synthesizer

A frequency synthesizer is usually the most convenient equipment to generate a signal with arbitrary frequency and phase. A typical application in telecommunications is when, having a 10-MHz frequency standard, we want to measure the stability of a slave clock that accepts only a 2.048-MHz signal as reference frequency.

The core of a *frequency synthesizer* is a system of programmable frequency multipliers and dividers. This equipment allows thus to synthesize an output signal having specified waveform, amplitude, phase and frequency, based on a reference timing signal at its

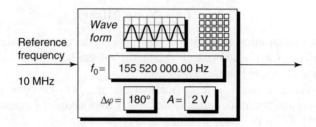

Figure 7.11. Example of digital frequency synthesizer

input. Typical input frequencies may be 5 or 10 MHz. Common output waveforms are sinusoidal, square, triangular, etc. The output frequency can be set in a wide range from 1 mHz to 150 MHz and above, with 10, 12 or more significant digits, as shown in the pictorial example of Figure 7.11.

In planning a measurement campaign on a clock, special care should be taken in characterizing the noise performance of all instruments used. However, common high-quality synthesizers supply timing signals with negligible phase noise, compared to the requirements of telecommunications-clock stability measurement.

7.2.3 Analogue and Digital Spectrum Analysers

The spectrum analyser is an instrument of considerable utility in radio-frequency electronics. It allows one to observe a signal in the frequency domain, plotting its power versus its Fourier frequency. Spectrum analysers come in two basic varieties: swept-tuned and real-time instruments [7.26]. The former ones are mostly analogue, the latter ones mostly digital.

Swept-tuned spectrum analysers, for a long time the most common type, are analogue instruments based on the principle of the super-heterodyne receiver and work as shown in the basic scheme of Figure 7.12. A local oscillator, whose frequency is controlled by a ramp waveform generator, outputs a signal with frequency slowly varying over a

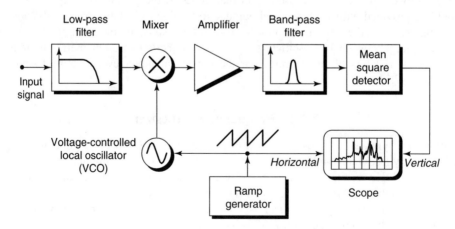

Figure 7.12. Scheme of principle of a swept-tuned spectrum analyser

given interval. The signal to analyse is mixed with this signal and then filtered by a narrow band-pass filter, centred on a fixed frequency. While the local oscillator is swept through its range of frequencies, different input frequencies are successively mixed to pass through the filter. Typically, the input frequency may vary in the interval from 1 to 10^9 Hz, while the interval of Fourier frequencies that can be analysed may range from 1 to 10^6 Hz.

Note that swept-tuned spectrum analysers allow one to observe only one Fourier frequency at a time. The complete spectrum is plotted cyclically in time, by sweeping the frequency range of interest. In some applications, this can be a real disadvantage, since it does not permit to observe transient events. In addition, the narrower the scanning bandwidth, the slower the sweep rate must be.

Hence, the use in many applications of *real-time spectrum analysers*, which allow one to observe a signal as a whole and to output the signal power measurement on all frequency bands simultaneously.

The clumsy approach to make a real-time spectrum analyser consists of employing a set of narrow band-pass filters to look at a range of frequencies simultaneously. Today, real-time spectrum analysers are based exclusively on numerical Fourier analysis (*digital spectrum analysers*), owing to the availability of inexpensive integrated circuits that compute instantly the FFT of vectors made of thousands of samples. The FFT approach is particularly attractive to analyse transients and low-frequency signals, where swept-tuned instruments would be too slow. Moreover, digital processing makes it natural and inexpensive to add many additional functions, such as data correlation, averaging, etc.

For a long time, spectrum analysers have been the main instrumentation for frequency-stability analogue measurement. Unfortunately, in telecommunications applications the most interesting part to observe in a spectrum lies at lowest frequencies, in the order of mHz or even μHz. This is the reason why time-domain measurements are normally preferred today.

7.2.4 Digital Time Counter

The digital time counter is the core instrument in modern set-ups for clock stability measurement, both in the time and in the frequency domains. Its basic function is to measure the elapsed time between two events (*start* and *stop trigger events*), by incrementing a counter register according to a base reference frequency.

As explained already in Section 5.9.3, a trigger event typically consists of detecting an electrical signal $V(t)$ going over some threshold on the input channel, as shown in Figure 7.13.

The time interval is measured by incrementing a counter register according to a very high and stable base frequency, supplied by an internal oscillator or by an external reference. The best measurement resolution, obviously, is attained by setting the base frequency as highest as possible: high-performance commercial counters use a local reference as high as 500 MHz, phase-locked to a very stable OCXO running at 5 or 10 MHz. This corresponds to a time-interval measurement resolution of 2 ns.

This resolution limit can be beaten by some interpolation tricks, without further increasing the base frequency [7.27]. Among them, the most common in commercial time counters is the so-called *Vernier interpolation*. This technique owes its name to the

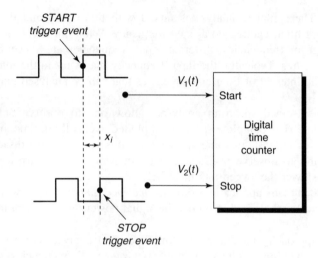

Figure 7.13. A digital time counter measures a Time Error sample x_i between two square timing signals $V_1(t)$ and $V_2(t)$

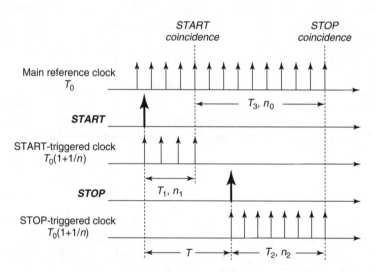

Figure 7.14. Principle of Vernier interpolation for increasing the resolution of time interval measurement

French mathematician P. Vernier (1580–1637), who invented it to increase the resolution of the *nonius* (hence known as *vernier*) for measuring fractions of divisions in a graduated circular or rectilinear scale.

This method allows to determine in what point of the base-clock cycle the input pulse occurred. As shown in Figure 7.14, three clock signals are involved:

- the main reference clock runs continuously and generates a timing signal with period T_0 (coarse measurement resolution);

- the input START pulse triggers a second clock, with slightly longer period $T_0(1 + 1/n)$;

- the input STOP pulse triggers a third clock, with same period as the other triggered clock $T_0(1 + 1/n)$.

Fast circuitry then looks for the first coincidence between the pulses of the two triggered clocks and the pulses of the main reference clock, while counting the number n_1, n_2 of pulses of each triggered clock before coincidence. Then, it is possible to determine the time interval T elapsed between START and STOP events as follows:

$$
\begin{aligned}
TI &= T_1 + T_3 - T_2 \\
&= n_1 T_0 \left(1 + \frac{1}{n}\right) + n_0 T_0 - n_2 T_0 \left(1 + \frac{1}{n}\right) \\
&= T_0 \left[n_0 + (n_1 - n_2)\left(1 + \frac{1}{n}\right)\right]
\end{aligned}
\tag{7.19}
$$

The resulting precise measurement resolution is equal to $1/n$ of the period T_0 of the main clock. High-performance time counters on the market exploit this technique, to improve the coarse measurement resolution attainable with the base frequency synthesized from the local reference oscillator. For example, the book [7.27] cites an excellent time counter that was on the market in the 1980s (the Hewlett-Packard 5370A): with a coarse measurement resolution $T_0 = 5$ ns (base frequency equal to 200 MHz) and a refining factor $n = 256$, it achieved a precise measurement resolution in the order of 20 ps.

Commercial time counters are able to measure not only time intervals, as described above, but also period and frequency, according to a similar scheme. To measure period, the number of internal cycles falling within a period of the input signal is counted. To measure frequency, the number of cycles of the input signal falling within an internal period is counted.

Time counters are today the standard choice to measure clock stability in several applications and in telecommunications in particular. A time-interval measurement resolution in the order of 20–200 ps is common in high-performance commercial time counters. This resolution, although perfectly adequate for example for making standard measurements on telecommunications equipment, is still too coarse to measure short-term instability in good quartz-crystal oscillators. However, several techniques have been conceived that enhance the time and frequency measurement sensitivity (cf. Section 7.3.4) and thus make the use of time counters attractive also in these cases.

7.3   DIRECT DIGITAL MEASUREMENT

The most straightforward method to measure the stability of a timing signal is to directly measure its Time Error, compared to a reference signal having nearly the same frequency, simply by means of a digital time counter (*direct digital measurement of Time Error*). Three different approaches can be adopted: to measure time intervals (TE) between the edges of the two signals, to measure the frequency of the timing signal under test and to measure its period.

7.3.1   Measurement of Time Intervals (Direct Digital Measurement of Time Error)

The technique described in this section has become widely used, since high-resolution time counters have been available. In fact, this is the simplest technique and therefore it is most commonly used today, except when special accuracy requirements are posed.

The TE is directly measured between a reference signal $s_1(t)$ and the signal under test $s_2(t)$, having actual frequencies ν_1 and ν_2 with nominal values ν_{1n} and ν_{2n} respectively. That is, the time intervals elapsed between two significant instants of the two timing signals are measured by the time counter. In the simplest mode, the time counter is started at some instant t_1, when the signal $s_1(t)$ is detected to cross positively the zero-voltage level on input channel 1. Then, the counter is stopped at the next instant t_2, as soon as the signal $s_2(t)$ is detected to cross positively the zero-voltage level on input channel 2. Same-direction zero crossings are normally chosen in order to provide immunity from changes in both the amplitude and symmetry of the waveform.

The time interval so measured is acquired as the ith TE sample x_i. The measurement is then repeated with period τ_0. This procedure is shown in Figure 7.15, for example in the case of two square timing signals.

Paper [7.4] remarks that, because of distortion, the phase of an oscillator is generally not well known except at zero-crossings. Thus, exactly speaking, the time interval $T_2(t_2) - T_1(t_1)$ is measured: the counter is started when the clock 1 states to be at instant t_1 and stopped when the clock 2 states to be at instant t_2. Nevertheless, all analysis techniques require the time difference at the same time: $T_2(t_1) - T_1(t_1)$. The translation requires a correction that takes into account the difference in frequency between the two oscillators. This correction, usually negligible, is the term between parentheses in the following expression:

$$T_2(t_1) - T_1(t_1) \cong -P\tau_c \left(1 + \frac{\nu_{2n} - \nu_{1n}}{\nu_{1n}}\right) \tag{7.20}$$

where the sign is a pure convention (time advance is positive), P is the value of the counter register, τ_c is the period of the counter base clock and hence the measurement resolution. It is important to notice that the units of the time difference measured in Equation (7.20) are [seconds of clock 1]. Therefore, aiming at a better accuracy over long measurement intervals, the counter base clock should be synchronized by the reference signal $s_1(t)$ itself (as in the scheme drawn forth in Figure 7.16).

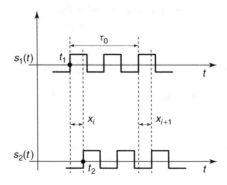

Figure 7.15.   Direct digital measurement of Time Error between two timing signals $s_1(t)$ e $s_2(t)$

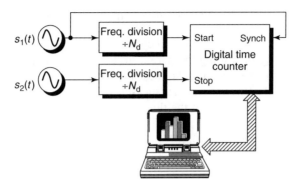

Figure 7.16.   Direct digital measurement of Time Error with frequency division

The measurement is repeated with period τ_0 and a sequence of N TE samples $\{x_i\}$ is acquired. Usually, the measurement is driven by a PC, which acquires and stores the measurement data for subsequent numerical processing and graphical plotting of stability quantities computed. Alternatively, a few modern time counters include some data storage, processing and graphical plotting capability.

The simple scheme described above measures a maximum accumulated TE not larger than one cycle $1/\nu_{2n}$ of the stop signal. When the phase difference between input signals exceeds this value, the counter readout exhibits a cycle slip and becomes periodic. More formally, the time counter provides TE values *modulo* $1/\nu_{2n}$. That is, given an instrument readout Z at time t_1, the exact value $TE(t_1)$ is known with the ambiguity of an integer number of cycles, as

$$\mathrm{TE}(t_1) = Z + \frac{k}{\nu_{2n}} \qquad k \in 0, \pm1, \pm2, \ldots \tag{7.21}$$

This ambiguity can be reduced by dividing the frequency of the signals from each oscillator by a factor N_d, before the time-interval measurement, as shown in the scheme of Figure 7.16. The effect of such frequency division is to increase the time interval before an ambiguity occurs to N_d/ν_{2n}. To measure long-term frequency deviations between atomic clocks, signals are usually frequency-divided down to 1 Hz.

7.3.2   Measurement of the Frequency

Average frequency can be measured directly by means of a time counter used as *frequency counter* [7.4]. This way, the counter determines the number of whole cycles M_c completed by the signal under test during a time interval τ given by the counter's time base. Hence, we have

$$\overline{\nu}(0; \tau) = \frac{M_c + \Delta M_c}{\tau} \cong \frac{M_c}{\tau} \tag{7.22}$$

where $\overline{\nu}(t_1; t_2)$ denotes the average frequency over the interval $[t_1, t_2]$ and ΔM_c is the fraction of cycle not measured by the counter. The quantization error $\Delta\nu_Q$ is limited by

$$\frac{\Delta\nu_Q}{\overline{\nu}(0; \tau)} < \frac{1}{M_c} \tag{7.23}$$

7.3.3 Measurement of the Period

For low frequencies, the number of counted cycles may be small and thus the quantization error becomes large. By measuring the period, instead of the frequency, it is possible to decrease the error without increasing the measurement duration.

A time counter used as period counter measures the duration of M_c whole cycles of the signal under test, as N_c cycles of the time base τ_c [7.4]. The fraction of cycle ΔN_c is not measured. Since

$$M_c = \bar{v}(0; M_c/v_n) \times (N_c + \Delta N_c)\tau_c \tag{7.24}$$

the measured period is

$$\overline{T}(0; M_c/v_n) = \frac{1}{\bar{v}(0; M_c v_n)} \cong \frac{N_c \tau_c}{M_c} \tag{7.25}$$

while the quantization error ΔT_Q in this case is limited by

$$\Delta T_Q < \frac{\bar{v}(0; M_c v_m)}{N_c} \tag{7.26}$$

Frequency measurements are seldom used to characterize precision oscillators, while period measurements are very common.

7.3.4 Techniques for Enhancing the Measurement Sensitivity

Even though the resolution[3] of modern digital time counters is continuously improving (typically it ranges from 20 to 200 ps), yet the short-term time fluctuations of a state-of-the-art crystal oscillator may be even smaller, for example in the order of 1 ps or less. Thus, the accuracy of test set-ups based on direct digital measurement of phase and frequency fluctuations may be not sufficient, if aiming at characterizing precision oscillators.

Several sensitivity-enhancement methods have been developed, aiming at distinguishing very small time and frequency fluctuations. These methods can be classified as:

- *heterodyne techniques*, which consist of mixing the timing signal under test with the reference signal, having *nearly* the same frequency, to measure the same *phase* fluctuations but on the resulting low-frequency beat signal (thus measuring larger *time* fluctuations);

- *homodyne techniques*, which are a limit of the previous method, occurring when the mixed reference signal has exactly the *same* average frequency as the signal under test (the reference signal may be obtained from the signal under test for example by means of a delay line or a PLL);

- *multiple conversion techniques*, in which the actual signal to measure is obtained from the signal under test through several mixing and frequency synthesis stages; among them, the Dual-Mixer Time-Difference (DMTD) technique may be considered the state of the art of time and frequency stability measurements.

---

[3] Do not let *resolution* be confused with *precision* or *accuracy*: the actual accuracy of any instrumentation has manifestly nothing to do with the number of digits displayed!

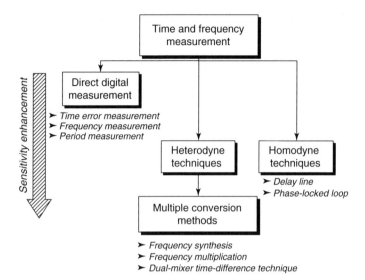

Figure 7.17.   Taxonomy of time and frequency measurement techniques. (Reproduced from [7.28], ©1997 IEEE, by permission of IEEE)

Examples of the above techniques are discussed in detail in the cited works [7.1]–[7.10]. Among these references, fundamental and thorough surveys are provided by [7.2]–[7.4].

The taxonomy of the main time and frequency measurement techniques is thus summarized in Figure 7.17.

7.4 HETERODYNE TECHNIQUES: THE BEAT-FREQUENCY METHOD

One of the most common techniques to enhance the instrumentation sensitivity in measuring time and frequency fluctuations consists of multiplying (mixing) the timing signal under test with the reference signal, having *nearly* the same frequency. Then, the measurement is carried out on the resulting beat signal, which has a much lower frequency. Doing so, the same *phase* and *frequency* fluctuations are measured, which result in larger *time* fluctuations in the beat signal.

Since the signal under measurement is obtained by mixing two signals having nearly the same, but different frequency, this technique has been named *heterodyne* (from the Greek etyma ἕτερος = *different* and δύναμις = *force*), or also *beat-frequency method*.

Then, let us consider two timing signals whose frequency difference is much less than the frequency of each oscillator [7.4]:

$$s_1(t) = A_{10} \sin\left[2\pi \nu_{10} t + \varphi_1(t) + \varphi_{10}\right]$$

$$s_2(t) = A_{20} \sin\left[2\pi \nu_{20} t + \varphi_2(t) + \varphi_{20}\right] \tag{7.27}$$

where $|\nu_{10} - \nu_{20}| \ll \nu_{10}$ and the constants φ_{10} and φ_{20} are the nominal phases of the two signals.

If the two signals are mixed in a balanced mixer and then low-pass filtered, so to make negligible the signal with sum-frequency $\nu_{10} + \nu_{20}$, the heterodyne signal results

$$s(t) \cong A_0 \cos\left[2\pi\left(\nu_{10} - \nu_{20}\right)t + \varphi_{10} - \varphi_{20} + \varphi_1(t) - \varphi_2(t)\right] \tag{7.28}$$

This signal can be then put under measurement, for example with a time counter as described in the previous section. Its frequency $|\nu_{10} - \nu_{20}| \ll \nu_{10}$ is called *beat frequency*. From the definition provided in Equation (5.11), for the heterodyne signal it follows

$$x_H(t) = \frac{\Delta\varphi(t)}{2\pi\nu_H} \tag{7.29}$$

where

$$\nu_H = |\nu_{10} - \nu_{20}| \tag{7.30}$$

$$\Delta\varphi(t) = \varphi_1(t) - \varphi_2(t) \tag{7.31}$$

We can then rewrite Equation (7.29) as

$$x_H(t) = \frac{\nu_0}{\nu_H} x(t) \tag{7.32}$$

from which we conclude that a given phase change corresponds to a larger time deviation for the heterodyne signal than for the original signal. We define *heterodyne improving factor* the ratio

$$\eta_H = \frac{\nu_0}{\nu_H} \tag{7.33}$$

Hence, the quantization error in a period measurement is reduced by the factor η_H. The heterodyne factor is often in the order of 10^2 in practical measurements.

This theory can be put in practice by setting up the test-bench illustrated in Figure 7.18. The signals from two independent oscillators are sent to the two input ports of a double balanced mixer. Its output is low-pass filtered to obtain the beat signal with frequency ν_H, amplified and eventually sent to the digital time counter, for example to measure its TE. Also in this case, to achieve a better measurement accuracy especially over long observation intervals, it is advisable to synchronize the time-counter internal oscillator to the reference signal, as shown in Figure 7.18. This system allows one to attain an excellent accuracy.

As a final remark, let us notice that beat frequencies, obtained by mixing two oscillating signals, are profitably used in several fields, also outside electronics. For instance, players of stringed musical instruments know the so-called beat-note tuning technique. Employed especially by electric-bass players, it consists of tuning a string by making it to vibrate while playing the corresponding harmonic note on the lower string at the same time. This harmonic note is used as a reference note for tuning. In this case, the beat note is not generated by the product of two oscillations, but by their sum (the notes played together). From basic trigonometry, we have

$$\cos\omega_1 t + \cos\omega_2 t = 2\cos\left(\frac{\omega_1 - \omega_2}{2}t\right) \cdot \cos\left(\frac{\omega_1 + \omega_2}{2}t\right) \tag{7.34}$$

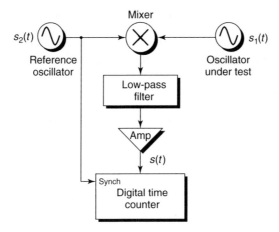

Figure 7.18.   Measurement set-up for the beat-frequency method

The result is a note at the average frequency $\omega_1 \cong \omega_2$, amplitude-modulated by a sine wave at half the difference of the frequencies (beat note). What is heard is a pulsating sound, with a pace as slower as the two strings are mutually better tuned. Therefore, to easily tune the string even without ear for music, it is enough to tighten it until the beat note stops pulsating.

7.5   HOMODYNE TECHNIQUES

The limit of the heterodyne method occurs when $v_{10} = v_{20}$ (that we will simply let v_0) [7.4]. Since the signal under measurement is now obtained by mixing two signals having the *same* frequency, this technique has been named *homodyne* (from the Greek etyma ὅμος = *same* and δύναμις = *force*). In this case, the signal output by the mixer and low-pass filtered results

$$s(t) \cong A_0 \cos \left[\varphi_{10} - \varphi_{20} + \varphi_1(t) - \varphi_2(t)\right] \tag{7.35}$$

By means of a phase shifter, it is easy to arrange the test set-up to have

$$\varphi_{10} - \varphi_{20} = \frac{\pi}{2} \tag{7.36}$$

so that

$$s(t) \cong -A_0 \sin \left[\varphi_1(t) - \varphi_2(t)\right] \cong A_0 \left[\varphi_2(t) - \varphi_1(t)\right] \tag{7.37}$$

Now, the problem is to produce a reference signal $s_2(t)$ that has exactly the same frequency as the signal under test v_0, without introducing significant correlation between $\varphi_1(t)$ and $\varphi_2(t)$. When a suitable method to produce this result is used, then it is possible to use $s(t)$ as a measure of $\varphi(t)$; the mixer, followed by the low-pass filter, works as *phase detector*. Two such methods are described in [7.4]: one based on a delay line and one on a PLL.

7.5.1   Delay-Line Methods

The homodyne method based on a delay line, to produce the reference signal $s_2(t)$ with exactly the same frequency v_0 as the signal under test $s_1(t)$, is illustrated in Figure 7.19 [7.4].

The delayed signal is given by

$$s_2(t) = \hat{s}_1(t - t_\mathrm{d}) = A_{20} \sin\left[2\pi v_0(t - t_\mathrm{d}) + \varphi_1(t - t_\mathrm{d}) + \varphi_{10} + \varphi_\mathrm{s}\right] \tag{7.38}$$

When the phase shifter is adjusted for quadrature, i.e. $\varphi_\mathrm{s} - 2\pi v_0 t_\mathrm{d} = \pi/2$, we have

$$s_2(t) = A_{20} \sin\left[2\pi v_0 t + \varphi_1(t - t_\mathrm{d}) + \varphi_{10} + \pi/2\right] \tag{7.39}$$

and the output of the mixer followed by the low-pass filter (phase detector) becomes

$$s(t) = A_0\left[\varphi_1(t - t_\mathrm{d}) - \varphi_1(t)\right] \tag{7.40}$$

Remembering Equation (5.81), we finally get

$$\overline{y}_{t-t_\mathrm{d}}(t_\mathrm{d}) = \langle y(t - t_\mathrm{d})\rangle_{t_\mathrm{d}} = \frac{1}{t_\mathrm{d}} \int_{t-t_\mathrm{d}}^{t} y(t)\mathrm{d}t = -\frac{s(t)}{2\pi v_0 A_0 t_\mathrm{d}} \tag{7.41}$$

From this relationship, we can see that the delay-line method can be used to produce samples $\overline{y}(n\tau_0)$, by varying the delay time t_d accordingly. However, this technique is used more frequently with a fixed delay by restricting its application to the range $\tau \gg t_\mathrm{d}$, so that $\overline{y}_{t-t_\mathrm{d}}(t_\mathrm{d})$ is a good approximation of the instantaneous frequency $y(t)$.

Under this assumption, the spectral analysis of the signal output by the mixer can be used to estimate the PSD of the frequency fluctuations

$$S_y(f) \cong \frac{1}{(2\pi v_0 t_\mathrm{d})^2} S_{s/s_0}(f) \quad \text{for } f \ll \frac{1}{\pi t_\mathrm{d}} \tag{7.42}$$

7.5.2   Phase-Locked Loop Methods

A general scheme of homodyne measurement based on a PLL, to produce the reference signal with exactly the same frequency v_0 as the signal under test, is illustrated in Figure 7.20 [7.3][7.4].

The signal from the oscillator under test $s_1(t)$ is mixed with that from the reference oscillator. The signals are kept in quadrature, that is a $\pi/2$ relative phase error between

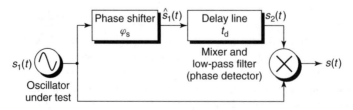

Figure 7.19.   Homodyne measurement set-up based on a delay line. (Adapted from [7.4] by permission of Academic Press)

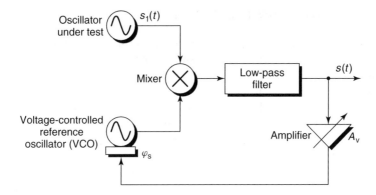

Figure 7.20. Homodyne measurement set-up based on a PLL

them is aimed at. Doing so, the signal $s(t)$ out of the mixer corresponds to phase fluctuations of the signal under test $s_1(t)$, with average nominally equal to zero, as yielded by Equation (7.37).

The reference oscillator is a voltage-controlled oscillator (VCO) held in phase lock with the oscillator under test (with constant relative phase difference $\varphi_s = \pi/2$) by a feedback circuit: the mixer output is low-pass filtered and then amplified to control the reference VCO frequency. The mixer is the key element of this system. The advent of the Schottky barrier diode was a significant breakthrough to make low-noise balanced mixers suitable for this kind of measure.

Two measurement methods have been conceived, based on this general scheme. They distinguish for the time constant of the loop, which controls the VCO frequency aimed at keeping it in quadrature with the reference oscillator.

The first method is called *loose phase-lock loop method* [7.3]. In this case, the system is adjusted to have a loop time constant (attack time[4]) τ_{PLL} very long. Therefore, the PLL tracks the phase fluctuations of the oscillator under test very slowly (loose phase lock). The system depicted in Figure 7.20 behaves then as a high-pass filter with cut-off frequency $B = 1/(2\pi\tau_{PLL})$ very small. Short-term phase fluctuations between the two oscillators appear as voltage fluctuations $s(t)$ out of the mixer. Thus, the system should be designed so that the time constant τ_{PLL} is much greater than the inverse of the lowest Fourier frequency of interest. For example, if we want to measure the spectrum of phase fluctuations down to 1 Hz, the loop time constant τ_{PLL} must be greater than $1/2\pi$ s.

The second method based on a PLL to produce the reference signal is the so-called *tight phase-lock loop* method [7.3]. The system scheme is exactly the same depicted in Figure 7.20, except that in this case the loop time constant is chosen very small, for example in the order of a few milliseconds, to keep the reference oscillator in tight phase-lock condition. In this case, the phase fluctuations are being integrated, for observation intervals longer than the attack time τ_{PLL} of the loop. Therefore, the voltage output $s(t)$ results proportional to the relative frequency fluctuations between the two oscillators.

A more detailed analysis of the behaviour of this homodyne measurement system, even according to different loop-filter transfer functions, is featured by [7.4].

---

[4] The PLL *attack time* is defined as the time needed for the PLL to recover on its output at least 70% of the amplitude of a phase step on its input.

7.6   MULTIPLE CONVERSION TECHNIQUES

The idea of mixing the signal from the oscillator under test with that from the reference one allows to achieve a measurement sensitivity inconceivable with the simple direct digital measurement. Yet, sometimes the resulting beat frequency is still unsuitable for frequency-stability measurement.

This frequency may be too high for available time counters or the heterodyne factor may be too small to achieve the desired enhancement. In this case, a second mixing stage, cascaded with the first one, may be used to produce the beat frequency desired.

Conversely, the direct beat frequency between two oscillators may be too small. For example, consider the beat frequency between two caesium frequency standards, which is in the order of one cycle per day (10^{-5} Hz): observing stability at shorter times is impossible. This limitation can be overcome by the use of two parallel mixing stages.

In the following, two examples of the above techniques will be outlined. They are summarized from papers [7.3][7.4], to which the reader is referred for further details.

7.6.1   Frequency Synthesis

A frequency synthesizer is usually the most convenient means to produce reference signals with arbitrary frequency. A mixing stage preceding the synthesizer may be used both to move the signal to a more convenient frequency range and to enhance the noise from the oscillator under test compared to the short-term noise from the synthesizer. An example of measurement configuration demonstrating both aspects was described in [7.4] and is shown in Figure 7.21.

The first mixing stage from the microwave to the radio frequency results in a substantial heterodyne improving factor (about 77). The output of the first conversion stage lies within the range of low-noise commercial frequency synthesizers, thus making it possible to obtain a fixed, low beat frequency over a wide range of input frequencies with a second mixing stage. The initial mixing stage also reduces the frequency synthesizer's contribution to the measurement-system noise.

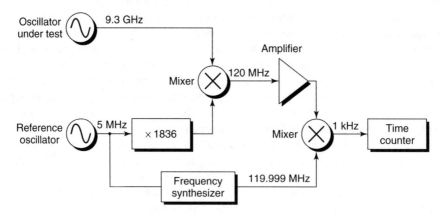

Figure 7.21.   Use of frequency synthesis for measuring oscillators whose frequency differs significantly from the available low-noise reference. (Adapted from [7.4] by permission of Academic Press)

7.6.2 The Dual-Mixer Time-Difference Technique

The *Dual-Mixer Time-Difference* (DMTD) method, originally proposed in [7.1][7.5], is often reported as the state of the art in frequency-stability measurements. In this section, we summarize its descriptions featured by [7.3][7.4].

The underlying idea consists of measuring the time error between the beat signals obtained by multiplying the signals from two oscillators with a common signal. In its original form, a transfer oscillator and two mixers in parallel are used, to permit short-term frequency-stability measurements between oscillators that have an inconveniently small frequency difference (e.g., in the case of atomic standards), as shown in Figure 7.22 [7.3].

In this figure, *oscillator 1* may be considered the oscillator under test and *oscillator 2* the reference oscillator. Their signals, with frequency ν_1 and ν_2 respectively, are sent to the ports of a pair of double balanced mixers. A third oscillator (*transfer oscillator*), with separate symmetric well-buffered outputs, is fed to the other two mixer ports. The frequency ν_0 of this common oscillator is offset by a small amount from that of oscillators 1 and 2, so to obtain the desired heterodyne improving factor.

Then, two different beat signals come out of the two mixers, out of phase by the same amount as oscillators 1 and 2. Therefore, their relative time error corresponds to the time difference between oscillators 1 and 2, amplified by the heterodyne improving factor. Further, the beat frequencies differ by an amount equal to the frequency difference between oscillators 1 and 2.

A phase shifter may be inserted to adjust the relative phase by an additional shift φ_s, so that the two beat rates are nominally in phase. This adjustment sets up the nice condition that the noise of the common oscillator tends to cancel (at least for certain types of noise) when the time difference is measured.

The time difference $x(t)$ between the signals from the two oscillators compared is thus given simply by

$$x(t) = \frac{x_H(t)}{\tau_b \nu_0} - \frac{\varphi_s}{2\pi \nu_0} + \frac{k}{\nu_0} \tag{7.43}$$

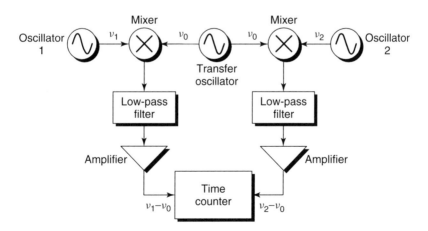

Figure 7.22. Basic set-up for the Dual-Mixer Time-Difference technique. (Adapted from [7.3], NIST. Not copyrightable)

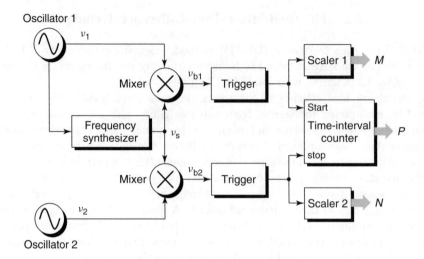

Figure 7.23. Set-up for the Dual-Mixer Time-Difference technique with synthesis of the common transfer frequency. (Adapted from [7.4] by permission of Academic Press)

where $x_H(t)$ is the time difference measured by the counter between the two beat signals, τ_b is the nominal beat period, ν_0 is the nominal carrierfrequency, φ_s is the phase shift added to the signal from oscillator 1 and k is an integer to be determined in order to remove the cycle ambiguity.

It is important to notice that knowing φ_s and k is needed only if the absolute time difference is desired. For frequency measurements and for time fluctuations measurements, k may be assumed zero and the only problem is to realize whether it went through a cycle during a set of measurements. Finally, we remark that the product $\tau_b \nu_0$ is the heterodyne improving factor of Equation (7.33).

The frequency of the common oscillator is most easily generated with a synthesizer locked to one of the two oscillators, for example to oscillator 1, as shown in Figure 7.23 [7.4].

By convention, let oscillator 1 be the one under test and oscillator 2 the reference one. Moreover, let the synthesizer output frequency be set lower than that of the oscillator under test. Then, we can express the synthesizer output frequency as

$$\nu_s = \nu_1 \left(1 - \frac{1}{R}\right) \tag{7.44}$$

The constant R is equal to the heterodyne improving factor, since the beat frequency between oscillator 1 and the synthesizer is

$$\nu_{b1} = \nu_1 - \nu_s = \frac{\nu_1}{R} \tag{7.45}$$

The ensemble oscillator-synthesizer-mixer works as a frequency divider. The triggers are plain zero-crossing detectors. The scalers are registers that work as system clocks, by counting the time in units of cycles of the input signals.

The phase difference between the two oscillators can be written in terms of the readings of the three counters:

$$\varphi_2(t_M) - \varphi_1(t_M) = 2(N - M)\pi - 2\pi \left[\bar{v}_{b2}\left(t_M; t_N\right)\right] \tau_c P \tag{7.46}$$

where M is the reading of scaler 1, N is the reading of scaler 2, t_M and t_N are the trigger instants respectively of scalers 1 and 2, $\bar{v}_{b2}(t_M; t_N)$ is the average beat frequency in the interval $[t_M, t_N]$, P is the reading of the time-interval counter and τ_c is the period of its time base. Comparison of Equation (7.46) to Equation (7.20) for the direct digital measurement reveals that the role of scalers is to accumulate the coarse phase difference between the oscillators, while the time-interval counter provides fine-grain resolution of the fractional cycle.

The main advantage of the DMTD technique over plain direct time-interval measurement is that the noise performance is improved by the large heterodyne factor, allowing sub-picosecond time resolution.

The average beat frequency $\bar{v}_{b2}(t_M; t_N)$ cannot be known exactly, but it can be estimated with sufficient precision if it changes slowly compared to the interval between measurements. By distinguishing the variables of two independent measurements with a prime, then

$$\bar{v}_{b2}\left(t_M : t_N\right) \cong \frac{N' - N}{R\dfrac{M' - M}{v_{10}} + \tau_c(P' - P)} \tag{7.47}$$

7.7 CLOCK STABILITY MEASUREMENT IN TELECOMMUNICATIONS

The methods outlined in the previous sections do provide the engineer with a wide set of powerful techniques to measure time and frequency while coping with the most various requirements, with both analogue and digital instrumentation.

Actually, clock stability measurement in telecommunications usually does not require the resolution enhancement featured by heterodyne and homodyne methods, so that the TE direct digital measurement has been adopted *de facto* as the standard technique. Nevertheless, telecommunications measurements pose somehow peculiar issues, for example in the measurement of MTIE. In this section, the whole subject is expounded in some detail, aiming at pointing out the distinctive features of stability measurement in telecommunications and the techniques adopted in this context, for conformance testing and for more advanced performance measures.

7.7.1 Distinctive Features

Time and frequency stability measurements are carried out in telecommunications mainly to three different purposes:

(1) *conformance testing* of clocks, i.e. for simply checking their compliance to relevant standards;

(2) *performance evaluation*, aiming at a deeper insight into the actual clock behaviour and noise characterization;

(3)   *in-service performance monitoring* of timing equipment deployed in office buildings, to ensure proper operation of the network, by means of field personnel or built-in hardware.

Aside from the purpose, clock stability measurement in telecommunications poses several peculiar issues and requirements with respect to traditional laboratory measurements on oscillators. Its most distinctive features may be summarized as follows.

- The range of interest of the observation interval τ is most often centred on the range 10^{-1} s $\leq \tau \leq 10^4$ s, over which the international standards specify the limit values of stability quantities (MTIE and TDEV in particular). Wider ranges are sometimes characterized for performance evaluation and for scientific purposes.

- Several SASE, SEC and other equipment clocks, although built around a very low-noise crystal or atomic oscillator, control their phase locking by means of digital electronics and numerical algorithms. These DPLLs may produce a considerable short-term noise in the output timing signal (in some cases, an amplitude even up to *tens* of nanoseconds was measured, for example as it has been shown in Figures 5.34 through 5.36).

- The standard masks specifying telecommunications clock stability allow rather high limits for the phase noise under measurement (e.g., even more than 10 or 100 ns of TIErms or MTIE) [7.29]–[7.31], well above the resolution of commercial time counters.

- Time and frequency measurements in telecommunications applications do not necessarily take place in a laboratory under strictly controlled conditions, but are often accomplished in the field.

All the above considerations make the TE direct digital measurement, although not adequate to resolve the short-term time fluctuations of precision oscillators (i.e., for $\tau \ll 1$ s), well suited to telecommunications clock stability testing. On the one hand, the resolution of a good time counter (in the order of 200 ps or less) is suitable for measuring the stability of most telecommunications clocks in the range of interest, also beyond mere conformance testing. On the other hand, the robustness of this set-up is essential when performing field measurements, where more sophisticated sensitivity-enhancement methods easily suffer mechanical or electromagnetic interference.

7.7.2   Practical Measurement Procedure by Acquisition of Sequences of Time Error Samples

According to what was stated in the previous section, the practical measurement procedure outlined in the following steps has consolidated *de facto* as the standard method to characterize the stability of telecommunication clocks, also beyond mere conformance testing.

Step 1: Acquisition of a TE Data Sequence

Acquire and store a sequence $\{x_i\}$ of N samples of TE, with sampling period τ_0 and over a measurement interval $T = (N - 1)\tau_0$.

Typical values of the measurement parameters are τ_0 in the order of a few milliseconds and T up to $10^3 - 10^4$ s. Thus, typical sequence lengths are in the order of N from 10^4 to 10^6.

Latest ETSI and ITU-T standards [7.14][7.29]–[7.31] recommend to measure the TE data with $\tau_0 \leq 33$ ms, $T \geq 12\tau$ and *'through an equivalent 10 Hz, first-order, low-pass measurement filter'*, for stability measures on observation intervals in the most common range from 0.1 s to 1000 s. For observation intervals in the range from 10 s to 10^5 s, ETSI and ITU-T recommend to measure data through a 0.1 Hz low-pass filter, with sampling time $\tau_0 \leq 3.3$ s and measurement period $T \geq 12\tau$. For any other range of observation intervals, the maximum sampling time τ_0 and the filter cut-off frequency f_c are recommended to have the same ratio to the minimum observation interval τ_{min} as above: i.e., $\tau_0 \leq \tau_{min}/3$ and $f_c = 1/\tau_{min}$.

Actually, the specifications above may be considered as general guidelines in most cases. Performing measurements with sampling time longer than $\tau_{min}/3$ or collecting data for a measurement period shorter than $12 \tau_{max}$ is not really an issue, yielding just some lack in the confidence of results. Several measurement results that we provided in this book were measured without strictly complying with the limits above (anyway, they were collected before those ETSI and ITU-T specifications were made).

On the other hand, the low-pass measurement filter deserves some more specific comment. Its rationale seems to have been the wish of getting rid of the spectrum aliasing, which is supposed to occur by sub-sampling the TE function in the presence of broadband noise. Nevertheless, in the opinion of the author of this book, these arguments are questionable indeed [7.32]. Some discussion of this aspect will be provided in Section 7.7.3.

The samples x_i are usually collected through direct digital measurement of the TE between the CUT and the RC, as shown in Figures 7.13 and 7.15 for square-wave timing signals. Thus, a typical test set-up is shown in Figure 7.24, in the case of the synchronized-clock measurement configuration. An atomic frequency standard is a good choice for the RC, especially when performing measurements in the independent-clock configuration. The RC should also supply the time base to the time counter. Data acquisition, storing, numerical processing and graphic plotting are accomplished by a computer interfaced to the time counter.

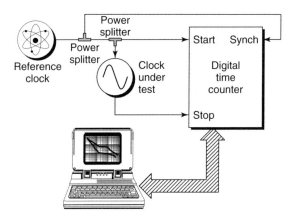

Figure 7.24. Measurement set-up for TE acquisition, storing and post-processing (synchronized-clock configuration). (Reproduced from [7.28], ©1997 IEEE, by permission of IEEE)

Alternatively, some commercial instruments include in one box the time counter and a built-in computer, equipped with all the software. Such instruments, therefore, merge the functions of TE data acquisition, storing, numerical processing and graphic plotting. Obviously, they have some appeal, because they are sold as ready-to-use black boxes that are capable of doing easily everything needed by telecommunications engineers involved in synchronization testing. When purchasing one of those instruments, nevertheless, special care should be taken in order to ascertain indeed what algorithms are used to compute the stability quantities and what all the relevant measurement parameters are. Moreover, there is some lack in flexibility. We expect that a researcher who carries out measurements to scientific purposes would enjoy adjusting parameters, trying new algorithms, etc. In this case, a set-up like that depicted in Figure 7.24 is the only solution.

Finally, when special sensitivity is required, e.g. for laboratory characterization of the short-term noise of very good clocks, TE measurement may be accomplished through more sophisticated methods, such as heterodyne techniques.

Step 2: Numerical Processing and Computation of Stability Quantities

Compute the stability quantities of interest (viz. ADEV, MADEV, TDEV, MTIE, TIErms) with the standard estimators of Equations (7.4) through (7.8) based on the acquired sequence $\{x_i\}$.

Other meaningful quantities may be computed as well with suitable numerical algorithms: for example PSDs, autocovariance function, other variances, etc. (cf. for example Appendix 7A).

Numerical processing may be carried out both off-line and in real time, while acquiring TE data. In the latter case, special implementations of estimators (7.4)–(7.8) must be used, to allow progressive computation.

It is worthwhile noticing that MTIE computation, owing to its distinctive nature of peak measurement and to the number of operations nested in the estimator (7.8), may be quite difficult and troublesome in most practical cases. This issue is widely discussed in Section 7.7.3.

Step 3: Measurement of Test-Bench Background Noise

Do the same without the CUT depicted in Figure 7.24, by directly feeding both time-counter input ports with the timing signal split from the RC, for assessing the test-bench background noise floor.

As remarked previously, the resolution of commercial time counters is definitely adequate for conformance testing of telecommunications clocks. In the short term, nevertheless, the oscillator output noise may be cloaked by the test-bench background noise. Therefore, it is good practice to characterize this background noise for any measurement campaign of some importance, in order to discriminate what parts of the measurement results are really due to the CUT and what are added by instrumentation instead.

The test-bench background noise can be characterized by splitting the timing signal from the reference clock and by feeding it into the time-counter input ports, according to the scheme depicted in Figure 7.25. For splitting the signal under test, a power splitter or even a simple T-junction may be used. Obviously, the twin signals sent to the counter input ports should have more or less the same physical characteristics (viz. amplitude, waveform, impedance, etc.) as the signals sent in the real measurements, in order to allow meaningful comparison between the results obtained in the two cases. The experimental results shown in Figures 5.26 through 5.28 have been obtained in this way, on characterizing the background noise of a TE direct digital measurement set-up.

Then, the results so obtained must be compared to that obtained by measuring the clock under test. A quick comparison of the stability curves obtained in the two cases allows one to distinguish where the clock noise was actually measured and where the test equipment noise dominates instead. Where the measured stability curves coincide, as shown in the example plot of Figure 7.25, it is evident that the values measured on the clock under test are not meaningful.

Measurement Results: Example of Clock Experimental Characterization

As final example of application of the measurement procedure outlined above, here below we present a complete set of graphs of stability quantities measured on a SEC [7.28]. These results were measured on an ADM-4 of the same supplier as the ADM-1 dealt with in Section 5.10.3 (measurement results provided in Figures 5.34–5.36), which we denoted as supplier B. Actually, this ADM-4 clock was designed more recently, and it is supposed to be an improved version of the previous model equipped on the ADM-1.

The measurement procedure is fully detailed in Appendix 7A. The scheme of the experimental set-up, in synchronized-clock configuration, is shown in Figure 7.27. A TE sequence of $N = 96\,750$ samples was acquired, with sampling period $\tau_0 \cong 37.5$ ms and over a total measurement interval $T = 3600$ s. Then, all stability quantities were computed with the usual standard estimators or as described in Appendix 7A.

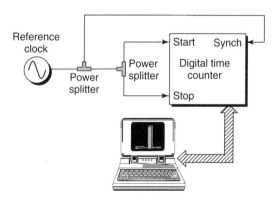

Figure 7.25.   Experimental set-up for the characterization of the test-bench background noise (direct digital measurement of TE)

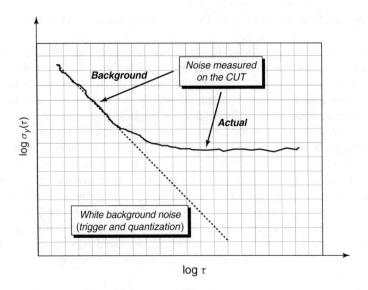

Figure 7.26. Comparison of the stability curves measured with the set-ups of Figures 7.24 and 7.25, in order to distinguish where the clock noise was actually measured and where the background noise dominates instead

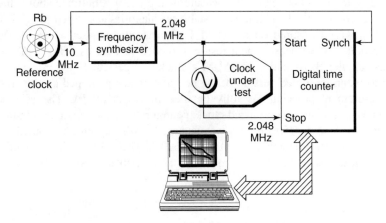

Figure 7.27. Experimental set-up used for the characterization of the ADM-4 SEC (measurement results shown in Figures 7.28–7.33)

The graphs shown in Figures 7.28 through 7.33 summarize in one nice experimental example some of the concepts expounded previously. In particular, Figures 7.28 and 7.29 plot the TE sequence measured, using different decimation ratios to cope with the limitations of the graphic plotting package. The former graph plots the whole sequence over the total measurement interval, for $0 \leq t \leq 3600$ s. The latter, instead, zooms on the interval $800 \leq t \leq 1400$ s, to give a glance on the TE trend on a shorter time scale. The subsequent Figures 7.30, 7.31, 7.32 and 7.33 plot the spectrum, the Allan deviations, the time deviation and TIErms curves and the MTIE curve respectively. From inspection of these graphs, a few comments can be drawn.

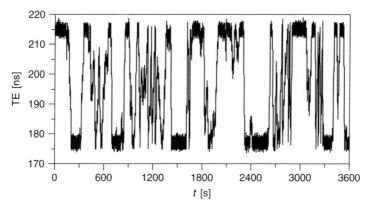

Figure 7.28. TE sequence $\{x_i\}$ measured on the ADM-4 SEC from supplier B ($N = 96\,750$, $\tau_0 \cong 37.5$ ms, $T = 3600$ s) plotted for $0 \leq t \leq 3600$ s. (Reproduced from [7.28], ©1997 IEEE, by permission of IEEE)

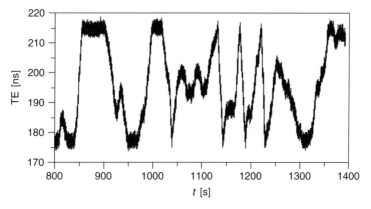

Figure 7.29. TE sequence $\{x_i\}$ measured on the ADM-4 SEC from supplier B ($N = 96\,750$, $\tau_0 \cong 37.5$ ms, $T = 3600$ s) plotted for $800 \leq t \leq 1400$ s [7.28]

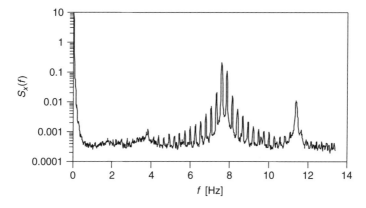

Figure 7.30. PSD estimate $S_x(f)$ measured on the ADM-4 SEC from supplier B ($N = 96\,750$, $\tau_0 \cong 37.5$ ms, $T = 3600$ s). (Reproduced from [7.28], ©1997 IEEE, by permission of IEEE)

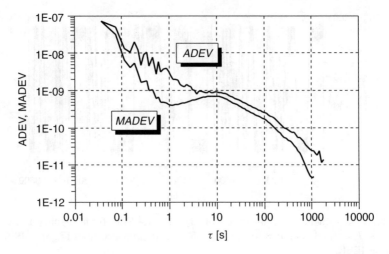

Figure 7.31.   ADEV(τ) and MADEV(τ) curves measured on the ADM-4 SEC from supplier B ($N = 96\,750$, $\tau_0 \cong 37.5$ ms, $T = 3600$ s). (Reproduced from [7.28], ©1997 IEEE, by permission of IEEE)

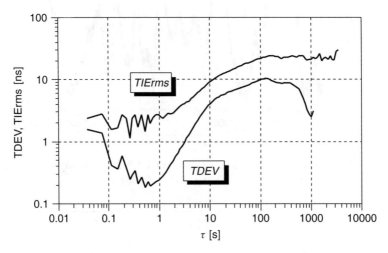

Figure 7.32.   TDEV(τ) and TIErms(τ) curves measured on the ADM-4 SEC from supplier B ($N = 96\,750$, $\tau_0 \cong 37.5$ ms, $T = 3600$ s). (Reproduced from [7.28], ©1997 IEEE, by permission of IEEE)

- Although the DPLL's frequency quantization error of the ADM-1 SEC (the 26-ns short-term saw-toothed noise plotted in Figure 5.34) appears to have been eliminated in this new ADM-4 design, a slower wander appears in the TE process of this clock. As shown by the graph in Figures 7.28 and 7.29, these TE fluctuations have amplitude even exceeding 45 ns. This pseudo-periodic component does not appear in the PSD of Figure 7.30, because it is located at a very low Fourier frequency.

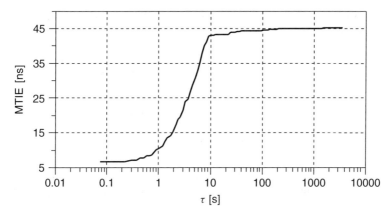

Figure 7.33. MTIE(τ) curve measured on the ADM-4 SEC ($N = 96\,750$, $\tau_0 \cong 37.5$ ms, $T = 3600$ s). (Reproduced from [7.28], ©1997 IEEE, by permission of IEEE)

- The PSD in Figure 7.30 (evaluated neglecting a multiplicative factor) exhibits several spikes at harmonic frequencies centred on 3.8 Hz, 7.6 Hz and 11.4 Hz. These periodic components appear as ripples in the time-domain plots of ADEV, MADEV, TDEV and TIErms (Figures 7.31 and 7.32).

- The ADEV and MADEV curves in Figure 7.31 exhibit the typical inflection point of the measurements in the synchronized-clock configuration (cf. Figures 7.5 and 7.6).

- The MTIE plot of Figure 7.33 confirms that the 45-ns output phase shifts of the CUT may be sharp: the MTIE curve rises from 7 to 45 ns in the narrow interval $1\text{s} \leq \tau \leq 10$ s.

Finally, it must be pointed out that repeated measurements on this SEC yielded similar results. Therefore, the behaviour herein shown is not accidental but should be considered typical of this clock.

7.7.3 Measurement of Maximum Time Interval Error

The measurement of MTIE features some peculiar issues, due to its nature of raw peak-to-peak measure and to the heavy computational weight of the estimator (7.8). In this section, this topic is addressed in detail, discussing several possible approaches from the crude computation of the standard estimator to other faster ones. A non-standard, but significant technique for assessing clock performance in terms of MTIE is also described. Finally, the impact of the so-called anti-aliasing low-pass filter recommended by ETSI and ITU-T for MTIE measurements is discussed.

7.7.3.1 *Direct Approach: Crude Computation*

Starting from the TE measured sequence $\{x_i\}$, the most straightforward way to compute MTIE(τ,T) is to directly compute the estimator (7.8). Letting $N_T = T/\tau_0 + 1$ be the total number of available TE samples and $N_\tau = \tau/\tau_0 + 1$ be the number of samples available in a window (observation interval) of span τ, to obtain a single value $\text{MTIE}(\tau, T)$ the following expression has to be computed:

$$\text{MTIE}(\tau, T) = \max_{j=1}^{N_T - N_\tau + 1} \left[\max_{i=j}^{N_\tau + j - 1} (x_i) - \min_{i=j}^{N_\tau + j - 1} (x_i) \right] \qquad (7.48)$$

As pointed out in Appendix 5B, the plain computation of this formula is unadvisable and quickly tends to be unmanageable on TE sequences of practical length, due to the number of operations nested in evaluation loops.

Moreover, MTIE has proved very useful for characterizing clocks focusing on short-term noise too, beyond clock conformance testing. In this case, it should be computed on data collected at the highest possible rate, in order to capture the fastest phase fluctuations, and thus on very long TE sequences.

7.7.3.2 Suitable Approaches to Practical Measurement

Different approaches can be envisaged to face up to the issue of storing and processing this huge amount of data, and therefore to accomplish the practical measurement of MTIE without making use of a supercomputer.

(1) Contriving a suitable efficient algorithm alternative to the crude computation of (7.48). A fast algorithm, based on binary decomposition, has been described in Appendix 5B. Any computational trick (except real-time data processing), however, does not let us avoid storing all those data.

(2) Drastically reducing — trivial but effective! — the number N_T of samples x_i to process. This is achievable through shortening the measurement period T and/or lengthening the sampling period τ_0 (which is equivalent to sample decimation).

(3) Following an alternative approach, such as the one outlined next (*measurement on disjointed intervals*) [7.34].

7.7.3.3 Practical Approaches: Techniques Based on Estimator Computation

The evaluation of MTIE curves through a plain implementation of estimator (7.48) is feasible only on sequences of few samples, acquired over a short measurement period T and/or over a long sampling period τ_0. More efficiently, the binary-decomposition algorithm described in Appendix 5B allows to crunch longest sequences, but it still needs them stored on the computer hard disk.

7.7.3.4 Practical Approaches: Measurement on Disjointed Intervals

The underlying idea is not to consider *all* the sliding windows of width τ over the measurement interval T, but to perform a sequence of M consecutive independent measurements of $\text{MTIE}(\tau)$, as shown in Figure 7.34 [7.34], each one taking into account disjointed sets of samples x_i collected at the *maximum rate* allowed by the measurement instrumentation. Each $\text{MTIE}(\tau)$ measurement is carried out by letting $T = \tau$ in Equation (7.48). Thus

$$\text{MTIE}(\tau) = \max_{i=1}^{N_\tau}(x_i) - \min_{i=1}^{N_\tau}(x_i) \qquad (7.49)$$

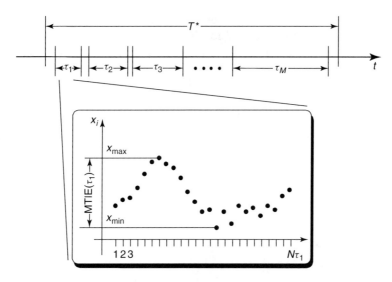

Figure 7.34.   MTIE(τ) measurement technique on disjointed intervals. (Reproduced from [7.34], ©1996 IEEE, by permission of IEEE)

In Figure 7.34, the overall measurement interval has been marked with T^*, since it has nothing to do with T of MTIE definition and Equation (7.48). Here, it is simply the period during which the clock is under test including dead time between measurements.

Moreover, M different measurements of MTIE(τ) can be accomplished in sequence by varying τ step by step from a minimum value $\tau_1 = \tau_{MIN}$ up to a maximum $\tau_M = \tau_{MAX}$, as a geometric progression of ratio

$$\rho = \sqrt[M-1]{\frac{\tau_{MAX}}{\tau_{MIN}}} \tag{7.50}$$

The spirit is to collect M snapshots of the clock peak-to-peak noise, by evenly sweeping the interval of interest [τ_{MIN}, τ_{MAX}]. Since the expression (7.49) is very simple, it can be evaluated in real time (e.g., in hardware) with virtually no limits on the number of samples x_i to process. Thus, it becomes possible to attain the maximum instrumentation sampling rate without worrying about data storage and, what's more, the time needed to compute (7.48).

The output of this test procedure is a scatter diagram, which typically shows a cloud of points (measurement results) of coordinates (τ, MTIE(τ)). The cloud represents the CUT behaviour during the whole measurement period T^*.

A *necessary condition* for the clock compliance with standards would be that *all* the measured points be below the specified mask, if adopting the classical definition of MTIE(τ, T) (i.e., not the percentile definition). Obviously, the test becomes stricter as the number M of subsequent MTIE measurements becomes larger.

7.7.3.5   *Experimental Results*

Both the standard technique based on estimator computation and the one on disjointed intervals have been extensively applied by the author of this book in testing several telecommunications clocks, for conformance testing and for research purposes. In all cases, the scheme of the experimental set-up, in synchronized-clock configuration, was the same already shown in Figure 7.27 for the ADM-4.

As far as the *estimator-computation technique* is concerned, a first example of MTIE curve, measured on the ADM-4 SEC from supplier B, was shown in Figure 7.33. A second example is now provided in Figures 7.35 and 7.36 [7.34], which depict respectively the TE and MTIE measured on the same ADM-1 from supplier B already dealt with in Section 5.10.3 (cf. Figures 5.34–5.36). In this case, a TE sequence of $N = 47\,450$ samples was acquired, with sampling period $\tau_0 \cong 37.5$ ms and over a total measurement interval $T = 1800$ s.

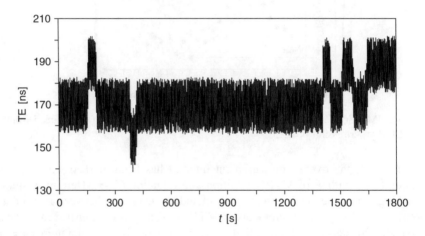

Figure 7.35. TE sequence $\{x_i\}$ measured on the ADM-1 SEC from supplier B ($N = 47\,450$, $\tau_0 \cong 37.5$ ms, $T = 1800$ s). (Reproduced from [7.34], ©1996 IEEE, by permission of IEEE)

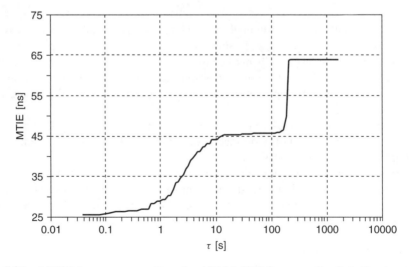

Figure 7.36. MTIE(τ) curve measured on the ADM-1 SEC from supplier B (estimator computation technique, $N = 47\,450$, $\tau_0 \cong 37.5$ ms, $T = 1800$ s). (Reproduced from [7.34], ©1996 IEEE, by permission of IEEE)

As far as the *measurement on disjointed intervals* is concerned, we notice that this technique allows exploiting the maximum TE sampling rate achievable by the time counter. A special hardware function of the time counter that we used (Hewlett-Packard HP5372A) allows, in fact, real-time evaluation of (7.49) on up to 2×10^9 TE samples x_i measured on *every* edge of the 2.048 MHz timing signals (i.e., $\tau_0 = 488$ ns).

Some measurement results are provided in Figures 7.37 through 7.42 [7.34]. They were obtained with this technique by testing a SASE and the clocks of various telecommunications equipment, namely a PSTN digital switching exchange, two ADM-1s, an ADM-4 and a DXC 4/3/1. We denoted their suppliers according to the same notation used so far, since we already showed experimental results for all of them except the DXC. For ease of reference, the actual values of the parameters τ_{MIN}, τ_{MAX}, M and T^* for each test, together with the CUT type and supplier label, are summarized in Table 7.1.

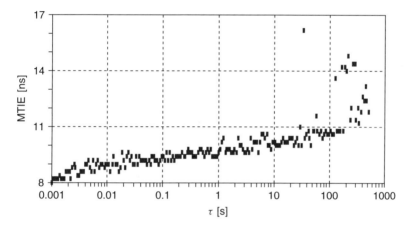

Figure 7.37. MTIE values measured on the clock of the PSTN digital switch from supplier D (measurement on disjointed intervals). (Reproduced from [7.34], ©1996 IEEE, by permission of IEEE)

Table 7.1 Actual measurement parameters for the results shown in Figures 7.37 through 7.42 (measurement on disjointed intervals)

Figure	CUT Type	Supplier	τ_{MIN}	τ_{MAX}	M	T^*
7.37	PSTN digital switch clock (OCXO)	D	1 ms	500 s	200	~2.5 h
7.38	ADM-1 SEC (TCXO)	B	1 ms	500 s	300	~3.5 h
7.39	ADM-4 SEC (TCXO)	B	1 ms	1000 s	500	~10.5 h
7.40	ADM-1 SEC (TCXO)	A	1 ms	1000 s	100	~2.5 h
7.41	DXC 4/3/1 SEC (TCXO)	E	1 ms	1000 s	500	~10.5 h
7.42	SASE (OCXO)	C	1 ms	1000 s	1000	~21 h

MTIE graphs regarding SECs (Figures 7.38 through 7.41) depict also the limits to not exceed as specified by ITU-T Rec. G.813 [7.29] and ETSI EN 300 462 [7.14]. A solid line is used to plot the current mask (G.813-2) while the dashed line (G.813-1) denotes the old mask, for a long time in force, specified in earlier versions of ITU-T and ETSI standards.

As shown by the graphs provided, some SECs exhibit higher noise than other CUTs, as they are implemented with lower cost oscillators. In particular, let us focus on the SEC of ADM-1 from supplier B, whose measurement results have been already shown in Figures 5.34–5.36 and in Figures 7.35–7.36. The MTIE measurements plotted in Figure 7.38 show that it is not compliant with the original mask G.813-1, since it exhibits a 26-ns noise floor for $\tau \le 13$ ms and some measured MTIE values are above the limits

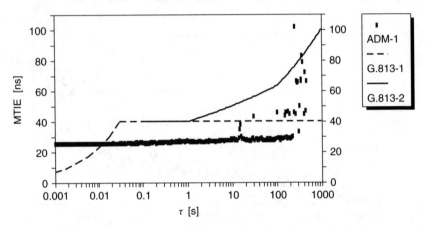

Figure 7.38. MTIE values measured on the SEC of the ADM-1 from supplier B vs ITU-T G.813 old and newer masks (measurement on disjointed intervals). (Reproduced from [7.34], ©1996 IEEE, by permission of IEEE)

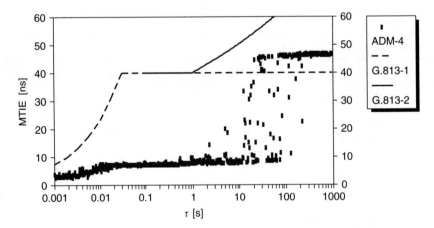

Figure 7.39. MTIE values measured on the SEC of the ADM-4 from supplier B vs ITU-T G.813 old and newer masks (measurement on disjointed intervals). (Reproduced from [7.34], ©1996 IEEE, by permission of IEEE)

in the range 10 s $\leq \tau \leq$ 1000 s. The new mask G.813-2, on the contrary, is much more permissive, so that almost all the measured values are below those limits.

Analogous considerations apply for the ADM-4 SEC of the same supplier B, whose measurement results have been already shown in Figures 7.28–7.33. As shown by the graph in Figure 7.39, while the measured values for $\tau > 10$ s are well above the original mask G.813-1, the new mask G.813-2 has granted 'amnesty', allowing under the limits also the 46-ns values.

Not surprisingly, finally, the clock featuring the best performance is the SASE of Figure 7.42.

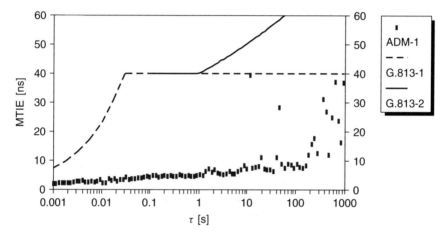

Figure 7.40. MTIE values measured on the SEC of the ADM-1 from supplier A vs ITU-T G.813 old and newer masks (measurement on disjointed intervals). (Reproduced from [7.34], ©1996 IEEE, by permission of IEEE)

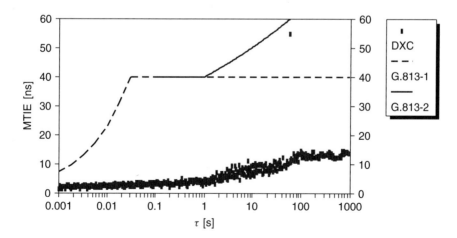

Figure 7.41. MTIE values measured on the SEC of the DXC 4/3/1 from supplier E vs ITU-T G.813 old and newer masks (measurement on disjointed intervals). (Reproduced from [7.34], ©1996 IEEE, by permission of IEEE)

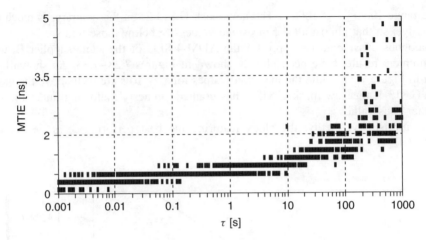

Figure 7.42.   MTIE values measured on the SASE from supplier C (measurement on disjointed intervals). (Reproduced from [7.34], ©1996 IEEE, by permission of IEEE)

7.7.3.6   Comparison between the Estimator-Computation Technique and the Measurement on Disjointed Intervals

While the technique based on the estimator computation is more in line with the formal definition of MTIE than the measurement on disjointed intervals, yet it implies the issue of its computational weight exploding with the number of TE stored samples.

The method of measurement on disjointed intervals, on the other hand, distinguishes itself for its ability in capturing the fastest phase fluctuations, thus allowing very small values of τ in characterizing clocks with MTIE. Moreover, scatter diagrams are very useful in rendering the statistics of measured values — clouds thicken where values are more likely — and thus help in checking percentile masks at a glance.

We want to point out, moreover, the excellent agreement between the results measured with the two techniques on the SECs of the ADM-1 and the ADM-4 of supplier B. To this aim, Figures 7.43 and 7.44 merge in one graph the results shown previously for these clocks.

Although measurements were accomplished in very different times, in both cases the MTIE curves evaluated through estimator computation do match with measurements on disjointed intervals, thus confirming on the one hand the validity of the latter method, on the other that those graphs really express the typical behaviour of CUTs and not fortuitous misbehaviours.

Finally, measurements accomplished on disjointed intervals appear to be stricter conformance tests as they yield slightly worse results (i.e., higher MTIE(τ) values being equal τ), compared to those obtained through estimator computation on *short* sequences $\{x_i\}$ sampled at a much lower rate. This is quite natural, since the peak-to-peak deviation of a set of N_τ TE samples, spanning an observation interval τ, is greater for larger N_τ (keeping constant τ, i.e. by decreasing the sampling period τ_0). We remark, again, that in the graphs shown above MTIE measurements for $\tau = 1000$ s are based on 2×10^9 samples.

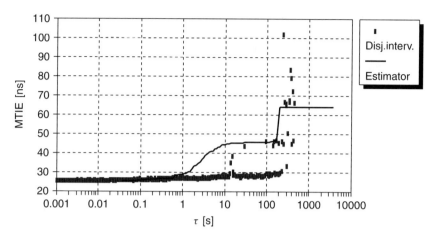

Figure 7.43. Comparison of MTIE results collected by estimator-computation technique (cf. Figure 7.36) and by measurement on disjointed intervals (cf. Figure 7.38) on the SEC of the ADM-1 from supplier B

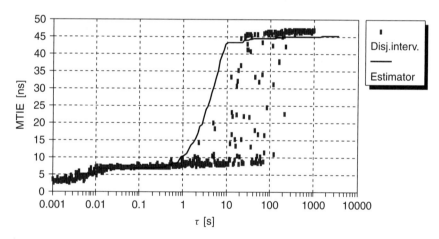

Figure 7.44. Comparison of MTIE results collected by estimator-computation technique (cf. Figure 7.33) and by measurement on disjointed intervals (cf. Figure 7.39) on the SEC of the ADM-4 from supplier B

7.7.3.7 *Impact of the Anti-Aliasing Low-Pass Pre-Filtering*

As stated previously, latest ETSI and ITU-T standards [7.14][7.30][7.31][7.29] recommend to evaluate MTIE based on TE data measured, for observation intervals in the range from 0.1 s to 1000 s, *'through an equivalent 10 Hz, first-order, low-pass measurement filter'*. The rationale of this choice seems to have been the wish of getting rid of the spectrum aliasing, which is supposed to occur by sub-sampling the TE function in the presence of broadband noise, and of achieving consistency among measurements carried out with different set-ups.

Nevertheless, the actual impact and the rationale itself of such an anti-aliasing pre-filtering should be investigated thoroughly in order to support this choice or not. In paper [7.32], the impact of such pre-filtering was thoroughly studied, on simulated noise and on TE sequences measured on real clocks.

A first reflection is that the 'anti-aliasing' name itself is not appropriate. Aliasing must be counteracted, by low-pass pre-filtering, when we sample a signal and we want to reconstruct it back from the sample sequence. Stability measures, such as the Allan variance, are not affected by aliasing issues.

Second, in a bit synchronizer or an SDH pointer processor, TE fast fluctuations are obviously as important as slower ones and *may cause buffer threshold overflows as well*, depending on their peak amplitude. Of course, no 10-Hz low-pass filtering is performed on the TE of digital signals at equipment input ports before writing bits in elastic buffers. Therefore, it is possible that measuring MTIE on the timing interfaces through a 10-Hz low-pass filter may yield too optimistic results, which seem to ensure that buffer thresholds

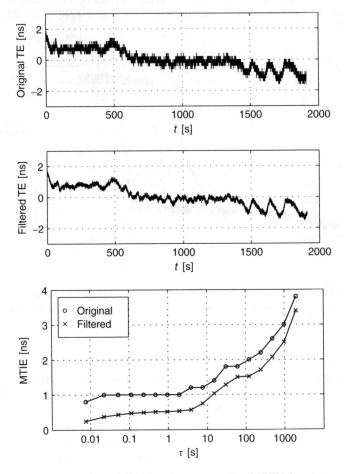

Figure 7.45. Original TE(t), filtered TE(t) and corresponding MTIE(τ) curves measured on the SASE from supplier C ($N = 262\,144$, $\tau_0 \cong 7.5$ ms, $T = 1966$ s). (Reproduced from [7.32], ©1997 IEEE, by permission of IEEE)

cannot be exceeded, while the true peak-to-peak TE fluctuations are much larger and thus may cause slips or pointer justifications

Obviously, the higher the noise frequency, the lower the MTIE values measured after low-pass filtering. In particular, the impact of the anti-aliasing filter is substantial for WPM and FPM noises, owing to their broadband spectrum.

Moreover, paper [7.32] pointed out the impact that such anti-aliasing pre-filtering can have in practical cases by providing some real measurement results, which are reported here below. These results were measured in synchronized-clock configuration by means of the same test set-up described in Section 7.7.2 (see Appendix 7A for details). The CUTs are two SASE clocks (from suppliers C and F), a DXC 4/3/1 SEC (supplier G) and an ADM1 SEC (supplier B). We denoted the suppliers with the same notation used so far, since we already showed experimental results for all these CUTs except the SASE F and the DXC G. The sequence length N and duration T, together with the sampling period τ_0, are specified in the respective figure captions.

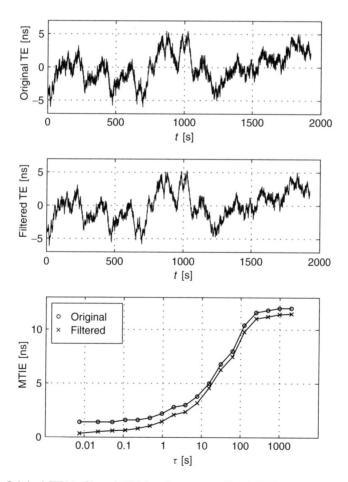

Figure 7.46. Original TE(t), filtered TE(t) and corresponding MTIE(τ) curves measured on the SASE from supplier F ($N = 262\,144$, $\tau_0 \cong 7.5$ ms, $T = 1966$ s). (Reproduced from [7.32], ©1997 IEEE, by permission of IEEE)

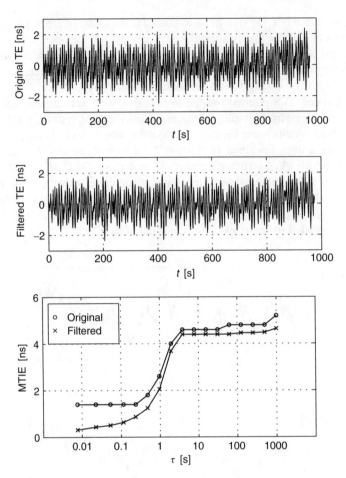

Figure 7.47. Original TE(t), filtered TE(t) and corresponding MTIE(τ) curves measured on the DXC SEC from supplier G ($N = 131\,072$, $\tau_0 \cong 7.6$ ms, $T = 998$ s). (Reproduced from [7.32], ©1997 IEEE, by permission of IEEE)

The graphs shown exhibit a different behaviour according to the underlying noise spectrum. The impact of the 10-Hz low-pass pre-filtering is always evident, but is substantial in particular in the case of the ADM-1 SEC B (Figures 7.48 and 7.49). This clock exhibits a wide and very fast periodic noise, a saw-toothed TE waveform due to frequency-quantization error of peak-to-peak amplitude 25 ns and period 32 μs, as made evident in Figures 5.34 and 7.35. In the measurement accomplished with sampling period $\tau_0 = 488$ ns (Figure 7.48), this noise gets completely cancelled by the 10-Hz low-pass filter, so that, while the filtered MTIE reports a reassuring 0 ns, the true MTIE is over 25 ns.

Considering the measurement accomplished with sampling period $\tau_0 = 7.6$ ms (Figure 7.49), on the other hand, it is interesting to notice that the filtered MTIE curve does not start from the 0-ns level for $\tau = \tau_0$, as it might be expected on the basis of Figure 7.48. The reason lies in the spectrum aliasing due to subsampling the saw-toothed noise: the aliased noise is not fully filtered out due to its lower frequency. In spite of this, the filtered MTIE still reports a noise level three times smaller than the actual one.

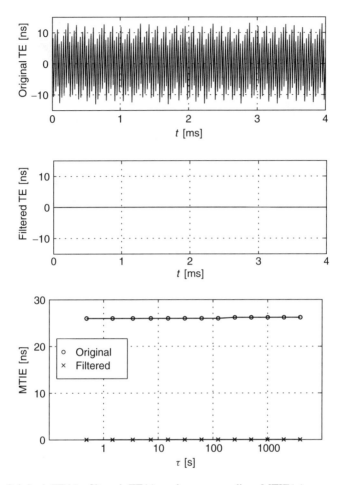

Figure 7.48.   Original TE(t), filtered TE(t) and corresponding MTIE(τ) curves measured on the ADM-1 SEC from supplier B ($N = 8196$, $\tau_0 \cong 488$ ns, $T = 4$ ms). (Reproduced from [7.32], ©1997 IEEE, by permission of IEEE)

In conclusion, the impact that such anti-aliasing pre-filtering can have in practice may be substantial and even misleading. Measurements accomplished through such a filter may yield too optimistic results, hiding the *true* amplitude of TE fluctuations at timing interfaces, and therefore may yield errors inbuffer-size design or in assessing the pointer justification rate in SDH transmission chains.

7.8   OTHER MEASUREMENTS ON EQUIPMENT CLOCKS

The stability measurements that were described in the previous sections are not the only ones that are carried out on equipment timing interfaces for a complete clock characterization. Other important quantities and parameters to measure on equipment clocks as well as on SASE are, for instance: the output jitter, the jitter tolerance, the jitter transfer function, the operational ranges and the various output phase transients occurring for example

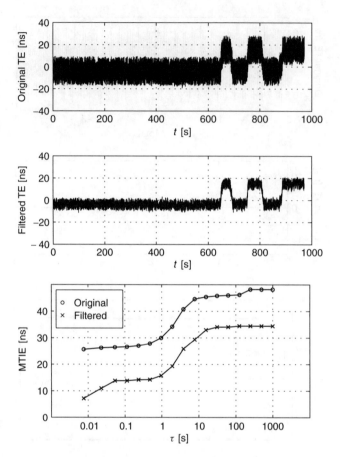

Figure 7.49. Original TE(t), filtered TE(t) and corresponding MTIE(τ) curves measured on the ADM-1 SEC from supplier B ($N = 131\,072$, $\tau_0 \cong 7.6$ ms, $T = 998$ s). (Reproduced from [7.32], ©1997 IEEE, by permission of IEEE)

under reference switch or microinterruptions. In this section, some details will be given on measurement techniques for all these items.

Other measurements are also required for a thorough conformance testing of clocks to international standards. For the full list of possible measurements, the reader is referred directly to the relevant standards specifying clock characteristics (e.g., ITU-T Recs. G.811 [7.30], G.812 [7.31] and G.813 [7.29]).

7.8.1   Output Jitter

The expression *output jitter* in this section denotes, in a wide sense, all phase impairments that are generated by the clock on its output timing interfaces in absence of input jitter or other external disturbances. That may include, for example, fortuitous phase discontinuities due to infrequent internal testing.

Several ITU-T Recommendations, cited throughout the book, specify jitter and wander limits for all digital and timing interfaces. The stability measurements in

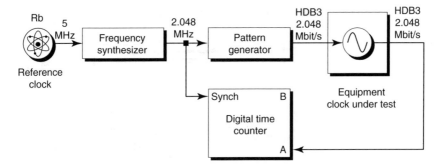

Figure 7.50.   Experimental set-up for measuring the output jitter of a telecommunications clock

synchronized-clock configuration described in the previous section can thus be ascribed to the group of output jitter measurements, in a particular sense.

More qualitative measurements of the output jitter may be carried out simply by means of a time counter, used to collect samples of phase or time error with respect to the expected values. The resulting phase or time error sequence plots describe the clock behaviour along the measurement interval.

An example of test set-up based on this principle for measuring the output jitter of a clock is depicted in Figure 7.50. There, the CUT is the clock of some telecommunications equipment that accepts, as synchronization reference, a 2.048 Mbit/s signal, HDB3-coded according to ITU-T Rec. G.703 [7.35]. For instance, several PSTN digital switching exchanges, deployed in the last 20 years, can be synchronized only via an E1 signal. In this case, moreover, the time counter is not triggered by two start and stop signals, but it measures the phase deviation of the input signal with respect to the reference.

Experimental Results

Examples of output jitter, measured with the test-bench depicted in Figure 7.50, are shown in Figures 7.51 and 7.52. In this and in all other cases in this section, a Hewlett-Packard HP5372A time counter was used. The graphs plot sequences of phase error samples measured on two widely deployed PSTN digital switching exchanges from different suppliers, say A and B[5]. Units on the Y-axis are [UI] and on the X-axis are [s]. These graphs are analogous to those plotting TE sequences shown in previous sections (e.g., Figures 7.28 and 7.35).

The measurement in Figure 7.51 ($T = 16$ s) revealed a short-term noise with peak-to-peak amplitude of about 6 mUI. The long-term measurement ($T = 50\,400$ s), on the other hand, captured a wider wander (160 mUI), probably due to some environmental disturbance.

7.8.2   Jitter Tolerance

The jitter tolerance of a clock is defined as the minimum phase noise level that the clock should be able to tolerate at its input, whilst staying within prescribed performance limits

---

[5] In this section, the suppliers of PSTN digital switching equipment will be denoted with different letters (namely A, B and C) than as done previously in the book.

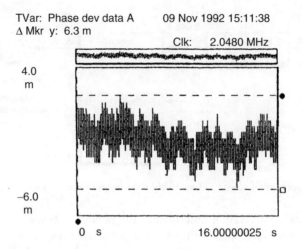

Figure 7.51. Output jitter [UI] measured in the short term on the clock of a PSTN digital switching exchange (supplier A)

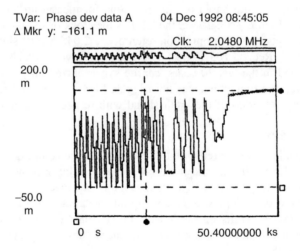

Figure 7.52. Output jitter [UI] measured in the long term on the clock of a PSTN digital switching exchange (supplier B)

in locked mode of operation, not causing any alarms or reference switch or entering hold-over. Several ITU-T Recommendations, cited throughout the book, specify jitter tolerance for all digital and timing interfaces.

In this case, a time counter may be used to check whether the CUT stays locked or not. An example of test bench for measuring the jitter tolerance of a clock is thus depicted in Figure 7.53, on a CUT of the same kind as in the previous measurement. The reference signal is phase-modulated by a sine wave (jitter modulation), with given frequency and amplitude. The jitter tolerance measured, for a given jitter frequency, will be the maximum jitter amplitude that will not make the CUT to lose lock.

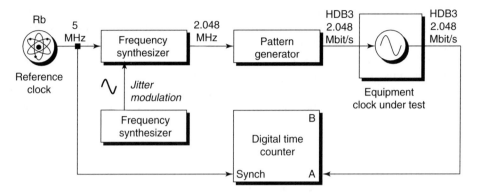

Figure 7.53.   Experimental set-up for measuring the jitter tolerance of a telecommunications clock

7.8.3   Jitter Transfer Function

The jitter transfer function is the ratio between the output jitter amplitude and the input jitter amplitude over a jitter frequency range. Obviously, the underlying assumption is that the CUT is working according to a linear model (i.e., the input jitter amplitude is small).

This test is accomplished by feeding the CUT input with a reference signal affected by sinusoidal jitter at increasing frequencies, with small fixed amplitude. In this case, a time counter may be used to measure the phase deviation of the CUT output, along a time interval greater than the jitter period. The test bench is the same as depicted in Figure 7.53.

If the CUT is working according to a linear model, then the phase deviation measured by the time counter is a sine wave at the same frequency as the input jitter, but having different amplitude. The ratio of the two peak amplitudes, plotted versus the jitter frequency, is the jitter transfer function measured (cf. Figure 5.10).

Actually, ITU-T and ETSI standards [7.14][7.29][7.31] address the issue of noise transfer characteristic by stating generally that it determines the slave clock properties with regard to the transfer of excursions of the input phase relative to the phase modulation. These standards provide noise transfer specifications in terms of a linear jitter transfer function, as we did above, and of a TDEV mask that should not be exceeded on the clock output when the reference signal is affected by noise shaped accordingly to the TDEV input tolerance specification.

The two most interesting parameters to check in a jitter transfer function measured are the clock *bandwidth* and *maximum gain* (cf. Section 5.5.2). The ITU-T and ETSI standards cited above specify that a SASE should not exceed a maximum bandwidth of 3 mHz and a maximum gain of 0.2 dB. A SEC, on the other hand, should have bandwidth between 1 Hz and 10 Hz and a maximum gain of 0.2 dB.

7.8.4   Hold-in, Pull-out, Lock-in and Pull-in Ranges

The measurement of all four hold-in, pull-out, lock-in and pull-in ranges is difficult and error prone. In particular the definition of lock-in range (as provided in Section 5.5.6 and encoding to ITU-T and ETSI standards) does not allow to derive unambiguously a measurement procedure.

International standard bodies do not give limits for all four ranges: ETSI addresses only the pull-out and pull-in ranges, while ITU-T only the pull-in, hold-in, and pull-out ranges. Moreover, most limit values of the parameters above have not been specified yet but are still reported as 'under study'.

The measurement of the hold range in static and dynamic conditions (hold-in and pull-out ranges) is of particular interest. It can be carried out with the same test bench depicted in Figure 7.50, by adjusting the synthesizer output frequency — very slowly or abruptly according to what parameter is being measured — until the time counter does not detect that the CUT is not locked anymore.

7.8.5 Output Phase Transients

The clock reference signal may be affected by disturbances or transmission failures of several different kinds, for instance micro-interruptions (e.g., interruptions lasting milliseconds or even microseconds) or switching between two alternate synchronization signals. Any impairment on the synchronization reference signal results in a phase transient (discontinuity) on the slave clock output. The ability of the slave clock to withstand such disturbances is a major aspect in clock quality, since transmission failures and disturbances of different kinds are common stress conditions in real telecommunications networks. Other possible causes of output phase discontinuities are infrequent internal testing or rearrangement operations within the slave clock (e.g., a switch of active internal oscillator). ITU-T Recommendations on clocks specify limits for output phase transients caused by all these possible disturbances.

An example of test bench for measuring output phase transients caused by micro-interruptions on the clock reference signal is depicted in Figure 7.54. The CUT is again of the same kind as in the previous examples and accepts at the input a HDB3 2.048 Mbit/s signal. The micro-interruption generator is a simple device able to physically interrupt the signal, for single or multiple intervals in series, each lasting a few microseconds or milliseconds.

A test bench for measuring output phase transients caused by reference switching is depicted in Figure 7.55. In this case, the equipment clock under test is made to switch between two alternate reference synchronization signals by a manual command. The phase

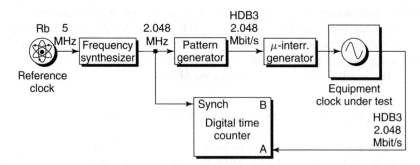

Figure 7.54. Experimental set-up for measuring the output phase transients, caused by micro-interruptions on the input reference signal, of a telecommunications clock

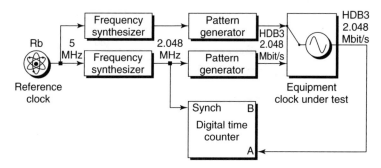

Figure 7.55. Experimental set-up for measuring the output phase transients, caused by switching between two alternate synchronization signals, of a telecommunications clock

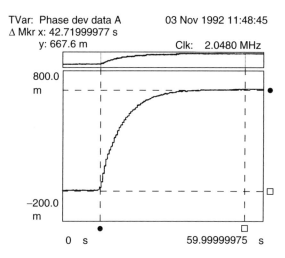

Figure 7.56. Output phase transient [UI] caused by a series of 10 micro-interruptions each of duration 10 ms on the input reference signal of a PSTN digital switching exchange (supplier C)

control of one synthesizer may be adjusted to feed the CUT inputs with two signals having substantial phase offset.

Finally, output phase transients due to active internal oscillator switching can be measured with the same test bench depicted in Figure 7.50. Also in this case, the equipment under test is made to switch between two alternate internal oscillators by a manual command.

7.8.5.1 *Experimental Results: Micro-Interruptions*

An example of output phase transient caused by micro-interruptions on the input reference signal is shown in Figure 7.56. It was measured with the test-bench depicted in Figure 7.54 on a widely deployed PSTN digital switching exchange from supplier C, by applying a series of 10 interruptions each of duration 10 ms on the CUT reference.

The micro-interruptions on the reference signal, in this case, caused the clock to exhibit a phase hit of amplitude 0.667 UI and with exponential shape. The transient had duration in the order of 20 or 40 s.

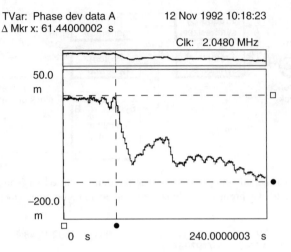

Figure 7.57. Output phase transient [UI] caused by reference switching on a PSTN digital switching exchange (supplier A)

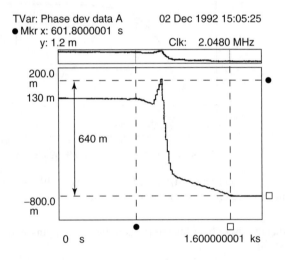

Figure 7.58. Output phase transient [UI] caused by reference switching on a PSTN digital switching exchange (supplier B)

7.8.5.2   *Experimental Results: Reference Switching*

Examples of output phase transients caused by reference switching are shown in Figures 7.57–7.59. They were measured with the test-bench depicted in Figure 7.55 on widely deployed PSTN digital switching exchanges, from suppliers A, B and C, by commanding the equipment clock under test to switch between the two reference synchronization signals.

The reference switching caused these clocks to exhibit phase hits of different shape and duration. In the case of the equipment from supplier A, the phase transient had a duration of somehow longer than 250 s, complex shape and amplitude limited to 140 mUI. On

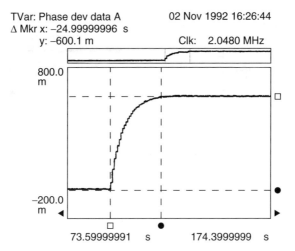

Figure 7.59.   Output phase transient [UI] caused by reference switching on a PSTN digital switching exchange (supplier C)

the equipment from supplier B, the phase transient lasted 600 s, had complex shape as well but with larger amplitude than in the previous case (600 mUI). The equipment from supplier C, finally, exhibited an exponential-shape phase hit almost identical to that shown previously in Figure 7.56.

7.8.5.3   *Experimental Results: Active Internal Oscillator Switching*

Examples of output phase transients caused by switching of the active internal oscillator are shown in Figures 7.60–7.62. They were measured with the test-bench depicted in Figure 7.50 on widely deployed PSTN digital switching exchanges, from suppliers A, B and C, by commanding the equipment under test to switch between the two internal

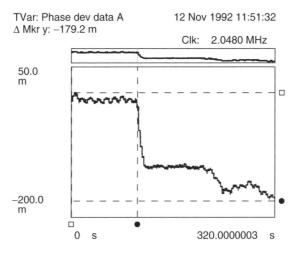

Figure 7.60.   Output phase transient [UI] caused by active internal oscillator switching on a PSTN digital switching exchange (supplier A)

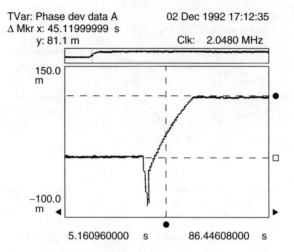

Figure 7.61. Output phase transient [UI] caused by active internal oscillator switching on a PSTN digital switching exchange (supplier B)

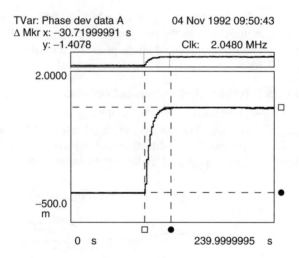

Figure 7.62. Output phase transient [UI] caused by active internal oscillator switching on a PSTN digital switching exchange (supplier C)

oscillators. The active oscillator switching caused these clocks to exhibit phase hits of similar shape and duration as measured under reference switching (cf. Figures 7.57–7.59), with the exception of the equipment from supplier B, which in this case featured a shorter transient.

7.9   NETWORK MEASUREMENTS

Network measurements aim at the characterization of the quality of network synchronization, as implemented in the digital telecommunications network under test. They are usually carried out by comparing the timing signals coming from different, remote network nodes. Thus, their results include the impact of the quality of transmission links as well as of the performance of clocks and network synchronization distribution.

The main issues affecting the quality of transmission links are the added jitter and wander and the micro-interruptions. Even periods of unavailability may be expected.

Possible measurements that can be carried out to characterize the quality of network synchronization are the measurement of jitter and wander, by acquisition of long sequences of phase or time error as described in Section 7.8, and the measurement of standard stability quantities MTIE, TDEV, etc., as described in Section 7.7. These measurements can be performed either by comparing the timing of digital signals transported to the measurement location from remote clocks or by using a GPS reference directly at CUT place. It is evident how this subject is closely related to that already discussed in Section 4.6.

A general scheme of possible network synchronization measurements is depicted in Figure 7.63. A time counter, located for example at the PRC location, measures the phase or time error between a signal received directly from the PRC (measurement reference) and signals received from remote network clocks. The latter signals are carried to the measurement location across transport networks. PDH systems are the best long-distance transmission facilities to this purpose, as widely discussed in previous chapters.

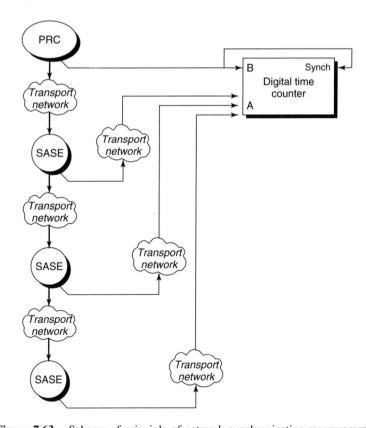

Figure 7.63.   Scheme of principle of network synchronization measurements

As obvious, the main drawback of this scheme is that transport links add impairments of various types on the signals under measurement. Alternatively, one could perform measurements at network clock locations, by comparing the output of network clocks with an absolute timing reference from a GPS receiver. This measurement configuration was already introduced in Figure 7.4(b), as independent-clock standard configuration for the characterization of network clock interfaces.

7.10   SUMMARY

The RF-spectrum is definitely not a good tool to characterize clock frequency stability: given $S_s^{RF}(f)$, it is not possible to determine unambiguously whether the power at various Fourier frequencies is the result of amplitude rather than phase fluctuations in the timing signal $s(t)$.

By the mid-1960s, the IEEE convened a technical committee that recommended the basic quantities $y(t)$, $x(t)$, $v(t)$ and the measures of frequency stability $S_y(f)$ and $\sigma_y^2(\tau)$. By the mid-1990s, the international standard bodies ETSI and ITU-T recommended five time-domain quantities for the specification of timing interface requirements: ADEV, MADEV, TDEV, TIErms and MTIE.

Allan variances are not influenced by synchronization and syntonization errors x_0 and Δv, because they are based on the second difference of TE data. A linear frequency drift D, on the contrary, causes a quadratic ramp in TE(t) and a trend of AVAR and MAVAR proportional to τ^2. A practical procedure for estimating the linear frequency drift D consists of averaging the second difference of the TE sequence.

The Allan variance is defined as infinite-time average of the square difference of two consecutive frequency samples $\overline{y}_k$. In practice, nevertheless, we can have only finite sequences of N measurement samples x_k. Hence, it is possible to compute only an *estimate* s_y^2 of the Allan variance σ_y^2. The estimate s_y^2 is a random variable. To model its distribution, the so-called χ^2 distribution with l degrees of freedom is used. Knowing the exact χ^2 distribution of the estimate s_y^2 is useful to determine its *confidence interval*.

Two standard configurations for measuring clock stability have been defined by international bodies ITU-T and ETSI: the *synchronized-clock* measurement configuration and the *independent-clock* measurement configuration. The measurement configuration substantially affects the behaviour of the stability quantities. The curves measured in the two cases coincide in the short term (i.e., for observation intervals τ much shorter than the PLL time constant), but exhibit a substantially different trend in the long term. The reason is that the CUT internal noise is high-pass filtered in the synchronized-clock measurement configuration.

An issue to consider carefully when performing stability measurements is the possible impact of the TE sampling period τ_0 (for a given observation interval τ) on the behaviour of the stability quantities. While ADEV, MTIE and TIErms do not feature any significantly different behaviour as the sampling period is varied for all the power-law noise types, the behaviour of MADEV and TDEV is substantially dependent on the chosen measurement sampling period in the observation intervals where WPM or FPM noises dominate. In such cases, it would not be possible to juxtapose curves measured with different τ_0, because extremes would not meet.

A *mixer* is a device that accepts two analogue pseudo-periodic signals at its inputs and forms an output signal with components at the sum and difference frequencies. For example, a four-quadrant multiplier is a mixer.

The core of a *frequency synthesizer* is a system of programmable frequency multipliers and dividers. This equipment allows thus to synthesize an output signal having specified waveform, amplitude, phase and frequency, based on a reference timing signal at its input.

Spectrum analysers allow to observe a signal in the frequency domain, plotting its power versus its Fourier frequency. Spectrum analysers come in two basic varieties: swept-tuned and real-time instruments. The former ones are mostly analogue, the latter ones mostly digital.

The *digital time counter* measures the elapsed time between two start and stop trigger events, by incrementing a counter register according to a base reference frequency. A trigger event typically consists of detecting an electrical signal going over some threshold on the input channel. A time-interval measurement resolution in the order of 20–200 ps is common in high-performance commercial time counters.

The most straightforward method to measure the stability of a timing signal is to directly measure its Time Error, compared to a reference signal having nearly the same frequency, simply by means of a digital time counter (*direct digital measurement of TE*). Three different approaches can be adopted: to measure the time intervals between the edges of the two signals, to measure the frequency of the timing signal under test and to measure its period.

Sensitivity-enhancement methods have been developed, aiming at distinguishing very small time and frequency fluctuations. These methods can be classified as:

- *heterodyne techniques*, which consist of mixing the timing signal under test with the reference signal having nearly the same frequency;

- *homodyne techniques*, which are a limit of the previous method, occurring when the mixed reference signal has exactly the same average frequency as the signal under test; examples are methods based on a delay line and on a PLL to generate the reference signal;

- *multiple conversion techniques*, in which the actual signal to measure is obtained from the signal under test through several mixing and frequency synthesis stages.

The TE direct digital measurement has consolidated as standard to characterize the stability of telecommunication clocks. The practical measurement procedure is outlined in the following steps:

- acquire and store a sequence $\{x_i\}$ of N samples of TE, with sampling period τ_0 and over a measurement interval $T = (N - 1)\tau_0$;

- compute the stability quantities of interest (viz. ADEV, MADEV, TDEV, MTIE, TIErms) with the standard estimators;

- do the same without the CUT, by directly feeding both time-counter input ports with the timing signal split from the RC, for assessing the test-bench background noise floor.

The measurement of MTIE features some peculiar issues, due to its nature of raw peak-to-peak measure and to the heavy computational weight of the estimator.

Other important quantities and parameters to measure on equipment clocks as well as on SASE are, for instance: the output jitter, the jitter tolerance, the jitter transfer function, the operational ranges and the various output phase transients occurring for example under reference switch or microinterruptions.

Network measurements aim at characterizing the quality of network synchronization, as implemented in the digital telecommunications network under test. They are usually carried out by comparing the timing signals coming from different, remote network nodes. Thus, their results include the impact of the quality of transmission links as well as of the performance of clocks and network synchronization distribution.

7.11   REFERENCES

[7.1] L. S. Cutler, C. L. Searle. Some aspects of the theory and measurement of frequency fluctuations in frequency standards. *Proceedings of the IEEE*, vol. 54, no. 2, Feb. 1966.

[7.2] P. Lesage, C. Audoin. Characterization and measurement of time and frequency stability. *Radio Science* 1979; **4**, (4). © American Geophysical Union.

[7.3] D. A. Howe, D. W. Allan, J. A. Barnes. Properties of signal sources and measurement methods. *Proceedings of the 35th Annual Frequency Control Symposium*, 1981.

[7.4] S. R. Stein. Frequency and time — their measurement and characterization. In *Precision Frequency Control*. Edited by E. A. Gerber and A. Ballato, vol. 2, ch. 2, pp. 191–232. New York: Academic Press, 1985.

[7.5] D. W. Allan, H. Daams. Picoseconds time difference measurement system. *Proceedings of the 29th Annual Frequency Control Symposium*, 1975.

[7.6] F. L. Walls, D. W. Allan. Measurement of frequency stability. *Proceedings of the IEEE*, vol. 74, no. 1, Jan. 1986.

[7.7] C. A. Greenhall. A method for using a time interval counter to measure frequency stability. *Proceedings of the 41st Annual Frequency Control Symposium*, 1987.

[7.8] F. L. Walls, A. J. D. Clements, C. M. Felton, M. A. Lombardi, M. D. Vanek. Extending the range and accuracy of phase noise measurements. *Proceedings of the 42th Annual Frequency Control Symposium*, 1988.

[7.9] D. W. Allan. Time and frequency metrology: current status and future considerations. *Proceedings of the 5th European Frequency and Time Forum*, 1991.

[7.10] C. Hackman, D. B. Sullivan. Time and frequency measurement. *American Journal of Physics* 1995; **63** (4): 306–317.

[7.11] IEEE Std. 1139. *IEEE Standard Definitions of Physical Quantities for Fundamental Frequency and Time Metrology*. Revised version approved Oct. 20, 1988.

[7.12] J. A. Barnes, A. R. Chi, L. S. Cutler, D. J. Healey, D. B. Leeson, T. E. McGunigal, J. A. Mullen Jr., W. L. Smith, R. L. Sydnor, R. F. C. Vessot, G. M. R. Winkler. Characterization of frequency stability. *IEEE Transactions on Instrumentation and Measurements* **IM-20** (2).

[7.13] ITU-T Rec. G.810 *Definitions and Terminology for Synchronisation Networks*. Geneva, August 1996.

[7.14] EN 300 462 *Transmission and Multiplexing (TM); Generic Requirements for Synchronization Networks*. Part 1-1: *Definitions and Terminology for Synchronization Networks*. Part 2-1: *Synchronization Network Architecture*. Part 3-1: *The Control of Jitter and Wander within Synchronization Networks*. Part 4-1: *Timing Characteristics of Slave Clocks Suitable for Synchronization Supply to Synchronous Digital Hierarchy (SDH) and*

Plesiochronous Digital Hierarchy (PDH) Equipment. Part 5-1: *Timing Characteristics of Slave Clocks Suitable for Operation in Synchronous Digital Hierarchy (SDH) Equipment*. Part 6-1: *Timing Characteristics of Primary Reference Clocks*.

[7.15] J. Rutman. Characterization of phase and frequency instabilities in precision frequency sources: fifteen years of progress. *Proceedings of the IEEE*, vol. 66, no. 9, Sept. 1978.

[7.16] P. Lesage, C. Audoin. Characterization of frequency stability: uncertainty due to the finite number of measurements. *IEEE Transactions on Instrumentation and Measurement* 1973; **IM-22** (2).

[7.17] J. A. Barnes. Atomic timekeeping and the statistics of precision signal generators. *Proceedings of the IEEE*, vol. 54, no. 2, Feb. 1966.

[7.18] S. Bregni, M. Carbonelli, M. D'Agrosa, D. De Seta, D. Perucchini. Different behaviour of frequency stability measures in independent and synchronized clocks: theoretical analysis and measurement results. *Proceedings of IEEE ICC '94*, New Orleans, USA, 1-5 May 1994.

[7.19] S. Bregni, M. Carbonelli, D. De Seta, D. Perucchini. Impact of the time error sampling period on telecommunications clock stability measures. *European Transactions on Telecommunications* 1997; **8** (6).

[7.20] P. Lesage, T. Ayi. Characterization of frequency stability: analysis of the modified Allan variance and properties of its estimate. *IEEE Transactions on Instrumentation and Measurement* 1984; **IM-33** (4).

[7.21] L. G. Bernier. Theoretical analysis of the modified allan variance. *Proceedings of the 41st Annual Frequency Control Symposium*, 1987.

[7.22] J. Rutman, F. L. Walls. Characterization of frequency stability in precision frequency sources. *Proceedings of the IEEE*, vol. 79, no. 7, July 1991.

[7.23] D. E. Knuth. *The Art of Computer Programming*, vol. 2, p. 118. London: Addison-Wesley, 1981.

[7.24] J. A. Barnes, D. W. Allan. A statistical model of flicker noise. *Proceedings of the IEEE*, vol. 54, no. 2, pp. 176–178, Feb. 1966.

[7.25] P. Horowitz, W. Hill. The art of electronics Sec. 13.12. *Radiofrequency Circuit Elements*. Cambridge: Cambridge University Press, 1980.

[7.26] P. Horowitz, W. Hill. The art of electronics, Sec. 14.18. *Spectrum Analyzers*. Cambridge: Cambridge University Press, 1980.

[7.27] P. Horowitz, W. Hill. The art of electronics. Sec. 14.10. *Frequency, Period, and Time-Interval Measurements*. Cambridge: Cambridge University Press, 1980.

[7.28] S. Bregni. Clock stability characterization and measurement in telecommunications. *IEEE Transactions on Instrumentation and Measurement* **IM-46** (6).

[7.29] ITU-T Rec. G.813 *Timing Characteristics of SDH Equipment Slave Clocks (SEC)*. Geneva, August 1996.

[7.30] ITU-T Rec. G.811 *Timing Characteristics of Primary Reference Clocks*. Geneva, Sept. 1997.

[7.31] ITU-T Rec. G.812 *Timing Requirements of Slave Clocks Suitable for Use as Node Clocks in Synchronization Networks*. Geneva, June 1998.

[7.32] S. Bregni, F. Setti. Impact of the anti-aliasing pre-filtering on the measurement of maximum time interval error. *Proceedings of IEEE GLOBECOM '97*, Phoenix, AZ, USA, 1997.

[7.33] W. H. Press, B. P. Flannery, S. A. Teukolsky, W. T. Vettering. *Numerical Recipes in C—The Art of Scientific Computing*. Cambridge: Cambridge University Press (also available for other programming languages).

[7.34] S. Bregni. Measurement of maximum time interval error for telecommunications clock stability characterization. *IEEE Transactions on Instrumentation and Measurement* 1996; **IM-45** (5).

[7.35] ITU-T Rec. G.703 *Physical/Electrical Characteristics of Hierarchical Digital Interfaces*. Blue Book, Geneva, Oct. 1998.

APPENDIX 7A
EXPERIMENTAL SET-UP AND PROCEDURE
FOLLOWED FOR MEASUREMENT RESULTS SHOWN
IN CHAPTERS 5 AND 7

The measurement procedure outlined in Section 7.7.2 has been extensively applied throughout the 1990s in testing several telecommunications clocks, for conformance testing and for research purposes, within the national activities for the development of the new synchronization network of the public operator Telecom Italia. With this regard, the author is indebted to all members of the National Study Group on Synchronization, established by Telecom Italia and joined by SIRTI, Fondazione Ugo Bordoni and CSELT (now Telecom Italia Lab) in the early 1990s.

All the experimental results shown in Chapters 5 and 7 were measured in the synchronized-clock configuration, according to the scheme in Figure 7.3(a), except those in the independent-clock configuration provided in Section 7.1.8, those provided in Section 5.9.3 (background noise of the measurement set-up) and most of those provided in Section 7.8 (other measurements on equipment clocks).

As shown in Figure 7.24, a digital time counter (Hewlett-Packard HP5372A), with a resolution of 200 ps, measured the TE between the output timing signal of the CUT and its input reference (both G.703 [7.35] signals at 2.048 MHz unless otherwise stated). The latter was synthesized from the 10 MHz output signal of the RC, a rubidium frequency standard, which also supplied the time base to the time counter.

In the independent-clock configuration, on the other hand, the TE between the timing signal at the output of the CUT and that one from the independent reference clock (rubidium frequency standard) was measured.

The background noise of the measurement set-up, finally, was measured by splitting the 2.048 MHz reference timing signal, synthesized from the rubidium oscillator output, and by feeding it directly into the time counter input ports (see Figure 7.25).

The time counter was driven via a GPIB IEEE488.2 interface by a laptop computer, which controlled data acquisition and then accomplished their numerical processing and the visualization of the results.

The hardware bandwidth of the time counter is in the order of 500 MHz. Moreover, measuring the TE between two 2.048 MHz square signals imposes a maximum bandwidth $\hat{f}_h$ of the measured phase fluctuations that cannot be larger than 1 MHz (sampling theorem). However, the natural cut-off frequency f_h of the clock output noise is expected to be not larger than 100 kHz. Since $\hat{f}_h > f_h$, it is then not necessary to specify exactly the actual value of $\hat{f}_h$, as it does not affect substantially the total phase noise power measured (cf. Section 5.9.1).

All sets of the stability curves shown in Chapters 5 and 7 were evaluated based on TE sequences made of N samples, measured with sampling period τ_0 and over a measurement interval T.

The ADEV, MADEV, TDEV, TIErms and MTIE curves were computed for up to 24 points per decade, to achieve an excellent rendering of actual trends, with the standard estimators of Equations (7.4) through (7.8).

Moreover, the PSD $S_x(f)$ was computed (always neglecting a multiplicative factor) on 1024 points through the Fast Fourier Transform (FFT) periodogram technique with Parzen data windowing (triangular shape) [7.33], while the autocovariance function $C_x(\tau)$ was evaluated (again neglecting a multiplicative factor) on 2047 points, as inverse transform of the PSD for $\tau \geq 0$ and mirrored copy for $\tau < 0$.

The actual values of measurement parameters N, τ_0 and T together with other particular information have been specified in each case when providing measurement results.

GLOSSARY OF TERMS

Accounting management. Area of network management functions that deal with the identification of costs for the use of hardware and software resources.

Accuracy of a clock. The maximum time or frequency error, which may be measured in general *over the whole clock life* unless differently specified. The *time accuracy* means how well a clock agrees with UTC, while the *frequency accuracy* is the maximum frequency error $\Delta \nu_{max}$ compared to the nominal value ν_n.

Add-Drop Multiplexer. Multiplexer equipment with drop and insert capability. ADMs of level N (ADMs-N) are used in intermediate nodes of SDH transmission chains or rings to insert or drop tributaries from the STM-N line in transit. Tributaries may be PDH signals of any kind and hierarchical level as well as SDH STM-M signals ($M \leq N$).

Administrative Unit. SDH multiplexing element. An AU consists of a higher order VC-i ($i = 3, 4$) and of an AU pointer indicating the offset of the pointed VC within the AU itself.

Administrative Unit Group. SDH multiplexing element. The purpose of AUGs is similar to that of TUGs. They contain an homogeneous group of AU-3 byte-interleaved or one AU-4, located in fixed positions.

Analogue signal. A signal a characteristic quantity of which follows continuously the variations of another physical quantity representing information.

Anisochronous. A non-isochronous digital signal.

Annual wander. Wander caused by pseudo-periodical (seasonal) temperature variations on the transmission medium along the year.

Asynchronous. Two digital signals are so called if they are not synchronous.

Asynchronous digital multiplexing. The type of digital multiplexing where the tributary signals are asynchronous (usually plesiochronous) and the *bit justification* technique is adopted to synchronize tributaries into the multiplex signal.

Asynchronous transfer mode (1). Information transfer mode where the information sources are mutually asynchronous and the information is segmented in information units, which are sent by sources in independent instants, with intervals dependent on source demand.

Asynchronous Transfer Mode (2). The term ATM denotes in particular the cell-switched technique chosen by international standard bodies for the implementation of the B-ISDN.

Atomic clock. Clock that uses as frequency reference the oscillation of an electromagnetic signal associated with a quantum transition between two energy levels in an atom.

Autonomous clock. A stand-alone device able to generate a timing signal, suitable for the measurement of time intervals, starting from some periodic physical phenomenon that runs independently of external influence.

Beat signal. Low-frequency signal at the difference frequency $|v_1 - v_2|$, obtained by mixing two signals with nearly same frequencies v_1 and v_2 in *heterodyne techniques.*

Bit justification. The technique used in asynchronous digital multiplexers to perform bit synchronization of tributary signals into the multiplex signal. Three types of bit justification exist: positive, negative and positive/negative bit justification. Positive justification has been standardized in North America and Europe to establish the PDH. The three types of bit justification are not conceptually different: they all consist of transmitting either tributary or dummy bits at certain bit positions within the multiplex signal (justification opportunity bits), adapting to the variable input frequency of the tributaries.

Bit leaking. Technique used in SDH *desynchronizers* for reducing pointer adjustment jitter. In the case of AU-4 pointer justification, for example, it consists of spreading each of the incoming 24-bit phase steps into 24 one-bit phase steps in some way (phase spreading).

Bit synchronization. The synchronization of an asynchronous bit stream according to a local equipment clock. For example, the bit and frame aligning of PCM signals at the inputs of a digital switching exchange or the synchronization of tributaries into the multiplex signal in a digital multiplexer (possibly by bit justification).

Building Integrated Timing Supply. The ANSI equivalent of the ITU-T and ETSI SSU/SASE. It represents the building clock.

Byte justification. Term sometimes used to refer to *pointer justification.*

Carrier supply. Equipment generating all the necessary carriers to be used by FDM multiplexers and demultiplexers for all FDM hierarchical levels and using PLLs to generate reference frequencies synchronous with a pilot input frequency.

Carrier synchronization. The extraction, from an amplitude-modulated signal, of a signal coherent with the carrier in frequency and phase for coherent demodulation.

Cell. Fixed-size packet.

Chronosignal. From the greek etymon $\chi\rho\acute{o}\nu o\varsigma$ = time. A periodic or pseudo-periodic signal used to control the timing of actions (e.g., symbol decision at reception) on digital signals. Also called *timing signal.*

Clock. Device able to supply a *timing signal,* i.e. a pseudo-periodic (ideally periodic) signal usable to control the timing of actions. From a theoretical point of view, the operation principle of any kind of clock consists of a generator of oscillations and in an automatic counter of these oscillations.

Clock recovery. Same as *symbol synchronization*: the extraction of the symbol timing from a signal.

Configuration management. Area of network management functions that deal with equipment configuration setting and identification.

Container. A Container (C-*i*) is the basic SDH synchronous element mapping the tributary signal transported between the access points of the SDH network. The mapping of PDH signals into a structured, fixed-size set of bytes C-*i* may be *asynchronous*, where accommodation (synchronization) of a plesiochronous signal into the synchronous C-*i* is done through bit justification, or *bit/byte-synchronous*, where bit justification is not necessary as the mapped signal is synchronous with the timing signal of the C-*i*.

Decision circuit. A circuit that decides the most probable value of a signal element of a received digital signal.

Demapping. The inverse process than mapping, i.e. the extraction of a tributary signal from a SDH Container (C-*i*).

Desynchronizer. The block performing the demapping of the tributary from the multiplex signal, in a digital demultiplexer, thus comprising mainly the buffer and the read and write controls. It fulfils the complementary function of the synchronizer: it smoothes the gaps in the output tributary clock signal, due to mapping, thus returning a bit stream with regular clock signal.

Digital Cross-Connect. A DXC takes at the input ports both PDH and SDH signals. It allows flexible cross-connection of the signals and of the VCs from any input port *i* to any output port *j*, according to a cross-connection matrix that can be set by the management system.

Digital multiplexing. (In the proper sense of the word) it means specifically to TDM-multiplex *N* digital signals (tributaries), which usually have the same (but in general different) nominal bit rate. The distinctive feature of digital multiplexing, compared to PCM multiplexing, is that the tributary signals to multiplex are already in digital form.

Digital signal. A discretely timed signal in which information is represented by a number of well-defined discrete values that one of its characteristic quantities may take in time. A digital signal is therefore a sequence of numbers or binary digits and may be obtained through the joint processes of sampling and quantization on a continuous-time analogue signal.

Discretely timed signal. A signal composed of successive elements in time, each element having one or more characteristics which can convey information, for example its duration, its waveform or its amplitude.

Dithering noise. In this book, waveform modulating the *synchronizer* stuff thresholds to reduce *waiting time jitter* (cf. *stuff threshold modulation*).

Diurnal wander. Wander caused by pseudo-periodical (diurnal) temperature variations on the transmission medium along the day.

Equivalent slip rate. Between two timing interfaces, it is the controlled slip rate that would be detected should the two timing signals at such interfaces be used to drive the write and read processes in a bit synchronizer buffer.

Fault management. Area of network management functions that deal with the detection, isolation and correction of abnormal operation.

Frequency Division Multiplexing. Analogue multiplexing technique, consisting of shifting every tributary channel in the frequency domain to different locations in the spectrum of the multiplex signal, so that there is no channel interference and it is thus still possible to separate the single channels from the multiplex signal by band-pass filtering.

Frame. A cyclic set of consecutive time slots in which the relative position of each time slot can be identified.

Frame alignment. Same as *frame synchronization.*

Frame synchronization (1). The process of determining the start and the end of groups of code words, i.e. of delineating the frames in the raw uniform stream of received bits.

Frame synchronization (2). The process of aligning in time the incoming frames at the different inputs of a digital switching exchange.

Free-run mode. Mode of operation of a slave clock when it has to work autonomously, supplying the free-run frequency ω_F of the internal oscillator (VCO) working with null control tension, because the reference signal failed, or moved outside the hold-in range, or fast outside the pull-out range, and keeps outside the pull-in range so that the clock cannot re-lock.

Frequency. Number of times for which a periodical phenomenon repeats in the time unit. Ellyptical term used with two different meanings that should not be confused. The symbol $v(t)$ denotes the time-dependent *instantaneous frequency* of the clock. The symbol f indicates the time-independent *Fourier frequency* that is the argument of spectral densities and Fourier transforms: in base-band representation, positive values of f denote frequencies above the central frequency v_n, negative values denote frequencies below av_n.

Frequency source. A source that supplies a (pseudo-)periodic signal whose meaningful information is frequency, usable as reference for frequency synchronization.

Frequency synthesizer. A system of programmable frequency multipliers and dividers, which allows one to synthesize an output signal having a specified waveform, amplitude, phase and frequency, based on a reference timing signal at its input.

Frequency-Locked Loop. Feedback control system where the *frequency* of a locally generated pseudo-periodic signal is automatically adjusted to the frequency of an incoming pseudo-periodic signal.

Gapped clock (signal). An isochronous timing signal with pulses not uniformly spaced, obtained from a *regular clock signal* by inhibiting some unwanted pulses.

Global Positioning System. Satellite radio system providing continuously and in real-time three-dimensional position, velocity and time information to suitably equipped land, sea and airborne users anywhere on Earth. Born essentially as a navigation and positioning tool, it is used also as pure time reference to disseminate precise time, time intervals and frequency. The NAVSTAR system, operated by the US DoD, is the first GPS system made available to civilian users.

Heterochronous. Two *heterochronous* (from the Greek etyma $\xi\tau\varepsilon\rho\sigma\varsigma = different$, $\chi\rho\acute{o}\nu\sigma\varsigma = time$) digital signals are isochronous, asynchronous digital signals, whose respective timing signals have different nominal frequencies.

Heterodyne techniques. From the greek etyma $\xi\tau\varepsilon\rho\sigma\varsigma = different$, and $\delta\acute{u}\nu\alpha\mu\iota\varsigma = force$. Time and frequency measurement techniques that consist of mixing the timing signal under test with the reference signal, having nearly the same frequency, in order to measure the same phase fluctuations but on the resulting low-frequency *beat signal* (thus measuring larger time fluctuations).

Hierarchical Master–Slave synchronization. Master–slave network synchronization strategy in which clocks are organized in two or more hierarchical levels and alternate synchronization routes are allowed for protection: if the master fails, another clock takes its place according to a hierarchical plan.

Hold-in range. The largest offset between the frequency ω_{in} of the input reference signal $s_{in}(t)$ and a specified nominal frequency (approximately the VCO free-run frequency ω_F), within which the slave clock maintains lock as the frequency varies slowly (rigorously speaking with $d\omega_{in}/dt \rightarrow 0$) over the frequency range.

Hold-over mode. Mode of operation of a slave clock when it has to work autonomously, supplying the frequency generated by the VCO, because the reference signal failed, or moved outside the hold-in range, or fast outside the pull-out range, and keeps outside the pull-in range so that the clock cannot re-lock. Compared to the free-run mode, nevertheless, the last control tension value at the input of the VCO, before the reference failure, is maintained, so to hold the last output frequency value over.

Homodyne techniques. From the greek etyma ὅμος = *same* and δύναμις = *force*. Time and frequency measurement techniques that are a limit of the heterodyne method and that occur when the mixed reference signal has exactly the *same* average frequency as the signal under test. The reference signal may be obtained from the signal under test, for example, by means of a delay line or a PLL.

Homogeneous chain. Chain of slave clocks characterized by the same parameter values.

Hypermedia. Multimedia hypertext.

Independent-clock measurement configuration. Standard measurement configuration on clocks, where there is no common master clock controlling the timing signals between which the TE is measured. On a single clock, it consists of measuring the output phase noise $\phi_{out}(t)$ compared to some 'good' external reference clock.

Inter-node distribution. Distribution of the timing reference among offices. Mostly, HMS architectures are adopted.

Intra-node distribution. Distribution of the timing reference within an office. Mostly, a star topology from the local SSU/BITS is used.

Isochronous. A digital signal is so called when the time intervals between its significant instants have, at least on the average, the same duration or durations which are integer multiples of the shortest duration (from the Greek etyma ἴσος = *equal*, χρονός = *time*).

Jitter. The phase/time deviation of the received digital signal, at every time $t_k = kT$, from the expected pulse times $t_k = kT$ of the ideal signal.

Jitter tolerance. The minimum jitter amplitude that equipment should be able to accept at its input, whilst staying within prescribed performance limits (e.g., in the case of a digital equipment interface, with error-free reception).

Jitter transfer function. The ratio between the output jitter amplitude and the input jitter amplitude over a jitter frequency range.

Justification demand. Difference between the actual read and write frequencies in the elastic store of a synchronizer of an asynchronous digital multiplexer.

Justification frequency. The rate with which bit justification takes place in an asynchronous digital multiplexer. In case of positive justification, that is the frequency at which justification opportunity bits are not assigned to tributary bits. The opposite holds for negative justification, which happens when justification opportunity bits are assigned to tributary bits.

Justification ratio. The ratio of the actual justification frequency to the maximum justification frequency. Similarly, the *nominal justification ratio* is the ratio of the nominal justification frequency to the maximum justification frequency.

Line Terminal Multiplexer. SDH equipment. An LTM of level N (LTM-N) is used at the termination of an SDH transmission chain to multiplex/demultiplex several tributaries on one STM-N line. Tributaries are usually PDH signals of any kind and hierarchical level, as well as SDH STM-M signals ($M \le N$).

Lock-In Range. The largest offset between the input frequency ω_{in} and a specified nominal frequency (approximately the VCO free-run frequency ω_F), within which the PLL can lock fast (i.e., in a time in the order of $1/\omega_n$ seconds for a second-order PLL) to the new input frequency.

Loss Of Frame. The alarm state of out-of-alignment in SDH system frame aligners.

Loss Of Pointer. The alarm state in SDH equipment due to loss of pointer alignment.

Loss Of Signal. The alarm state in telecommunication equipment due to loss of the input signal.

Mapping. One of the possible relationships in the SDH multiplexing rules is the mapping of a tributary signal into a synchronous Container (C-i). Such mapping rules may include bit justification, to accommodate a plesiochronous signal with variable frequency into a synchronous C-i having fixed size (asynchronous mapping, analogously to what discussed for asynchronous digital multiplexing). Otherwise, if the PDH tributary is locked to the same reference frequency as the STM signal, the mapping can be synchronous.

Master clock. An autonomous clock that is used to control the frequency of other clocks.

Master–Slave synchronization. Network synchronization strategy based on the distribution of the timing reference from one clock (*master clock*) to all the other clocks of the network (*slave clocks*), directly or indirectly, according to a star (two-level despotism) or tree (multi-level despotism) topology.

Maximum Time Interval Error. The maximum peak-to-peak variation of Time Error in all the possible observation intervals τ within a measurement period T.

Mesochronous. Two mesochronous (from the Greek etyma $\mu\acute{\epsilon}\sigma o\varsigma = medium$, $\chi\rho\acute{o}\nu o\varsigma = time$) digital signals are isochronous, asynchronous digital signals, whose respective timing signals have the same frequency, at least on the average, but no control on the phase relationship.

Mixer. A device that accepts two analogue pseudo-periodic signals at its inputs and forms an output signal with components at the sum and difference frequencies. For example, a four-quadrant multiplier is a mixer.

Multi-frame synchronization. Further delineation of wider frames spanning more basic frames already delineated.

Multiframe. A cyclic set of consecutive frames in which the relative position of each frame can be identified.

Multimedia synchronization. The orchestration of time-dependent heterogeneous elements (images, text, audio, video, etc.) in a multimedia communication at different (e.g. physical and human interface) levels of integration.

Multiplexer. Equipment that performs the multiplexing task in transmission and the inverse task in reception (demultiplexing).

Multiplexer Section OverHead. The rows 5 through 9 of the SOH in SDH frames.

Multiplexing. The action of transmitting several signals together on a single shared transmission channel.

Mutual synchronization. Network synchronization strategy based on the direct, mutual control among the clocks, so that the output frequency of each one is function of the frequencies of the others.

Network synchronization. The distribution of time and frequency over a network of clocks, spread over an even wider geographical area, by using the communications capacity of links interconnecting them.

Out Of Frame. A pre-alarm state (prior to LOF) in SDH system frame aligners.

Output jitter. In a wide sense, all phase impairments that are generated by the clock on its output timing interfaces in absence of input jitter or other external disturbances. That may include, for example, fortuitous phase discontinuities due to infrequent internal testing.

Oven-Controlled Crystal Oscillator. Quartz-crystal oscillator where the resonator and the other temperature-sensitive elements are placed in a controlled oven whose temperature is set as closely as possible to a point where the resonator frequency does not depend on temperature, so to minimize the effect of residual temperature variations.

Packet synchronization. The equalization of packet random delays across a packet-switched network for CBR circuit emulation.

PAM-TDM signal. Analogue multiplex signal where N PAM signals are multiplexed in TDM.

PCM multiplexing. It consists of TDM multiplexing N analogue telephone signals, having converted them to PCM digital form.

PCM primary multiplex. The basic level of digital multiplexing. The European 2.048 Mbit/s PCM frame (E1) is made of 32 octets (time slots), 30 of which carry single 64 kbit/s telephone channels. The North American 1.544 Mbit/s PCM frame (DS1) is divided in 24 time slots, carrying 24 telephone channels, and one additional overhead bit.

Performance management. Area of network management functions that deal with Quality of Service (QoS) data logging.

Phase alignment. One of the possible relationships among synchronous multiplexing elements in the SDH multiplexing rules is the phase alignment of one multiplexing element into the containing one. In this case, the phase relationship between the contained element and the containing one may vary and is coded by the pointer word, added to the former to yield the latter.

Phase-Locked Loop. Feedback control system where the *phase* of a locally generated pseudo-periodic signal is automatically adjusted to the phase of an incoming pseudo-periodic signal.

Plesiochronous. Two plesiochronous (from the Greek etyma $\pi\lambda\eta\sigma\iota o\varsigma = close$, $\chi\rho\acute{o}\nu o\varsigma = time$) digital signals are isochronous, asynchronous digital signals, whose respective timing signals have same frequency values only nominally, but actually different within a given tolerance range.

Plesiochronous Digital Hierarchy. Standard digital multiplexing hierarchy defined by CCITT (now ITU-T) for transmission in digital telephone networks and based on asynchronous digital multiplexing, i.e. on a bit justification technique, which allows digital multiplexing of asynchronous tributaries with substantial frequency offsets.

Plesiochrony. Network synchronization strategy in which all clocks are autonomous (anarchy).

Pointer. An SDH pointer is a word allocated in a determined position within STM-N frames (e.g., the AU pointer is placed in the fourth row of the OH submatrix) which identifies the position of the first byte of the pointed VC within the STM-N frame or the containing VC.

Pointer justification. The mechanism of incrementing or decrementing the pointer value and phase-shifting the pointed VC. It allows the dynamic phase alignment of VCs within the containing multiplexing element.

Pointer processor. Part of a SDH network element where pointers are processed, in order to re-synchronize incoming VCs into the output STM-N stream. This re-synchronization process is based on an elastic store, where VC bits are written from the input STM-N signal and read out according to the local equipment clock.

Primary frequency standard. A source of standard frequency, which does not need to be steered by any other external reference. Examples of primary frequency standards are the caesium-beam and hydrogen-MASER atomic standards.

Primary Reference Clock. The function that represents either an autonomous clock or a clock that accepts reference synchronization from some radio or satellite signal (e.g., GPS) and performs filtering (ITU-T and ETSI standards). The PRC represents thus the network master clock. The same expression PRC denotes also the physical implementation of the logical function (stand-alone clock). Its ANSI equivalent is the PRS.

Primary Reference Source. The ANSI equivalent of the ITU-T and ETSI PRC. It represents the network master clock. It may be based on a stratum-1 clock (e.g., a cesium primary frequency standard) or on a LORAN-C or GPS receiver.

Pull-in range. The largest offset between the input frequency ω_{in} and a specified nominal frequency (approximately the VCO free-run frequency ω_F), within which the slave clock achieves locked mode, irrespective of how long it takes to lock.

Pull-out range. The largest offset between the input frequency ω_{in} and a specified nominal frequency (approximately the VCO free-run frequency ω_F) within which the slave clock stays in the locked mode and outside of which the slave clock cannot maintain locked mode, irrespective of the rate of the frequency change.

Pulse Amplitude Modulation. Technique in which a sequence of analogue samples is coded by a train of pulses whose amplitude is proportional to the sample values.

Pulse Code Modulation. A process in which a signal is sampled and each sample is quantized, independently of other samples, and encoded to digital form.

Pulse stuffing. Same as *bit justification* (ITU-T Rec. G.701 recommends the use of the term *justification* and deprecates the term *stuffing* in the same sense).

Quantization. The process in which a continuous range of values is divided into a number of adjacent intervals, and any value within a given interval is represented by a single predetermined value within the interval (in other terms, the process of making the amplitude axis discrete on an analogue signal).

Real-time clock synchronization. A special kind of network synchronization, in which the distribution of the absolute time is concerned (e.g., the national standard time), mainly to network management purposes.

Real-time spectrum analysers. Spectrum analysers that allow to observe a signal as a whole and to output the signal power measurement on all frequency bands simultaneously.

Regeneration. The process of reconstructing a received digital signal so that the amplitudes, waveforms and timing of its signal elements are constrained within specified limits.

Regenerator Section OverHead. The rows 1 through 3 of the SOH of SDH frames.

Regenerator. Equipment that performs digital regeneration. Regenerators can be cascaded to build very long transmission lines. While traditional digital line regenerators mainly perform clock recovery, bit decision and then retransmission of a 'clean' digital signal, SDH regenerators also process the RSOH bytes to simple maintenance and management purposes.

Regular clock (signal). An isochronous timing signal with all its expected significant instants evenly spaced in time, at least on the average (cf. Figure 2.2). Inhibiting some unwanted pulses yields a *gapped clock* signal.

Return synchronization signal. Timing signal transported back from the far end of a synchronization trail originating from an office.

Sample. A representative value of a signal at a chosen instant, derived from a portion of that signal.

Sampling. The process of taking samples of a signal, usually at equal time intervals (in other terms, the process of making the time axis discrete on a continuous-time signal).

Synchronous Digital Hierarchy. Standard digital multiplexing hierarchy defined by CCITT (now ITU-T) for transmission in digital telephone and broadband telecommunications networks. It is based on synchronous digital multiplexing and exploits a byte justification technique (pointer justification) to cope with clock frequency offsets between network nodes.

SDH Equipment Clock. The function that, in an SDH network element, accepts synchronization inputs from external sources, selecting one of them, and filters the timing signal derived from this selected source (ITU-T and ETSI standards). It may use an internal timing source should all external synchronization references fail or degrade. The same expression SEC denotes also the physical implementation of the logical function (internal clock of a SDH network element).

Secondary frequency standard. A source of standard frequency, which can be, and usually is, steered by some primary standard source. Examples of secondary frequency standards are the high-quality quartz-crystal oscillators and the rubidium-gas-cell atomic standard.

Security management. Area of network management functions that deal with the procedures enabling the manager to operate the functions that secure the managed resources from user misbehaviour and unauthorized access.

Signal. A physical phenomenon whose one or more characteristics may vary to represent information.

Significant instants. Special instants when a timing signal triggers the controlled process. Suitable significant instants can be identified, for example, at the signal zero-crossing instants, for ease of implementation.

Slave clock. A device able to generate a timing signal, suitable for the measurement of time intervals, having phase (or much less frequently frequency) controlled by a reference timing signal at its input. The Phase-Locked Loop (PLL) and the Frequency-Locked Loop (FLL) are examples of slave clocks.

Slip. The repetition or loss of a set of consecutive bits in a digital signal, to enable the signal to accord with a rate different from its own. It is due to buffer overflow or underflow, when accomplishing bit and frame synchronization by means of an elastic store (e.g. at the input of synchronous digital multiplexers and of digital switching exchanges). In *controlled slips*, both the magnitude and the instant of bit loss or repetition are controlled (for example, entire PCM frames are lost or repeated).

Slip buffering. The technique used in synchronous digital multiplexers and digital switching exchanges to perform bit synchronization of tributary signals.

Smoothed clock (signal). A *gapped clock* signal whose instantaneous phase has been low-pass filtered, in order to redistribute pulses the most uniformly as possible along the time axis, according to its average frequency.

Spectrum analysers. Instruments that allow to observe a signal in the frequency domain, plotting its power versus its Fourier frequency. Spectrum analysers come in two basic varieties: swept-tuned and real-time instruments.

Stability of a clock. The measurement of variations of its instantaneous frequency (or of the time generated), compared to the nominal value (i.e., in practice, to that of a reference clock), *over a given observation interval.*

Stand-Alone Synchronization Equipment. Physical implementation of the SSU as a stand-alone piece of equipment (ITU-T and ETSI standards).

Stratum. Performance level of the hierarchy defined by ANSI to classify telecommunication clocks.

Stuff threshold modulation. Technique used in *synchronizers* for *waiting time jitter* reduction, based on modulating the buffer thresholds with a suitable *dithering* waveform, in order to shift the waiting time jitter out of the pass-band of the PLL filter in the *desynchronizer* on the receiver side.

Swept-tuned spectrum analysers. Spectrum analysers that allow to observe one Fourier frequency at a time. The complete spectrum is plotted cyclically in time, by sweeping the frequency range of interest.

Switching. The function of connecting a given input–output pair in nodes where multiple transmission links are terminated. It deals thus with the dynamic assignment of the transmission channels available in a network, on the basis of user connection requests.

Symbol synchronization. The recovery of the symbol sequence timing from a received signal, to identify the sampling and decision times in order to extract the logical information (also known as *clock recovery*).

Synchronization. The act of synchronizing, i.e. making synchronous (cf. the Greek etymon σύγχρονος) the operation of different devices or the evolving of different processes by aligning their time scales.

Synchronization link. A link between two synchronization nodes over which a timing signal is transmitted. The facilities of a transport network can be used (e.g., 2.048 Mbit/s links).

Synchronization Status Message. Quality marker, embedded in synchronization signals (e.g., STM-N/STS-N signals or E1/DS1 signals), which provides an indication of the quality of the timing carried, by denoting the type of synchronization source to which the signal is traceable. They have been defined in order to allow automatic protection procedures of the network synchronization distribution.

Synchronization Supply Unit. The function that represents the building clock (ITU-T and ETSI standards). In a network node, it accepts synchronization inputs from external sources, selecting one of them, filters the timing signal derived from this selected source and distributes the filtered timing signal to other elements within the node. It may use an internal timing source should all external synchronization references fail or degrade. The physical implementation of the logical function SSU may be integrated within a SDH network element or within a PSTN digital exchange, but most commonly is a piece of *Stand-Alone Synchronization Equipment* (SASE). Its ANSI equivalent is the BITS.

Synchronization trail. The logical representation of one or several synchronization links, i.e. the path along which timing is transferred in inter-node distribution. Synchronization trails may contain SECs.

Synchronized-clock measurement configuration. Standard measurement configuration on clocks, which consists in measuring the phase error between output and input signals $\phi_{out}(t) - \phi_{in}(t)$, based on $\phi_{in}(t)$ itself to establish the reference time. This configuration is of particular interest because it allows to observe directly the impairments added by the slave clock, without having to deal with the frequency offset between the clock under test and a reference clock.

Synchronizer. The block performing the mapping of the tributary into the multiplex signal, in a digital multiplexer, thus comprising mainly the buffer and the read and write controls. It accomplishes the bit-synchronization of the tributary bits into the multiplex frame available positions.

Synchronous digital multiplexing. The type of digital multiplexing where the tributary signals are synchronous and the *slip buffering* technique is used to synchronize tributaries into the multiplex signal.

Synchronous multiplexing elements. Building blocks of SDH and SONET frames. They are structured, fixed-size sets of bytes, which are variedly byte-interleaved or mapped one into the other to eventually form STM or STS frames.

Synchronous transfer mode. Information transfer mode where the information sources are mutually synchronous and can start sending their information units only in preassigned time-slots. A special case of STM occurs when fixed-size words are sent out periodically like in PCM multiplexes.

Synchronous Transport Module of level N. The frames of the various standard hierarchical levels of SDH. The same name is used to refer extensively to the signals themselves.

Synchronous Transport Signal of level N. The signals of the various standard hierarchical levels of SONET.

Synchronous. Two synchronous (from the Greek etymon $\sigma\acute{v}\gamma\chi\rho\sigma\nu\sigma\varsigma$, built by $\sigma\acute{v}\nu = with$ and $\chi\rho\acute{\sigma}\nu\sigma\varsigma = time$) digital signals are isochronous digital signals whose respective timing signals have the same frequency, at least on the average, and a phase relationship controlled precisely (i.e., with phase offset $\Delta\Phi = $ constant).

Temperature-Compensated Crystal Oscillator. Quartz-crystal oscillator that implements an automatic control on the oscillation frequency based on the measurement of the crystal temperature, in order to reduce its sensitivity on environmental temperature.

Terminated synchronization signal. Timing signal terminating a synchronization trail in an office.

Time counter. Instrument whose basic function is to measure the elapsed time between two events (start and stop trigger events), by incrementing a counter register according to a fast base reference frequency.

Time Division Multiplexing. Multiplexing technique (dual of FDM) in which several signals are interleaved in time for transmission over a common channel. It consists of allocating tributary channels to different time slots in the time domain, so that there are no channel interference and it is thus still possible to separate the single channels by sampling the multiplex signal at proper times.

Time Error. Difference between the time generated by a clock under test and a reference time (the absolute time in the ideal case).

Time Interval. The measure of a time interval τ, starting at time t, accomplished by a clock under test. It expresses how long the clock under test perceives a time interval of ideal length τ starting at ideal time t.

Time Interval Error. The error committed by a clock under test in measuring an interval τ, starting at time t, with respect to the reference clock.

Time slot. Any cyclic time interval that can be recognized and defined uniquely. In TDM multiplexing, one of the time intervals, one per telephone channel, into which the sampling period is allotted.

Time source. It supplies a (pseudo-)periodic signal that carries absolute or relative time information, usable as reference for time synchronization. The most remarkable time sources are the UTC sources.

Timing loop. Configuration that occurs when a clock is synchronized by a timing reference that is traceable to the clock itself.

Timing recovery. Same as *clock recovery.*

Timing signal. Same as *chronosignal.*

Transmission. The action of conveying information point-to-point, for example from one node in a network to another one directly linked to it by a physical channel. Moreover, transmission can be also from one point to multiple points (*multicast*) or even from one point to all listeners on the medium (*broadcast*).

Tributary. One of the signals multiplexed in a multiplex signal.

Tributary Unit Group. SDH multiplexing element. TUGs have been defined in order to limit the large number of possible ways to combine lower order VCs into higher order VCs. They contain homogeneous groups of TUs, byte-interleaved and located in fixed positions.

Tributary Unit. SDH multiplexing element. A TU consists of a lower order VC-i ($i = 11, 12, 2, 3$) and of a TU pointer indicating the offset of the pointed VC within the TU itself.

True jitter. The difference between the occurrence time of the nth clock pulse (*significant instant*) and the occurrence time of the nth pulse of an ideal clock at the same average frequency. The true jitter expresses the deviation from the average rate of a clock, rather than from the nominal rate.

Unit Interval. The nominal time interval between consecutive significant instants of an isochronous signal, or rather the shortest interval if intervals are integer multiples of it.

Universal Time Coordinated. The world official time scale, maintained by the Bureau International des Poids et Mesures and the International Earth Rotation Service, which forms the basis of a coordinated dissemination of standard frequencies and time signal.

Virtual Container. The basic SDH multiplexing element. A VC is a structured set of bytes and maps (i.e., contains) a payload which can be a plesiochronous PDH signal as well as other synchronous multiplexing elements. VCs are combined to form STM-N frames and are *individually* and *independently* accessible through a *pointer* information, directly associated to them on multiplexing. A VC is not disassembled nor modified during its travel through the SDH network, whereas it may experience only phase shifts (pointer action).

Voltage-Controlled Oscillator. An oscillator whose output frequency is controlled, within given limits, by an external tension.

Waiting time jitter. Low-frequency jitter exhibited by signals demultiplexed from an asynchronous digital multiplex signal. Its origin consists in the fact that justification in the synchronizer can only take place in particular instants, due to the determined position of justification control and opportunity bits in the output multiplex signal, and not exactly when it is demanded.

Wander. Low-frequency jitter (for $f_j < 10$ Hz).

Word synchronization. The process of determining the start and the end of code words, i.e. of delineating the words (e.g. bytes) in the raw and uniform stream of received bits.

INDEX